QH75 .C6819 1997
Conserving peatlands

5·11·98

Conserving Peatlands

Conserving Peatlands

Edited by

L. Parkyn
R.E. Stoneman
Scottish Wildlife Trust
Edinburgh
UK

and

H.A.P. Ingram
University of Dundee
Dundee
UK

D. HIDEN RAMSEY LIBRARY
U.N.C. AT ASHEVILLE
ASHEVILLE, NC 28804

CAB INTERNATIONAL
Wallingford
Oxon OX10 8DE
UK

Tel: +44 (0)1491 832111
Fax: +44 (0)1491 833508
E-mail: cabi@cabi.org

CAB INTERNATIONAL
198 Madison Avenue
New York, NY 10016-4341
USA

Tel: +1 212 726 6490
Fax: +1 212 686 7993
E-mail: cabi-nao@cabi.org

©CAB INTERNATIONAL 1997. All rights reserved. No part of this publication may be reproduced in any form or by any means, electronically, mechanically, by photocopying, recording or otherwise, without the prior permission of the copyright owners.

Chapter 35 only is ©Crown copyright.

A catalogue record for this book is available from the British Library, London, UK.
A catalogue record for this book is available from the Library of Congress, New York, USA.

ISBN 0 85198 998 5

Printed and bound in Great Britain by
Biddles Ltd, Guildford and King's Lynn

Contents

Contributors

Mr A.R. Anderson
Forestry Commission,
Northern Research Station,
Roslin,
Midlothian EH25 9SY,
UK.

Mr C. Bain
The Royal Society for the
Protection of Birds,
Scottish Headquarters,
17 Regent Terrace,
Edinburgh EH7 5BN,
UK.

Dr J.J. Blackford
Department of Geography,
University of Durham,
South Rd.,
Durham DH1 3LE,
UK.

Ms G. Boswijk
Deparrtment of Archaeology and
Prehistory,
University of Sheffield,
UK.

Mr S.J. Brooks
British Dragonfly Society and
Department of Entomology,
Natural History Museum,
London SW7 5BD,
UK.

Mr S.J. Brooks
Scottish Wildlife Trust,
Cramond House,
Kirk Cramond,
Cramond Glebe Road,
Edinburgh EH4 6NS,
UK.

Dr D.A. Brown
Department of Geological
Sciences,
University of Manitoba,
Winnipeg, Manitoba R3T 2N2,
Canada.

Dr P.C. Buckland
Department of Archaeology and
Prehistory,
University of Sheffield,
UK.

Mr B. Burlton
Forest Enterprise,
Kielder Forest District,
Eals Burn,
Bellingham,
Hexham,
UK.

Ms S. Campeau
Département de Phytologie,
Université Laval,
Sainte-Foy,
Québec G1K 7P4,
Canada.

Professor F.M. Chambers
Centre for Environmental Change and Quaternary Research,
Department of Geography and Geology,
C&GCHE,
Francis Close Hall, Swindon Road,
Cheltenham GL50 4AZ,
UK.

Dr D.J. Charman
Department of Geographical Sciences,
University of Plymouth,
Plymouth,
Devon PL4 8AA,
UK.

Professor R.S. Clymo
School of Biological Sciences,
Queen Mary and Westfield College,
London E1 4NS,
UK.

Professor B. Coles
Department of History and Archaeology,
University of Exeter,
Queens Building,
Queens Drive,
Exeter EX4 4QH,
UK.

Mr P.McM. Corbett
Environment Service,
Department of the Environment for Northern Ireland,
Commonwealth House,
35 Castle Street,
Belfast BT1 1GH,
UK.

Mr N.R. Doar
Scottish Wildlife Trust,
Cramond House,
Cramond Glebe Road,
Edinburgh EH4 6NS,
UK.

Mr G. Donald
Department of the Environment,
Room C15/18,
2, Marsham Street,
London SW1 3EP,
UK.

Dr G. Doyle
Department of Botany,
University College Dublin,
Belfield,
Dublin 4,
Ireland.

Dr P. Foss
Irish Peatland Conservation Council,
119 Capel Street,
Dublin 1,
Ireland.

Mr K. Gilman
Institute of Hydrology (Plynlimon),
Staylittle, Llanbrynmair,
Powys SY19 7DB,
UK.

Mr C. Ginzler
Department of Vegetation Ecology and Conservation Biology,
Vienna University,
A-1091 Vienna,
Box 285,
Austria.

Dr A. Grünig
Advisory Service for Mire Conservation,
Swiss Federal Institute for Forest, Snow and Landscape Research (WSL/FNP),
CH - 8903 Birmensdorf,
Switzerland.

Mr R. Hingley
Historic Scotland,
Longmore House,
Salisbury Place,
Edinburgh EH9 1SH,
UK.

Mr P. Immirzi
Research and Advisory Services Directorate,
Scottish Natural Heritage,
2 Anderson Place,
Edinburgh EH6 5NP,
UK.

Dr H.A.P. Ingram
Department of Biological Sciences,
University of Dundee,
Dundee DD1 4HN,
UK.

Dr B.R. Johnson
English Nature,
Lowland Peatland Programme,
Nature Conservancy Council for England,
Roughmoor, Bishop's Hull,
Taunton, Somerset TA1 5AA,
UK.

Mr J.H.J. Joosten
Ernst-Moritz-Arndt Universitat,
Fachnichtung Biologie,
Botanisches Institut und Botanischer Garten,
Grimmerstraße 88,
D-17487 Greifswald,
Germany.

Dr J.M. Jones
School of Biology & Earth Sciences,
Liverpool John Moores University,
Byrom Street,
Liverpool L3 3AF,
UK.

Mr R.A. Lindsay
Research and Advisory Services Directorate,
Scottish Natural Heritage,
2 Anderson Place,
Edinburgh EH6 5NP,
UK.

Dr M. Löfroth
Naturvardswerket,
S-106 48 Stockholm,
Sweden.

Dr D. Maas
Technical University of Munich,
Department of Vegetation Ecology,
D - 85350 Freising-Weihenstephan,
Germany.

Dr A.W. Mackay
Environmental Change Research
Centre,
University College London,
26 Bedford Way,
London WC1H 0AP,
UK.

Professor E. Maltby
Institute for Environmental
Research,
Royal Holloway,
University of London,
Huntersdale,
Callow Hill,
Virginia Water,
Surrey GU25 4LN,
UK.

Mr G. McNally
Land Development Manager,
Bord na Móna,
Newbridge,
Co. Kildare,
Ireland.

Mr H.P. Nowak
Bietenberg,
CH 6418,
Rothenthurm,
Switzerland.

Dr C. O'Connell
Irish Peatland Conservation
Council,
119 Capel Street,
Dublin 1,
Ireland.

Dr S.E. Page
Departments of Adult Education
and Zoology,
University of Leicester,
UK.

Ms L. Parkyn
Scottish Wildlife Trust,
Cramond House, Kirk Cramond,
Cramond Glebe Road,
Edinburgh EH4 6NS,
UK.

Professor J. Pfadenhauer
Technical University of Munich,
Department of Vegetation Ecology,
D - 85350 Freising-Weihenstephan,
Germany.

Dr G.L.M. Raeymaekers
Ecosystems Ltd.,
Beckerstraat B-1050,
Brussels,
Belgium.

Ms E. Reid
Scottish Natural Heritage,
Research and Advisory Services
Directorate,
2 Anderson Place,
Edinburgh EH6 5NP,
UK.

Dr J.O. Rieley
Department of Life Science,
University of Nottingham,
UK.

Dr L. Rochefort
Département de Phytologie,
Université Laval,
Sainte-Foy,
Québec G1K 7P4,
Canada.

Ms S.Y. Ross
Department of Environmental Science,
University of Stirling,
Stirling FK9 4LA,
UK.

Mr G.R. Seymour
Environment Service,
Department of the Environment for Northern Ireland,
Commonwealth House,
35 Castle Street,
Belfast BT1 1GH,
UK.

Mr A. Shaw
The Peat Producers Association,
41, High Street,
Glastonbury,
Somerset,
UK.

Dr P.A. Shepherd
East Midlands Environmental Consultants Ltd.,
310, Sneinton Dale,
Nottingham, NG3 7DN,
UK.

Dr T. Simpson
Department of the Environment,
Room C15/18,
2, Marsham Street,
London SW1 3EP,
UK.

Dipl Ing J. Sliva
Technical University of Munich,
Dept. of Vegetation Ecology,
D - 85350 Freising-Weihenstephan,
Germany.

Dr C.A. Smith
Environmental Sciences Dept.,
The Scottish Agricultural College,
Auchincruive, Ayr KA6 5HW,
UK.

Professor T.C. Smout
Institute for Environmental History,
University of St. Andrews,
Fife KY16 9QW,
UK.

Mr K. Stanfield
Environment and Heritage Service,
Peatlands Park,
33 Derryhubbert Road,
Dungannon BT71 6NW,
UK.

Ms C. Steel
Senior Conservation Officer,
The Wildlife Trusts,
The Green,
Waterside South,
Lincoln LN5 7JR,
UK.

Dr G.M. Steiner
Department of Vegetation Ecology and Conservation Biology,
Vienna University,
Box 285,
Althanstrasse 14,
A-1091 Vienna,
Austria.

Dr R.E. Stoneman
Scottish Wildlife Trust,
Cramond House, Kirk Cramond,
Cramond Glebe Road,
Edinburgh EH4 6NS,
UK.

Mr N. Temple-Heald
The Peat Producers Association,
41, High Street,
Glastonbury,
Somerset,
UK.

Dr D.B.A. Thompson
Scottish Natural Heritage,
Research and Advisory Services
Directorate,
2 Anderson Place,
Edinburgh EH6 5NP,
UK.

Ms A.E. Ward
Department of the Environment,
Room C15/18,
2, Marsham Street,
London SW1 3EP,
UK.

Mr D. Welch
Institute of Terrestrial Ecology,
Banchory Research Station,
Hill of Brathens,
Banchory,
Kincardineshire AB31 4BY,
UK.

Ms N.J. Whitehouse
Department of Archaeology and
Prehistory,
University of Sheffield,
UK.

Dr D.P. Whitfield
Scottish Natural Heritage,
Research and Advisory Services
Directorate,
2 Anderson Place,
Edinburgh EH6 5NP,
UK

Foreword

David Bellamy

Three per cent of the Earth's land surface is, or rather was, covered with peatlands - living systems that fix carbon from the global greenhouse, storing it away in the form of peat and producing oxygen into the bargain. Peatlands, bogs and fens are habitats for many unique plants and animal communities, themselves made up of a cross section of the main groups of the worlds' flora and fauna. As they grow, the peat preserves a detailed record of the vegetation of the peatland, its catchment and the surrounding region in the form of sub-fossils, pollen grains and spores. This record may include human artefacts, treasure trove and even the mummified bodies of wayward travellers or victims of sacrifice.

Unfortunately this unique heritage and resource is being destroyed at an ever increasing rate. The agents of destruction are many: drainage for agriculture and urban development, mining for power generation and for the horticultural trade and even to produce land fill sites to hide the products of our throw away society.

The Peatlands Convention, brainchild of the Scottish Wildlife Trust, held a gathering of the clans of peatland expertise from all over the world to lay the facts of the matter before the public, before it is too late. The destruction of the world's peatlands must stop now ... that is the poignant message of this book, a book that details the facts of their wonder, importance, conservation and rehabilitation.

If all the peatlands of the world were destroyed, there would be only one consoling fact, there would be no more living history books to take down the evidence of the arrogance and stupidity of humankind.

Acknowledgements

As organisers of the Convention, the Scottish Wildlife Trust would like to thank everyone who contributed to the proceedings, in particular the many speakers who were able to donate their time freely.

We gratefully acknowledge the financial contributions made towards the running of the Convention and publication of this book by:

European Commission
Historic Scotland
Lothian and Edinburgh Enterprise Trust
Mitchell Trust
Morton Fraser Milligan
Scottish Enterprise

and the UK Statutory Nature Conservation Agencies:

English Nature, Countryside Council for Wales and Scottish Natural Heritage.

The Peatlands Convention 1995

Introduction

Dr Rob Stoneman
Scottish Wildlife Trust, UK.

Follow the mighty Firth of Forth up-river and the great finger of sea which etches into eastern Scotland dwindles to a small stream, meandering across a broad flat valley. The Carse of Stirling lies sandwiched between the Highland Trossachs and the Campsie Fells and, in itself, is a scene of great beauty and rural tranquillity. Explore a little more, and the true richness of the scene transpires. In winter, pink-footed geese wheel overhead and hen harriers soar low to the ground. Mountain hare fit oddly into this lowland valley. The Carse is a botanical and entomological delight, the only known UK location of Labrador tea; a haven for the large heath butterfly; the site of the second only record of the rare jumping spider, *Heliophanus dampfii*.

Now follow another mighty river - the Barito - which is over one mile wide at its exit from the island of Borneo into the Sea of Java. Two hundred and fifty kilometres up-river and one is still only eight metres above sea-level. Surrounding the river in all directions is a vast forest stretching over millions of hectares. Inside, the ground is wet, soaking and walking is hard, hot and uncomfortable. But the rewards are bounteous. Crashing through the trees are orang-utan, gibbon and civets. Above the canopy, hornbills flap out a steady beating rhythm whilst the forest buzzes, whistles and screams in a majestic cacophony of sound.

Now switch back to Scotland and follow a youthful burn up into the wilderness of the Scottish Highlands and gaze across a sparkling pattern of land and water. In summer, golden plover emit a mournful whistle whilst red throated divers nest on lochans. The landscape is aglow with heather blossom and under foot, the multi-coloured bog mosses squelch to your step.

Across the world, this watery richness of life thrives in some remarkably extreme conditions. Surface waters are acidic, nutrient availability is severely limited whilst, most obvious of all, the land is wet. Very wet; for these are the peatlands of the world. Commonly over 95% water, the definition of 'land' is stretched to its edge. As the first sections of this volume show, in our peatlands, there is a resource of immense interest and value; an unusual slice of global biodiversity; an archaeological and palaeoecological treasure trove; a vast store of carbon regulating climate change; hydrological control systems; an economic asset; an outdoor classroom; a natural delight and, in some cases, some of the world's last and most beautiful wilderness areas.

In Britain, peatlands have been described as the jewels of the UK's nature conservation crown and yet, for our raised bogs at least, we stare into the abyss of total destruction. Today, only a few per cent of Britain's raised bogs can be described as natural. Most have been cut-away, 'reclaimed' for agriculture, drained, afforested, burnt or fragmented.

In central Kalimantan, one million hectares of the wondrous peat swamp forests are to be cleared for rice paddy despite the dire predictions of likely failure. In the Baltic states, bogs are being drained and cut-away for export in plastic bags to supply a western European horticultural industry which has inadvertently led to wholesale destruction of its own bogs. The Highland blanket bogs of Scotland are afforested for tax incentives; the Great Bog of Allen of central Ireland exists no more, having been sacrificed to feed Ireland's electricity demand; 60% of Finnish mires have been cut or afforested. In Holland, the once great Bourtangermoor has been reduced to just a few fragments now zealously protected as icons of a lost culture.

The scale of destruction is immense. Today public exposure of this ecological disaster has now led to mire protection initiatives. Regional, national and even supra-national legislation is slowing the destructive pace. An example is the European Union's Habitats and Species Directive which picks out peatlands as particularly worthy of conservation. Indeed, certain types of mires, raised and blanket bog, are earmarked as priority habitat upon which member governments have a particular duty to conserve.

The implications of this Directive are wide and on-going. One side to it, is the provision of financial aid (LIFE funding) for suitable nature projects which progress implementation of the Directive.

One of these, the Scottish Raised Bog Conservation Project (1993 to 1995) ran forward on four interrelated actions: planning for management, conservation management, rehabilitation and information exchange. The Project culminated in the Peatlands Convention and the production of the Scottish Raised Bog Conservation Strategy.

The aim of the Peatlands Convention was to channel the knowledge and expertise of peatland ecologists and conservationists to secure positive actions to conserve peatlands across the Union.

The conservation process can be broken down into three phases:

1. An understanding of the value, use and importance of what is to be conserved.
2. An assessment of the scale of conservation effort required.
3. A consideration of the methods of conservation action.

The conference opened with papers relating to the first phase. Biodiversity was examined in sessions relating to the *Peatlands of the World* and by considering the flora and fauna which peatlands support. The rich palaeoecological and archaeological *Peat Archive* was also reflected upon, whilst the session entitled *Functioning Peatlands* examined the role of peatlands in carbon cycling and hydrology. Peatlands also have an important economic value for horticulture, fuel and forestry.

This understanding of the value and role of peatlands feeds directly through to actions to conserve them. Indeed, looking at bogs through history challenges some of the present assumptions of bog conservation management. However, an assessment of conservation needs and the methods required should be made before action can be taken. Only when the problem is identified can actions be put in place to conserve the resource.

Conservation action is wide and varied. Protection of sites from direct threats, such as peat extraction, were considered with reference to legislative protection. Damaged peatlands (and many peatlands are) have to be positively managed to stop, slow or reverse degradation of sites. This issue was considered under the sessions: *Conservation Management, Rehabilitation and Monitoring.* Raising awareness is also vital to conservation efforts in order to enthuse people to expand conservation efforts wider. Awareness raising efforts involving campaigning, education, media and access are examined here.

The Peatlands Convention ended with some moral considerations of bog conservation issue and the presentation of the Scottish Raised Bog Conservation Strategy.

This volume, therefore, acts as a valuable reference book to guide, enthuse and educate peatland conservationists, although it can also act as a blueprint for other habitat conservation. If the enthusiasm and will displayed during the Convention can genuinely be translated into actions, then the future of our peatlands can only become more and more secure.

Part One

Peatlands of the World

An Introduction

Dr Hugh A.P. Ingram
University of Dundee, UK.

Why should anyone bother to be interested in peat bogs and mires in general? They are not places where people have ever been notably comfortable. Indeed, in various ways our civilisation finds then very alien indeed. They are remarkably inaccessible, sometimes difficult to traverse on foot and always presenting obstacles to wheeled vehicles. They used to have a reputation for harbouring disease, and certainly the fen-men of former centuries were far from healthy (Sheail and Wells, 1983). Poor trafficability renders them unfit for modern arable farming unless they are first drained, and the expense of drainage for this purpose is not justifiable unless yields are very high (as they have been in the former fenlands of the Isle of Ely and S.E. Lincolnshire).

One could argue that the very intractability of mires imparts interest. Elsewhere in this volume I have emphasised the prolonged concern of people with bogs as areas of rough pasture (Part five) and in our own century many of the raised mires of North West Germany have been converted to intensively managed pasture grassland called Hochmoorgrünland (Bartels, 1977). When lightly managed they present the appearance of wilderness and it is easy (and perilous) to forget that they are managed at all. All this leads to a special interest in their natural history and in the particular ecological processes they transact.

We are coming to understand these aspects rather better, especially some of the most fundamental ones that involve the water relations of peat bogs (Ivanov, 1981; Bragg, 1995), but there is a good deal that still eludes us. In particular, we do not yet have any clear understanding of how the living plants and animals of a mire function as a community.

In the circumboreal bogs it is possible to discern a certain range of plant life-forms that recur over large areas. These include:

- mosses, especially *Sphagnum*;
- liverworts, closely associated with the mosses;
- dwarf shrubs, often displaying xeromorphic features;
- dwarf sedges, also often displaying xeromorphic features;
- insectivorous mesophytes;
- non-insectivorous mesophytes (*Narthecium*, *Listera cordata*);
- lichens;
- trees, which are however often sparse or absent.

The regularity with which these elements associate to form the vegetation of peat bogs is itself remarkable. What are the 'assembly rules' which underlie this regularity? How can we investigate this? These, I suggest, are questions of clear and fundamental relevance not just to the science of mires but to their conservation and hence to their very survival.

One way forward is to compare mires in which the rules are different: different, that is to say, to judge from their outcome as the following examples suggest. First, in the North Island of New Zealand the water-retaining layer of some bogs, which in the circumboreal zone is formed by *Sphagnum*, is replaced by certain very peculiar flowering plants belonging to the family Restionaceae (Campbell, 1983), rush-like monocots best known from arid parts of southern Africa where, however, they also occur in peaty places (le Maitre *et al.*, 1996). What part does water retention play in the workings of the community?

Secondly, Walton Moss in Cumbria is a common grazing. It is a raised mire with an active and diverse *Sphagnum* carpet but with virtually no dwarf shrubs because these are browsed as soon as they appear. There is also scarcely any of the hummock-hollow microtopography which we normally associate with raised mires. So do the dwarf shrubs provide a scaffolding: a kind of climbing-frame for hummock-building *Sphagnum*? This hypothesis was suggested by Prof. Nils Malmer (of Lund) at a conference at Abisko convened in 1988. How can we test it?

Thirdly, the peripheral parts of intact raised bogs in Sarawak are covered by vast trees. Their gloomy interiors support few Sphagna (Anderson, 1983) and a sparse and infrequent presence of the sedge *Thoracostachyum bancanum* in the field layer. If one treads on their surface one sinks until one's feet meet the first buried log. Bearing in mind the large allocation of dry matter to the underground parts of our dwarf sedges, compared with that to their aerial shoots (the 'botanical iceberg' effect: Wein, 1973), are the buried parts of these plants the sustainers of what limited trafficability our bogs possess? How important is it to the associated components of the vegetation that the mire surface is cohesive? How many of these associates could survive on a mechanically unstable surface?

It is clear from all this, that insights come from a knowledge of mires that is wide as well as deep; or, if not insights, at least questions. And in science it is often more difficult to ask the right questions than to provide acceptable answers. That is one reason for the importance of a session on mire diversity worldwide: the extraordinary helps us to comprehend the familiar.

Chapter one:

The Bogs of Europe

Dr Gert Michael Steiner
Vienna University, Austria.

Introduction

Bog development is a function of features of the environment (e.g. climate, geology and geomorphology) combined with the intensity and history of human impact. Thus, every region has its own bog types with a vegetation reflecting both environmental conditions and degree of disturbance. These regional types are of high importance for conservation issues, because, without the protection of a representative number of sites in each region, the diversity of European bogs will vanish in the near future. Many regions have already lost their bogs completely and some of them put a large effort into the rehabilitation of the remaining peat layers. A better knowledge of what is there, and what is typical of every mire region, may help us to prevent money from being spent on restoration rather than conservation. And indeed, all efforts for mire protection are vital, as there are not enough sites left to lose a single one.

The distribution of bogs in Europe

This survey of bog provinces is part of a proposal carried out in 1983 by Prof. Hugo Sjörs (Uppsala, Sweden) as a basis for discussion by the regionality group of the International Peat Society in order to produce a worldwide mire regionality study. Unfortunately it was never published. As a member of this group the author was responsible for central European mires.

The zone of ombrogenous bogs

Central Irish lowland province - nemoral

Extensive oceanic domed bogs. Today most of these are exploited for fuel or peat production and there is no wholly natural and undisturbed example left. Much conservation work is done by the Irish Peatland Conservation Council and a Dutch group in protecting those few remaining sites with limited human impact.

British bog province - boreo-nemoral

Some examples of oceanic domed bog are found along the British west coast and in the Hebrides (Taylor, 1983). In England today there are no undamaged sites and industrial peat cutting is continuing. The situation in Scotland and Wales is marginally better. Although undisturbed bogs no longer exist, an acceptable number of sites in fairly good condition do remain. British conservationists are constrained by ongoing discussions regarding the benefit and sense of bog conservation. Thus, the situation for mire conservation in Britain is the worst in the whole of Europe; if there is no progress in the near future towards protecting and rehabilitating at least the most important examples in each region, Britain will turn into a country with a large peat resource, but without peat-forming bogs.

North Sea province - nemoral

Suboceanic raised bogs (plateau bogs) that occurred extensively in the lowlands of Denmark (western Jutland peninsular) the Netherlands and north-west Germany, where bog development was maximal. Today nearly all are excavated or turned over to arable land. Only a few examples of intact mire surface, with the typical hummock-hollow pattern, remain. A large number of rehabilitation projects in this province are trying to re-introduce bog vegetation and initiate peat formation. The characteristic species (e.g. *Erica tetralix, Narthecium ossifragum, Myrica gale, Sphagnum molle, S. imbricatum* and *S. pulchrum*) are a mixture of oceanic and suboceanic elements (Overbeck, 1975).

South Baltic province - nemoral

Suboceanic bogs along the eastern coast of Jutland and the southern coast of the Baltic sea as far as Gdansk. Although the prevailing mire types in this area are fens, some extensive bogs have developed as well. Most of them are now destroyed and used as pastures or agricultural land. One remaining example exists in Denmark, no more than a handful in Poland and in the German part some rehabilitation projects are underway (with little success). Characteristic species are *Rubus chamaemorus, Ledum palustre* and *Chamaedaphne calyculata. Vaccinium uliginosum* is more frequent on hummocks than in the North Sea province and *Pinus sylvestris* becomes more important.

Southwest Swedish province - boreo-nemoral

Suboceanic concentric and eccentric domed bogs of which the majority were formed by paludification. The mire margins usually have pine stands but the mire expanse is almost treeless with any development of the hummock-hollow pattern at right angles to the slope. Many of the bogs in this area are well preserved or only slightly influenced by human impact. The majority of the bogs in this area are concentric (Granlund, 1932) but in those regions with the highest precipitation bogs tend to cover large areas and be non-centric, often slightly sloping unilaterally. As in the North Sea province the bogs have suboceanic traits in the flora and vegetation (e.g. *E. tetralix, N. ossifragum, S. imbricatum*). The famous bog of Komosse (Osvald, 1925) is located here (see also Granlund, 1932; Malmer, 1962; 1965).

West Baltic province - boreo-nemoral

This province contains the east Swedish lowlands (excluding islands but extending westward as far as the Oslo Fjord). Paludification has been less important in this province, with most of the bogs developing from shallow lakes. The bogs are domed or plateau-like, almost always concentric and lack most of the suboceanic traits. Mire margins with pine woods and laggs are well developed. Many undestroyed examples can still be found in this area (see Granlund, 1932; Backeus, 1972; 1978).

West and South Finnish province - south boreal

Domed bogs with well developed concentric hummock-strings (kermis) prevail in this province, but some plateau-shaped bogs occur in the south-west where the climate is boreo-nemoral. Scattered pines are present on most of the sites showing the boreal character (Eurola, 1962; Ruuhijärvi, 1983).

East Norwegian and central Swedish province - south boreal

Non-centric raised bogs become dominant in the lower uplands north of the boreo-nemoral zone, most of them having developed via paludification (see Sjörs, 1948).

Finnish lakeland province - south-boreal

This province corresponds to the preceding one, although the altitude is lower and the climate is somewhat more continental. The kermis are narrower than the large hollows (rimpis) and mostly covered with *Empetrum hermaphroditum* and, more eastwards, *C. calyculata*. There are bogs entirely covered with *S. fuscum* lacking hollows (see Paasio, 1940; Eurola, 1962; Ruuhijaervi, 1983). To the east, in Finnish North Karelia, this province merges into the following one.

South Karelian province - south boreal

Raised - mainly domed - bogs prevail especially in the southern part. In the north, mixed mires (aapa mires) take over dominance (Botch and Masing, 1983).

Baltic east coast province - boreo-nemoral

This province is a narrow belt along the coast of the Baltic Sea. It is similar to the west Baltic province but with some suboceanic features (e.g. *C. calyculata* is absent). Plateau-shaped bogs prevail, but there are also some pine bogs. Two classical bogs are located within this province, the Moor von Augstumal (Weber, 1902) and the Zehlaubruch (Gams and Ruoff, 1929).

East Baltic province - boreo-nemoral

This province covers the eastern parts of the three Baltic states and much of adjacent western Russia (from St. Petersburg and Novgorod south to the headwaters of the Dniepr). Large mire complexes with raised - mainly domed - bogs prevail. The bogs have well developed hummocks (with *P. sylvestris* and *C. calyculata*) hollows and pools as well as pine-wooded margins. Many are located in ancient lake basins and are thus formed by terrestrialisation (Botch and Masing, 1983).

White Sea coast province - mid-boreal

This, and parts of the following province, have a remarkably northern extension for an area with raised bog dominance. This lowland area was submerged in the early post-glacial and the age of the mires depends upon their altitude. The oldest mires are raised bogs with pools, followed by bogs with hollows instead of pools, and the youngest mires are fens or reed-swamps (Botch and Masing, 1983).

Northeast European province - boreal

Unlike all other provinces this one has a large extension in the N-S direction as well as from the west (edge of the Fennoscandian Shield) to the east (Ural). The bogs are low, almost flat, and have a well developed microtopography with hummocks and large hollows. In the south they are more domed than in the north. The mire margins are usually wooded with pines and the laggs are often large fens (Botch and Masing, 1983).

Extrazonal bog provinces with mountain climates

Central European upland province

This province contains the upland region mainly of central Europe reaching from the Harz in Germany, the Ardennes in Belgium and the Massif Central in France in the west to the Bohemian Mass in the east. The bogs of this province developed via paludification mainly in basins, on slopes, saddles or watersheds. The bog vegetation is characterised by a more-or-less dense canopy of *P. mugo* s.l. Although the climatic gradient seems rather steep on first view, the predominating mountain climate causes a surprising uniformity across the bogs. Apart from rather oceanic bog types in the west European part of the province (Massif Central and Ardennes) the west-eastern gradient is only indicated by the decline of hollows. The north-south gradient is even less striking, except in the Harz where a plant community of the subalpine belt - the Eriophoro-Trichophoretum - dominates at a relatively low altitude.

Alpine bog province

This province reaches from the Jura mountains at the French-Swiss border to the Carpathian mountains in the east. Again, the mountain climate overlies the west-east gradient, but the influence of elevation becomes more important. The changes in vegetation along the altitudinal gradient are described later. The bogs developed mainly via paludification and terrestrialisation and all bog types (valley bogs, basin bogs, sloping bogs, saddle bogs etc.) can be found. In lower altitudes most of the sites have been damaged by peat extraction and few examples remain in fairly good condition. In higher altitudes the situation is much better, but many bogs are heavily disturbed by sheep or cattle grazing. Other bogs have been used as litter meadows after clearance of *P. mugo*. Although the original aspect has been lost, this kind of use has its benefits too, because *Sphagnum* growth is enhanced by mowing.

One outstanding type of bog where the water resource is vapour rather than rain is worth describing in detail. In some of the Alpine land-slide regions **condensation mires** occur on very steep slopes (35-40°) growing directly on the scree. The scree of a land-slide is unsorted and contains boulders of all sizes. Hence there is a system of cavities inside the scree where the air temperature is the annual average of the locality (5-6°C) and the relative air humidity is 100%. On a sunny day when the air at the base of the slope warms (20-25°C) and therefore rises, it is replaced by cooler air out of the scree. To maintain equilibrium, warm air is sucked back into the scree in the upper parts of the slope. In the cave-system of the scree the warm air from outside gets cooler, loses water and falls. Thus, a cyclic process - run by the sun's energy - begins. This circulation process, whereby warm surface air rises, is replaced by cold air out of the scree and warm air is sucked back into the scree, becomes quicker the warmer it is outside. When the circulation gets fast enough, the air-stream causes evaporation in the cavities (where the air is already over-saturated with water). Evaporation requires a lot of energy and, in the absence of the sun, this energy is taken from the air inside the scree, which therefore gets much colder (+/- 0°C) and begins to fall faster (5 km h^{-1}). The cold emerging air cools down the air on the surface of the scree causing condensation there. This water is now the resource for *S. capillifolium* growth and is enough for peat formation and the development of a hummock plant community, the Pino mugo-Sphagnetum magellanici (all data from Ellmauer and Steiner, 1992)

The zone of blanket bogs

Irish blanket bog province

Blanket bogs occur in the lowlands of the west coast, mostly in Connemara and Mayo where *Schoenus nigricans* is dominant in some of the plant communities. Blanket bogs also occur in the lowlands of south-western and north-western Ireland. Another type of blanket bog, often covering steep hillsides, is found all over the country (even south of Dublin) at altitudes above 400 m. Most of these bogs have been influenced by grazing, burning and domestic peat cutting. The present industrial extraction of peat is consuming more and more blanket bog areas. Afforestation and agriculture are less of a threat but the situation with regard to the conservation of these unique sites is serious.

British blanket bog province

Blanket bogs occur in Wales, northern England and all over Scotland including the Hebrides. In the south there are mainly upland blanket bogs, whilst, in the north, they almost reach sea level. Apart from the Irish province, northern Scotland is the centre of blanket bog distribution in Europe. The Flow Country is widely covered with high- and lowland blanket bogs (see Lindsay *et al.,* 1988). In addition to the damaging impact of burning and sheep-grazing, the last decade has seen the increase of afforestation by exotic conifers. Today most of the vast areas of blanket bog in this part of the United Kingdom are disturbed - at least in part - and it is nearly impossible to find a blanket mire complex free from human threat. The shape and pattern of hollows and bog-pools (a feature which is strongly linked to climatic conditions) is heavily endangered. Many of these pool systems vanished due to erosion following overgrazing, others were affected by the drainage for afforestation. It is not only hollow and pool systems which change with climatic conditions, vegetation aspect and species abundance (e.g. *Trichophorum cespitosum, Myrica gale, Pleurozia purpurea* etc.) also show a strong gradient from the west to the east of the province (Lindsay *et al.*, 1988).

North Atlantic blanket bog province

This province includes the Faeroes, the Shetlands and westernmost Norway. Due to steep slopes, blanket bogs are less common. Near the coast a blanket bog type occurs which is called Atlantic bog by the Norwegians. At higher altitudes, and on slopes, this type is replaced by blanket bogs (similar to Scottish ones), sloping bogs and fens. Sheep-grazing has a strong influence and grass-turf may replace real peat on places like the Faeroe Islands. The area is treeless except for parts of the Norwegian coast where sloping bogs with pines do occur in less exposed locations. *Racomitrium lanuginosum* plays a similar role as in Scottish and Irish bogs, forming large hummocks and giving some of the large bog areas a grey colour.

Northwest and North Norwegian blanket bog province

Somewhere near the mouth of the Trondheim Fjord the coastal blanket bogs become of a gradually more northern type. The climate is highly oceanic with outstandingly mild winters given the high latitude. The proportion of sloping fens increases, but there are still bogs in the few areas which are not too steep. *R. lanuginosum* is even more abundant, but other oceanic species decrease northwards.

Extrazonal blanket bogs

Subcontinental blanket bogs in Fennoscandia

In the Scandes mountain range, province of Västerbotten (Sweden) the Blaikfjället was identified as blanket bog during the third International Mire Conservation Group (IMCG) symposium in 1988. Due to its outstanding value, the Swedish government was asked to protect the site as the best example of suboceanic blanket bog. The group was successful and Blaikfjället is now protected.

The mires of the Risiitunturi, Finland, have been described as sloping fens by Havas (1961) but look very much like blanket bog and are an example of subcontinental blanket bog.

Alpine blanket bogs

Three big blanket bog complexes are known within the Central Alps. The Chaltenbrunnen bog in the canton Berne, Switzerland, the Wiege Moor in the Silvretta, Austria and the blanket bog of the Villanderer Alm, Sarntaler Alps,

South Tyrol, Italy. They all have a mosaic of *P. mugo* s.l. and *T. cespitosum* dominated vegetation, hollows and pools. As all three of them have developed at an altitude of about 2000 m the climatic conditions are rather humid; the average temperature is about the same as in northern Scotland and the annual precipitation is more than 2000 mm.

A short description of the bog plant communities in Europe

This survey is based on Dierssen (1982) for north-west European bogs and Steiner (1992) for central Europe (wherein the original tables are published). The relevé size for these tables was always 1 m^2 and the names used follow the code of the phytosociological nomenclature (Barkman *et al.*, 1986). Similar classifications for France and eastern Europe do not yet exist.

Class 1: Oxycocco-Sphagnetea Braun-Blanquet and Tüxen 1943

Characteristic species: *Andromeda polifolia, Drosera rotundifolia, Eriophorum vaginatum, V. oxycoccus, Cephalozia connivens, Mylia anomala, Polytrichum strictum, S. magellanicum, S. capillifolium, S. tenellum.*

Vegetation of peat-mounds, hummocks and ridges.

Order 1: Erico-Sphagnetalia Schwickerath 1940 emend. Braun-Blanquet 1949

Characteristic species: *E. tetralix, N. ossifragum, Cladonia impexa, Odontoschisma sphagni, S. subnitens.*

Differential species against the Sphagnetalia magellanici: *M. gale, Polygala serpyllifolia, Molinia caerulea, E. angustifolium, Rhynchospora alba, Hypnum ericetorum, Campylopus flexuosus.*

Oceanic raised bog, blanket bog and wet heath communities in the nemoral zone of western Europe.

Alliance 1: Oxycocco-Ericion (Nordhagen 1936) Tüxen 1937 emend. Moore 1968

Characteristic species: *E. tetralix, N. ossifragum, S. subnitens, T. cespitosum* ssp. *germanicum, O. sphagni, S. papillosum, S. magellanicum.*

Differential species for the Ericion tetralicis: *E. vaginatum, V. oxycoccus, S. imbricatum, S. papillosum, S. magellanicum, S. capillifolium, M. anomala.*

The alliance contains blanket bog, raised bog and poor fen communities in the oceanic and suboceanic part of the nemoral zone in Europe. In the euoceanic parts of the British Isles hollow species such as *E. angustifolium, Menyanthes trifoliata* and *R. alba* have a marked abundance in the otherwise dwarf shrub dominated vegetation types.

Association 1: Erico-Sphagnetum magellanici (Osvald, 1923) Moore 1968
Characteristic species: *E. tetralix*, *N. ossifragum*, *O. sphagni*, *S. imbricatum*.

Differential species for the Ericetum tetralicis: *E. vaginatum*, *S. papillosum*, *S. magellanicum*.

The Erico-Sphagnetum is the most important plant community of treeless bogs in the south boreal and nemoral zone of western and north-western Europe. Optimum conditions for the community are in the lowlands of northern Scotland where the association covers vast expanses of blanket bog.

Subassociations and variants
Although syntaxa are derived solely from their species composition, sub-associations of bog plant communities reflect distances to the average water table and the trophic conditions of their variants. In almost all mire plant communities, mosses or lichens are the differential species for sub-associations, whereas variants are determined by one or more differential species of any kind.

Most of the sub-associations in this class have a **typical variant** which is ombrotrophic and has no differential species and a **minerotrophic variant** where species like *M. caerulea* or *M. gale* occur with high constancy. The sequence of the sub-associations described in this survey always moves from wet through to dry conditions.

The term **phase** is used for a more detailed subdivision of important sub-associations with respect to the water regime.

The physiognomic dominance of a single species in a community is described as **facies** of this species.

The **sub-association of *S. angustifolium*** Dierssen 1982 occurs in the Pennines, north Wales, Devon and Cheshire. *E. nigrum*, *Aulacomnium palustre* and *S. papillosum* incline when the conditions get slightly drier.

The **sub-association of *S. pulchrum*** Dierssen 1982 tolerates higher water table fluctuations. In the British Isles *R. alba* has its main occurrence in this sub-association outside hollows.

The **sub-association *of S. cuspidatum*** Dierssen 1982 is the most frequent one in Britain, but less frequent in Norway. *S. cuspidatum* is accompanied by *S. papillosum* and *S. tenellum* which have their highest frequency in this sub-association. Although it grows close to hollows or pools, the species of the Rhynchosporion albae (alliance of hollow communities) are remarkably under-represented and dwarf shrubs like *E. tetralix* and *Calluna vulgaris* dominate.

The **typical sub-association** (Jonas, 1932) Tüxen 1937 is recorded to be the most common and characteristic vegetation type in the whole blanket bog area. On ridges it forms dense *Sphagnum* carpets scattered with dwarf shrubs. On the British Isles, *S. subnitens* accompanies *M. caerulea* and *M. gale* in the minerotrophic variant. In Norway the differences between the two variants are less marked. The abundance and wide distribution of the typical sub-association enables a subdivision into phases. Whereas the **typical phase** occupies the wettest sites close to the sub-association of *S. cuspidatum*, the **phase of *S.***

imbricatum abundant in Ireland and Norway but absent in Great Britain prefers drier sites. The **phase of *S. fuscum*** representing steep hummocks similar to *S. imbricatum* is less frequent except in the south-west of Ireland. The **phases of *P. strictum*** and ***Leucobryum glaucum*** both growing on drained sites are more or less restricted to Norway.

The sub-association of *C. uncialis* Birse 1975 was first described for the Shetlands and Scotland. It is typical of dry sites with steep surface morphology where peat formation was stopped by erosion processes. Along the hyperoceanic west coasts of the British Isles and Norway a phase of *R. lanuginosum* forms high and steep hummocks and in Norway a phase of *S. imbricatum* was identified similar to the corresponding phase in the previous sub-association.

Association 2: Pleurozio-Ericetum tetralicis Braun-Blanquet and Tüxen 1952 emend. Moore 1968

Characteristic species: *P. purpurea*,. *C. atrovirens*

Differential species against the Erico-Sphagnetum: *M. caerulea*, *S. nigricans*.

The Pleurozio-Ericetum replaces the Erico-Sphagnetum in the lowland blanket bogs of western Ireland and to a certain extent in western Scotland and northern Wales. Typical of this association is the high abundance of *S. nigricans* in western Ireland and *M. caerulea*. The characteristic species *P. purpurea* and *C. atrovirens* are less abundant and are absent in the sub-association of *S. imbricatum*. Important species are *D. rotundifolia*, *N. ossifragum*, *E. angustifolium* and *C. vulgaris*. *R. alba* is more common than in the Erico-Sphagnetum and becomes dominant in western Ireland. Another attribute of the community is the high proportion of minerotrophic species such as *Potentilla erecta*, *P. serpyllifolia* and *Carex panicea*. In contrast to the Erico-Sphagnetum *A. polifolia* and *V. oxycoccus* are absent and *E. vaginatum* becomes rare. Important moss components are *S. tenellum*, *S. subnitens*, *S. papillosum*, *S. capillifolium* and *S. cuspidatum*; *S. compactum* and *S. magellanicum* are less abundant. Compared with other ombrotrophic communities the number of species is high with a remarkably homogenous composition. The scattered *Sphagnum* patches give space for the growth of algae and liverworts, lichens are less important.

The association has two facies. Whereas the **facies of *S. nigricans*** is restricted to the hyperoceanic parts of Ireland, the **facies of *M. caerulea*** is common in the whole area of the community.

Subassociations and variants

The **sub-association of *S. auriculatum*** Dierssen 1982 prefers the wettest sites where *Eleocharis multicaulis* can become a typical species in western Ireland.

Although closely allied to the previous sub-association, the **typical sub-association** Dierssen 1982 is characterised by the absence of *S. auriculatum* and *E. multicaulis* due to somewhat drier conditions.

The **sub-association of *S. imbricatum*** Dierssen 1982 usually forms small secondary hummocks close to large pool systems. *S. tenellum* and *M. anomala* become important, whereas *S. magellanicum, P. purpurea, C. atrovirens* and *T. cespitosum* are absent. Although there is a similarity with the corresponding sub-association of the Erico-Sphagnetum, *E. vaginatum, A. polifolia, V. oxycoccus* and *Calypogeia sphagnicola* are very rare in the Pleurozio-Ericetum.

The **sub-association of *R. lanuginosum*** Dierssen 1982 has a much wider distribution than the previous and prefers drought conditions. Due to the ongoing drainage of bogs its distribution is becoming even wider. *R. lanuginosum* is usually accompanied by *C. arbuscula, T. cespitosum, E. vaginatum* and *H. ericetorum*. In western and north-western Scotland, *Pleurozia purpurea* and *C. atrovirens* indicate a slight minerotrophic influence, on true ombrotrophic sites the Pleurozio-Sphagnetum is replaced by the Erico-Sphagnetum there.

Association 3: Vaccinio-Ericetum tetralicis Moore 1962
Characteristic species: *E. nigrum, V. myrtillus, Juncus squarrosus.*

Differential species for the Pleurozio-Ericetum and the Erico-Sphagnetum: *Rhytidiadelphus loreus, Pleurozium schreberi, Diplophyllum albicans.*

In the upland blanket bogs of Britain this community intervenes between the Erico-Sphagnetum and the Empetro-Eriophoretum, whereas in Ireland it replaces both the Erico-Sphagnetum and the Pleurozio-Ericetum at higher altitudes. Dense dwarf shrub swards and scattered *Sphagnum* make the community look very homogenous. This physiognomy is allied to better drainage of the mires on steep slopes. Common species are *E. nigrum, C. vulgaris, V. myrtillus, E. vaginatum, T. cespitosum* and *E. tetralix*. In south-western Scotland, Wales and Devon *M. caerulea* can take over dominance. Under wet conditions *S. papillosum* is the dominant species in the moss layer. It is replaced by *S. capillifolium* and *S. subnitens* with increasing drought and by *R. lanuginosum* and *Cladonia* species in erosion complexes.

Subassociations and variants
The **typical subassociation** Moore 1962 is characteristic of sites without a pronounced surface topography. In Scotland, Wales and the Pennines *Deschampsia flexuosa* plays an important role in the vegetation aspect but is less important in Ireland. On hillsides with steeper slopes and better drainage *M. caerulea* can take over dominance. Chamaephytes are more or less absent under such conditions and mosses become rare.

The **subassociation of *C. arbuscula*** (Moore, 1962) Dierssen 1982 grows under much drier conditions. The **typical variant** can be found along erosion channels or on low ridges, the **variant of *R. lanuginosum*** on top of the hummocks only.

Alliance 2: Ericion tetralicis Schwickerath 1933

Characteristic species: *E. tetralix, S. compactum, S. molle, S. strictum, H. imponens, C. brevipilus.*

Differential species for the Oxycocco-Ericion: *J. squarrosus, Gentiana pneumonanthe.*

This alliance contains wet heath communities in the oceanic part of nemoral Europe and higher altitudes of western and central European uplands. The peat layer is usually less than 1 m deep and water table fluctuations are high. Therefore species associated with constant soil moisture conditions (e.g. *A. polifolia, E. vaginatum, V. oxycoccus, S. capillifolium* and *S. magellanicum*) are absent and species of heath communities become more abundant.

Association 1: Ericetum teralicis (Allorge, 1922) Jonas 1932
Characteristic species composition: *E. tetralix, T. cespitosum* ssp. *germanicum, J. squarrosus, S. compactum, S. molle.*

The Ericetum tetralicis replaces the communities of the Oxycocco-Ericion on gley-podsols and peaty soils (less than 30% organic matter). Sometimes it also grows on raised bogs or blanket bogs after drainage, burning and over-grazing. Vascular plants grow much denser as in Oxycocco-Ericion communities and mosses are less important. Common species are *E. tetralix, C. vulgaris, M. caerulea, T. cespitosum, N. ossifragum* (restricted to western Europe) *S. compactum, H. ericetorum, D. albicans, P. schreberi* and, especially in Britain, *Dicranum scoparium.*

Subassociations
The **typical subassociation** Tüxen 1937 can be found in Norway and Ireland where it grows in contact with the Erico-Sphagnetum.

The **subassociation of *C. impexa*** Jonas 1932 has a much wider distribution, prefers drier sites and mediates very often between blanket bog and heathland.

Order 2: Sphagnetalia magellanici Kästner and Flössner 1933

Characteristic species: *C. pauciflora, C. calyculata, L. palustre, V. uliginosum, P. strictum, S. fuscum.*

Differential species for the Erico-Sphagnetalia: *Calypogeia sphagnicola, S. russowii, Cetraria islandica, C. rangiferina.*

Sphagnum dominated communities of poor fens and bogs in the boreal zone and the mountains of central Europe.

Alliance 1: Oxycocco-Empetrion hermaphroditi Nordhagen ex Tüxen 1937

Characteristic species: *Betula nana, E. hermaphroditum, R. chamaemorus, V. microcarpum, Pinguicula villosa, C. bigelowii, Ochrolechia frigida, D.*

elongatum, Barbilophozia binsteadii, Lophozia marchica, Cephalozia leucantha, Sphenolobus minutus.

The Oxycocco-Empetrion contains the majority of bog communities in the mountains of the boreal and boreo-nemoral zone including some occurrences in the Alps as recorded by Krisai (1966) and Steiner (1992).

Association 1: Empetro hermaphroditi-Sphagnetum fusci DuRietz 1921 emend. Dierssen 1982 ex Steiner 1992

Characteristic species: *B. nana, E. hermaphroditum, R. chamaemorus, V. microcarpum.*

The Empetro-Sphagnetum fusci represents *Sphagnum* ridges in bogs and aapa mires throughout the boreal zone and in the Alps. The wettest places are occupied by *S. papillosum, S. rubellum, S. magellanicum* and *S. angustifolium* drier places by *S. capillifolium, S. fuscum* and *Cladonia* species. Frequent vascular plants are *A. polifolia, E. vaginatum, V. uliginosum, C. vulgaris, E. angustifolium* and *T. cespitosum* ssp. *cespitosum.*

Subassociations and variants

Due to the wide range of the community geographical races can be identified in some of the subassociations or phases respectively.

The **subassociation of *S. papillosum*** (Osvald, 1925) Dierssen 1982 represents the wettest part of the association with species of hollow communities like *S. lindbergii* and *S. balticum* and minerotrophic species such as *M. caerulea, E. angustifolium* and *Calliergon stramineum* as important components. The **variant of *S. lindbergii*** grows next to hollows or flarks where minerotrophic influence is high. A **phase of *S. magellanicum*** leads over to the next subassociation.

The **typical subassociation** (Osvald, 1925) Dierssen 1982 can be found on low and high ridges as well as hummocks. The decreasing soil moisture along this gradient is reflected by the **phases of *S. angustifolium*, *S. rubellum*** and ***P. schreberi***. The Empetro-Sphagnetum typicum is distributed all over Norway where the **race of *C. vulgaris*** tends to grow in low altitude near the coast and the **race of *B. nana*** in the Fjell, central and eastern Finmark and the Alps.

The **subassociation of *S. fuscum*** (DuRietz, 1921) Dierssen ex Steiner 1992 is the most common unit of the community. It represents high ridges, hummocks or strings in northern Scandinavia and the Alps. Compared with the two subassociations mentioned before minerotrophic species decline. *S. magellanicum* and *S. rubellum* prevail under wet, *S. capillifolium* and lichens under dry conditions. The two **races of *C. vulgaris*** and ***B. nana*** show a similar distribution pattern as described before.

The **subassociation of *C. arbuscula*** (Osvald, 1925) Dierssen ex Steiner 1992 represents the driest and northernmost occurrences of the community characterised by a bulk of frost resistant species, especially lichens. Three phases can be identified excluding each other geographically. The **phase of *R.***

lanuginosum growing on extremely acid hummocks in the coastal bogs of Norway occurs with two geographical races of which the **race of *E. hermaphroditum*** has a northern distribution. The **race of *C. stellaris*** has its distribution centre in the middle and south boreal zone, especially in the Troendelag area. The **typical phase** (with *C. arbuscula*) grows on inland bogs and also occurs with two geographical races. The **race of *C. vulgaris*** has a southern distribution reaching as far as the Alps, whereas the **typical race** (without *C. vulgaris*) tends to grow in northern and north-eastern Norway. The **phase of *D. elongatum*** is characteristic of the most continental parts of Norway from the mountains in the south up to the Finmark in the north.

Association 2: Ledo-Sphagnetum fusci DuRietz 1921
Differential species of the association and for the Empetro-Sphagnetum: *L. palustre, C. calyculata, V. vitis-idaea.*

The Ledo-Sphagnetum fusci replaces the Empetro-Sphagnetum fusci in eastern and north-eastern Fennoscandia. Due to the decline of oceanic species such as *C. pauciflora, E. vaginatum, D. rotundifolia, T. cespitosum, R. lanuginosum* and especially *C. vulgaris* and the increase of *L. palustre* and *C. calyculata* the hue of the community changes to white with the spring flowering. Dierssen (1982) identified three subassociations in eastern Norway but there are possibly more in Sweden and Finland.

Subassociations
The **subassociation of *S. russowii*** Dierssen 1982 grows in aapa mires along the base of the strings,

the **subassociation of *S. fuscum*** (DuRietz, 1921) Dierssen 1982 on high ridges, hummocks or strings and

the **subassociation of *C. pleurota*** Dierssen 1982 on dry and steep hummocks, strings, peat mounds or palsas.

Association 3: Empetro hermaphroditi-Eriophoretum McVean and Ratcliffe 1962 emend. Dierssen 1982
Characteristic species: *R. chamaemorus*, *E. hermaphroditum*, *B. nana*, *V. microcarpum*.

Differential species of the association: *V. myrtillus*, *V. vitis-idaea*, *R. loreus*, *E. nigrum*, *J. squarrosus*, *P. schreberi*.

This association is restricted to Great Britain and replaces the Empetro-Sphagnetum fusci in the blanket bogs of the subalpine belt in Scotland, the Pennines and Wales. Chamaephytes predominate in the community, mosses are of minor importance.

Subassociations and variants
The **subassociation of *S. papillosum*** Dierssen 1982 preferring waterlogged conditions is very rare.

The **typical subassociation** Dierssen 1982 is characterised by mosses with high drought resistance (e.g. *Hylocomium splendens, Plagiothecium undulatum* and *S. capillifolium*) and occurs in blanket bogs with low-relief patterns. In areas between erosion haggs it even grows in hollows forming secondary hummocks there.

The **subassociation of *S. fuscum*** (Birse and Robertson, 1976) Dierssen 1982 usually grows on dry secondary hummocks in otherwise unpatterned blanket bogs. *M. anomala* has a high constancy in this unit.

The **subassociation of *Cladonia arbuscula*** (McVean and Ratcliffe, 1962) Dierssen 1982 is the most common unit of the Empetro-Eriophoretum characterised by a high species richness. It occurs in drought areas drained by erosion channels and on top of hummocks. Typical species are *C. arbuscula, C. uncialis, C. impexa, C. chlorophaea, Coricularia aculeata, Anastrepta orcadensis, H. splendens, H. ericetorum* and *R. lanuginosum.*

Two variants can be identified, the first is the **typical variant** with the characteristics of the subassociation and the second is a **variant of *Racomitrium lanuginosum.*** They often grow together but the latter is more western in its distribution.

Association 4: Trichophoro cespitosi-Sphagnetum compacti Warén 1926 emend. Dierssen 1982 ex Steiner 1992

Characteristic species: *T. cespitosum* ssp. *cespitosum, S. compactum, C. delisei.*

The monotonous physiognomy of this community is caused by low species numbers and uniform ecological conditions. In the boreal zone it covers large areas of slightly sloping poor fen and aapa mire, whereas in its second distribution centre - the Central Alps of Switzerland and westernmost Austria - it tends to be a major component of alpine blanket bog or over-grazed bog with a high proportion of bare peat. In the boreal zone the characteristic species are accompanied by *A. polifolia, E. angustifolium, S. papillosum, S. balticum* and *S. lindbergii;* in the Alps the latter two boreal species are replaced by *S. tenellum* and *Gymnocolea inflata.*

Association 5: Vaccinio uliginosi-Sphagnetum papillosi Dierssen 1980

Characteristic species for the region only: *V. uliginosum, S. papillosum, S. subnitens.*

This plant community is restricted to the minerotrophic blanket mires of Iceland.

Alliance 2: Sphagnion magellanici Kästner and Flössner 1933 emend. Dierssen apud Oberdorfer et al. *1977*

Characteristic species: *A. polifolia, V. uliginosum, V. oxycoccus, P. mugo* s.l., *S. magellanicum, S. capillifolium, P. strictum.*

The Sphagnion magellanici contains hummock and ridge communities in the suboceanic and subcontinental parts of central Europe including uplands and mountains. The associations of the alliance usually lack distinct characteristic species but can easily be derived from the typical combination of differential species.

Association 1: Ledo-Sphagnetum magellanici Sukopp ex Steiner 1992
Differential species: *L. palustre, S. magellanicum.*

The Ledo-Sphagnetum magellanici is a common bog community in the subcontinental parts of the south boreal and nemoral zone as far as the northern border of Austria. Dense *L. palustre* and *V. uliginosum* swards grow beneath loose stands of *P. sylvestris.* The moss layer is represented by *S. magellanicum* and *S. capillifolium* (replaced by *S. fallax* or *S. girgensohnii* in the shade). Due to the rather continental climate drought conditions become common as indicated by the incline of *P. sylvestris* and the forest mosses *H. splendens, P. schreberi* and *D. scoparium.*

Association 2: Sphagnetum magellanici Kästner and Flössner ex Steiner 1992
Differential species: *A. polifolia, C. pauciflora, D. rotundifolia, E. vaginatum, V. oxycoccus, V. uliginosum, D. undulatum, P. strictum, S. angustifolium, S. capillifolium, S. magellanicum.*

The Sphagnetum magellanici is the most common plant community on hummocks and ridges throughout central Europe. The association is determined by a set of species that are characteristic of the alliance and the absence of typical species of the other communities within the alliance namely *L. palustre, T. cespitosum* and *P. mugo* s.l. Usually the Sphagnetum magellanici is a treeless community but minerotrophic influence in the marginal parts of bogs promotes the growth of *Picea abies*. There are examples with *P. abies* growing on truly ombrotrophic sites but only growing to a maximum height of 1 m. This phenomenon is described as the **facies of *P. abies***. Due to the wide distribution of the Sphagnetum magellanici right across central Europe, two geographical races can be described. In the **suboceanic race** dwarf shrubs are absent or scattered, whereas in the **subcontinental race** dwarf shrubs become an important part of the vegetation cover. From the west to the east the density of dwarf shrubs increases and the mire species *V. uliginosum* and *C. vulgaris* mix with the forest species *V. myrtillus* and *V. vitis-idaea.*

Subassociations and variants
The community occurs with six subassociations each with an ombrotrophic and a minerotrophic variant the latter represented by *C. nigra, M. caerulea, P. abies, P. erecta* and *P. commune*.

The **subassociation of *Scheuchzeria palustris*** (Kästner and Flössner, 1933) Dierssen apud Oberdorfer *et al.* ex Steiner 1992 growing under waterlogged conditions has a minerotrophic variant only. It usually occurs next to hollows in contact with the Caricetum limosae and is only represented in the suboceanic race, because in the subcontinental part of central Europe hollows become rare.

The **subassociation of *R. alba*** Dierssen apud Oberdorfer *et al.* ex Steiner 1992 is also restricted to the suboceanic race but prefers somewhat drier places. On sites with higher water table fluctuation (usually caused by drainage) it forms large unstructured carpets. Only a minerotrophic variant exists.

The **typical subassociation** Kästner and Flössner ex Steiner 1992 has the characteristics of the association and prefers high ridges or low hummocks. Both races and variants occur as well as the facies of *P. abies*.

The **subassociation of *S. fuscum*** (Kästner and Flössner, 1933) Dierssen apud Oberdorfer *et al.* ex Steiner 1992 grows on relatively dry places, where *S. fuscum* forms steep hummocks up to 1m.

The **subassociation of *P. schreberi*** Steiner 1992 with *H. splendens* and *D. scoparium* as constant associates indicates the end of peat growth, thus only a minerotrophic variant exists.

The **subassociation of *C. arbuscula*** Dierssen apud Oberdorfer *et al.* ex Steiner 1992 is restricted to the driest places, often indicating the end of peat growth, too. *C. arbuscula* with a more western distribution is replaced by *C. rangiferina* in the east. *C. islandica* is common all over the area. On a few places in the Alps and the eastern uplands *C. stellaris* was found. Again, only a minerotrophic variant exists.

Association 3: Eriophoro vaginati-Trichophoretum cespitosi Osvald 1923 emend. Steiner 1992

Differential species: T. cespitosum ssp. cespitosum, C. pauciflora.

The Eriophoro-Trichophoretum replaces the Sphagnetum magellanici in higher altitudes. It grows on low ridges, sometimes even in hollows or on peat areas with groundwater fluctuations.

Subassociations and variants

The **subassociation of *S. tenellum*** (Osvald, 1925) Dierssen apud Oberdorfer *et al.* ex Steiner 1992 grows in hollows or on wet bare peat often in contact with the Caricetum limosae or the Caricetum rostratae. *S. cuspidatum, S. majus, S. subsecundum* and *Gymnocolea inflata* are common species.

The **subassociation of *S. palustris*** Steiner 1992 is restricted to hollows of alpine bogs up to 1800m.

The **subassociation of *R. alba*** Steiner 1992 is very similar to the corresponding subassociation of the Sphagnetum magellanici. Again restricted to alpine bogs it prefers lower altitudes (max. 1200 m).

The **typical subassociation** (Osvald, 1923) Steiner 1992 can be found on sites where *S. magellanicum, S. capillifolium* and *S. papillosum* have their optimum growing conditions.

The **subassociation of *S. fuscum*** Dierssen apud Oberdorfer *et al.* ex Steiner 1992 forms small hummocks or ridges only and is restricted to undisturbed bogs.

The **subassociation of *C. arbuscula*** (Rudolph *et al.*, 1928) Dierssen apud Oberdorfer *et al.* ex Steiner 1992 is the driest unit of the community. It either caps hummocks in the mire expanse or replaces the typical subassociation in drained parts of the margin.

Association 4: Pino mugo-Sphagnetum magellanici (Kästner and Flössner, 1933) Neuhäusl ex Steiner 1992
Differential species: *P. mugo* s.str., *P. rotundata, P. uncinata.*

This association is the most common bog community in the subcontinental part of central Europe and is restricted to this area. *P. mugo* forms a more or less dense tree or shrub (Krummholz) canopy. These wooded bogs can be found from the plains of the Alpine foothills up to the subalpine belt in the uplands and mountains. Three bog species can be identified in the *P. mugo* group each of which represents a facies. *P. uncinata* is a monocormic tree with a western distribution, *P. rotundata* distributed in the central European uplands forms polycorm trees and *P. mugo* s.str. is a polycormic shrub with a more eastern distribution. The **facies of *P. uncinata*** grows in the Jura mountains, in parts of the central European uplands and in the Alps of Switzerland and western Austria. The **facies of *P. rotundata*** has its distribution centre in the central European uplands and the **facies of *P. mugo*** in the eastern Alps and the Carpathian mountains.

Subassociations and variants
The **typical subassociation** Kästner and Flössner ex Steiner 1992 is the central unit of the community and grows on the wettest sites. In the subcontinental part of central Europe, bogs usually consist of one big hummock only. The minerotrophic variant is indicated by *P. abies, Melampyrum pratense* ssp. *paludosum* and *S. angustifolium.*

The **subassociation of *S. fuscum*** Krisai ex Steiner 1992 is restricted to small hummocks in the wetter centres of large bogs.

The **subassociation of *P. schreberi*** Steiner 1992 usually grows in the dry marginal parts of the bogs but also indicates the end of peat growth.

The **subassociation of *C. arbuscula*** Dierssen apud Oberdorfer *et al.* ex Steiner 1992 is common on drier bogs in higher altitudes. Usually restricted to open patches in mountain pine thickets it also can indicate drainage or disturbance in lower altitudes.

In the wooded bogs of central Europe a plant community of the **class Vaccinio-Piceetea** forms marginal forests. This association is called **Bazzanio-Piceatum** Braun-Blanquet 1939 and is not only restricted to mire margins but grows frequently on drained peat or raw humus soils. Characteristic and differential species are *P. abies, Bazzania trilobata, S. girgensohnii* and *D. scoparium.*

Another association of the same class, the **Vaccinio uliginosi-Pinetum sylvestris** Kleist 1929 emend. Matuskievicz 1962, with *P. sylvestris, V. uliginosum, V. oxycoccus, B. pubescens, Frangula alnus, Sorbus aucuparia, S.*

magellanicum, S. angustifolium and *S. fallax* as typical species is a secondary bog community with an eastern distribution indicating recovery after drainage.

Class 2: Scheuchzerio-Caricetea nigrae (Nordhagen, 1936) Tüxen 1937

Characteristic species: *C. nigra, C. panicea, C. rostrata, C. lasiocarpa, E. angustifolium, P. palustris, M. trifoliata, S. fallax, S. lindbergii, S. riparium, S. subsecundum s.l., Drepanocladus exannulatus, D. revolvens, Bryum pseudotriquetrum, P. commune* etc.

The class contains small sedge communities of hollows, floating mats and fens.

Order 1: Scheuchzerietalia palustris Nordhagen 1936

Characteristic species: *C. limosa, D. anglica, R. alba, S. lindbergii, S. subsecundum* s.l.

This order contains the plant communities of hollows, floating mats and transitional mires.

Alliance 1: Rhynchosporion albae Koch 1926

Characteristic species: *D. intermedia, R. alba, R. fusca, S. palustris, D. fluitans, Cladopodiella fluitans, S. cuspidatum, S. majus, S. balticum, S. pulchrum, S. jensenii.*

Differential species for the Caricion lasiocarpae: Oxycocco-Sphagnetea species.

The Rhynchosporion albae is the alliance of hollow and flark communities. The communities usually grow in contact with bog communities and hence are enriched with species from hummocks or ridges. Three associations and the rank-less ***E. angustifolium*** **community** can be identified in this alliance, the latter growing along the margins of pools or flarks. They cover a wide range of nutrient conditions (base rich through acid) but are closely related to the waterlevel. This approach is subject to discussion especially amongst Scandinavian ecologists who prefer a classification on the basis of trophic conditions indicated by different mosses.

Association 1: Caricetum limosae Osvald 1923 emend. Dierssen 1982
Characteristic species: *C. limosa, C. livida, S. palustris.*

The Caricetum limosae is the most frequent plant community of *Sphagnum* and mud-bottom hollows as well as floating mats. It can be subdivided into several subassociations with regard to trophic conditions but only some of them are relevant for bogs. In Norway a **facies of *S. palustris*** and a **facies of *C. livida***, the latter on minerotrophic sites, are developed.

Subassociations

The **subassociation of *S. cuspidatum*** Krisai ex Steiner 1992 is the most common unit of the community in bog hollows all across Europe.

The **subassociation of *S. majus*** Krisai ex Steiner 1992 sometimes replaces the Caricetum limosae Sphagnetosum cuspidati in northern Scandinavia and in the subalpine belt of central Europe.

The **subassociation of *S. angustifolium*** Steiner 1992 has its distribution centre in the more continental parts of central Europe where hollows become rare.

The **subassociation of *S. auriculatum*** Dierssen ex Dierssen and Reichelt 1988 grows on a few sites in south-western Norway, but is common in the blanket bogs of the British Isles where it is restricted to lower altitudes in the nemoral zone.

The **typical subassociation** Osvald 1923 is characteristic of mud-bottom hollows. Mosses are usually absent but some floating specimens may sometimes occur.

Association 2: Sphagno tenelli-Rhynchosporetum albae Osvald 1923 emend. Dierssen 1982

Characteristic species: *R. alba, R. fusca, S. cuspidatum, S. tenellum.*

This association can be found along the margins of hollows usually in close contact with Oxycocco-Sphagnetea communities. *Drosera* species and some dwarf shrubs become more abundant. On eroded peat *Lycopodiella inundata* can take over dominance. Similar to the Caricetum limosae only a few subassociations are relevant for bogs.

Subassociations

The **subassociation of *S. tenellum*** (Osvald, 1923) Dierssen ex Dierssen and Reichelt 1988 is widely distributed in the oceanic and suboceanic parts of Europe including the lower altitudes of the Alps. *S. cuspidatum* replaces *S. tenellum* on wetter places and, in Norway, liverworts like *Cladopodiella fluitans* and *Gymnocolea inflata* represent the moss layer. On the British Isles *R. fusca* can replace *R. alba* in the hollows of raised and blanket bogs.

The **subassociation of *S. majus*** (Sjörs, 1948) Dierssen ex Dierssen and Reichelt 1988 is distributed in the south boreal zone and the Alps where it grows in hollows of raised bogs or on floating mats.

The **subassociation of *S. lindbergii*** Dierssen ex Dierssen and Reichelt 1988 grows in slightly minerotrophic bog hollows or flarks of aapa mires throughout the boreal zone.

The **typical subassociation** (Osvald, 1923) Dierssen and Reichelt 1988, very poor in species, forms small patches in the hollows of bogs and transitional mires.

Association 3: Caricetum rotundatae Fries 1913

Characteristic species: *C. rotundata, E. russeolum, E. medium, D. schulzei.*

This association is restricted to the northern boreal zone. It usually prefers the wet and acid conditions found along the margins of flarks and bog pools or at the base of palsas, although it can sometimes also be found in rich fens.

Other associations of the Scheuchzerio-Caricetea nigrae

There are other fen communities of the class growing in bogs as well. Two examples will be given in this survey. The **Caricetum rostratae** Osvald 1923 emend. Dierssen ex Steiner 1992 (alliance Caricion lasiocarpae in the same order) occurs in bog hollows and erosion channels and the **Caricetum magellanici** Osvald 1923 (order Caricetalia nigrae, alliance Caricion nigrae) sometimes replaces the Caricetum limosae in bog hollows of the subalpine belt.

Chapter two:

Blanket Bogs: An Interpretation Based on Irish Blanket Bogs

Dr Gerry Doyle
University College Dublin, Ireland.

Introduction

The features of European blanket bogs may be explored by reference to the euoceanic Atlantic blanket bogs that are a major feature of the western seaboard of Ireland (Doyle, 1982; 1990).

Blanket bogs - definition

Blanket bogs comprise deep, ombrotrophic peat deposits that cover major portions of the landscape in the oceanic sector of western Europe. Such blanket bogs are clearly separated from other extensive peatland complexes in northern continental Europe, such as aapa mires, which are relatively shallow and influenced to a large extent by relatively mineral-rich water that flows over the surface, especially at snow melt.

In Europe, blanket bogs are found in Ireland, Britain and Norway, where they cover entire landscapes wherever slopes remain less than 15°, and in places, such as parts of the south-west of Ireland, may cover slopes of up to 20°. Typically such blanket bogs occupy coastal plains and inter-montane valleys along the Atlantic seaboards and on plateau areas in the mountain regions subjected to oceanic climates. Blanket bogs in Ireland are of two distinct types (i) Atlantic or western blanket bogs confined to coastal plains below 200 m above sea level, and (ii) mountain blanket bogs found on plateau areas in mountain regions.

On flat terrain, Atlantic blanket bogs may have ombrotrophic domes, but lack the laggs typical of raised bogs (Osvald, 1949). The hummock/hollow complexes that are so characteristic of other European peatland types are poorly developed on the surfaces of these blanket bogs. Blanket bogs are, moreover, drained by shallow surface runnels which feed into drainage channels that dissect the blanket peat mass to varying depths, some flowing over peat, others cutting down into mineral soil. These drainage channels eventually connect with the major river systems in the peatland areas. Blanket bogs are also drained by subterranean systems that are fed through vertical shafts called swallow holes. The surface of these blanket peatlands may be festooned with a variety of pools and peat-based, or mineral soil-based, lakes. There is no uniformity in the distribution of these pools and lakes, reflecting the variation in topography of the underlying terrain and the absence of a predominant slope (which is responsible for the regular alignment of pools in other peatland types in Europe).

Blanket bog distribution in Ireland

Blanket bogs are major landscape features, covering extensive areas. In Ireland (Table 2.1) blanket bogs occupy some 11.6% of the entire landscape (Hammond, 1979), with Atlantic blanket bogs accounting for 4.3%, and mountain blanket bogs, the most extensive of the Irish peatland types, making up 7.3%. Other peatland types in Ireland are less extensive: minerotrophic fens, which are widely distributed and found wherever drainage conditions promote waterlogged conditions (Ó Críodáin and Doyle, 1994), account for 1.3% of the landscape; raised bogs, which typify the midland areas of Ireland and are characteristically formed in relatively low rainfall areas (>1200 mm), contribute 4.1% of the landscape.

Table 2.1. Extent of different peatland categories in Ireland.

	Percentage cover (%)	Extent (ha)
Total peatlands	***17.0***	***1,312,450***
Fens	1.3	101,810
Raised bogs	4.1	313,830
Atlantic blanket bogs	4.3	330,860
Mountain blanket bogs	7.3	565,950
Total blanket bogs	11.6	896,810

Source: Data extracted from Hammond (1979).

Climatic conditions in Irish Atlantic blanket bog regions

Blanket bogs are formed under the euoceanic climatic conditions of western Europe. In general terms such conditions may be summarised as mild, wet and windy (Table 2.2), with precipitation exceeding 1200 mm per annum and the number of rainy days exceeding 200 each year.

Table 2.2. Climate of an Atlantic blanket bog area at Glenamoy, north-west Mayo (cf. Doyle, 1982). The data is based on meteorological records for the period 1959 to 1973; open pan evaporation data refers to the period 1961 to 1973.

Temperatures	
Mean maximum	12.3°C
Mean minimum	6.1°C
Mean annual	9.2°C
Precipitation	1360 mm
Evaporation	603 m
Relative humidity	85%
Number of rainy days	264
Number of frosts	66
Days with persistent snow	7
Wind speed	
Mean	4.5 m s^{-1}
Maximum	44.0 m s^{-1}

The important features of climate demonstrated by these data include:
Mildness, as demonstrated in cool summer temperatures, relatively high winter temperatures, low number of frosts, and the virtual absence of snow cover.
Wetness, as indicated by high precipitation, high number of raindays, and constantly high relative humidity.
Excess of precipitation over evaporation, in this case in the ratio of 1360 mm:603 mm (or approximately 2:1) a ratio typical for blanket bog areas in general. Examination of meteorological data indicate that, in Ireland, blanket bogs occur in areas where precipitation exceeds evaporation by more than 700 mm. In Europe generally, blanket bogs are confined to those areas where precipitation deficits rarely occur (<1 in 10 years).

Soil development in Atlantic blanket bog areas

Atlantic blanket bogs in Ireland develop on acidic geological foundations, on gneisses, schists, quartzites, Carboniferous and Old Red Sandstones and granites. These basal rocks are extensively covered by thick deposits of glacial till. In north-west Mayo, which remained unglaciated during the most recent Midlandian Cold Stage, the landscape is covered by drift from the earlier Munsterian Cold Stage, which has been moulded by weathering and erosion to present a characteristically undulating landscape without obvious, newly-formed glacial features. As a result this region supports an apparently uniform cover of blanket peat that rolls over the landscape, uninterrupted by rocky outcrops. In contrast, in Connemara, County Galway, which was glaciated during the last or Midlandian Cold Stage, the relatively young landscape is more rugged, so that the blanket bog is less uniform with numerous rocks outcropping through the peat blanket.

Atlantic bog development in Ireland

The timing of blanket peat initiation in Ireland was not uniform, and the causal factors in initiation were varied. Peat initiation within the Atlantic blanket bog areas commenced in waterlogged depressions at the Boreal/Atlantic transition (Jessen, 1949), somewhere about 8000 BP, with *Phragmites* peat indicating minerotrophic conditions. The onset of mountain blanket peat in the Pennines was also put at the Boreal/Atlantic transition at 7500 BP (Conway, 1954).

Evidence from woody layers in blanket bog on sloped ground near Glenamoy in County Mayo indicates that the initiation of blanket peat proper probably commenced around 7000 BP as evidenced by preserved pine stumps (cf. Doyle, 1982; 1990) that are located at mineral soil/peat interfaces (e.g. at Bellanaboy Bridge, County Mayo ^{14}C dated at 7110 BP). Birks (1975) found analogous dates for pine stumps in mountain blanket peats in Scotland (7471 to 7165 BP). This pine period, found in both Ireland and Scotland, was contemporaneous with the Early Atlantic Pine Forest described from peatlands in continental Europe (Munaut, 1966; Vogel *et al.*, 1969; Munaut and Casparie, 1971; Casparie, 1972). Subsequent spread of blanket peat must have been promoted by increasing wetness of climate, since Lamb *et al.* (1966) estimated that average rainfall about 7000 BP was 11% greater than at the present day (cf. Taylor, 1983).

In places, there is a second layer of pine stumps separated from the mineral soil by more than 20 cm of peat. A stump from this upper layer at Bellanaboy bridge dated 4290 BP is contemporaneous with the Late Atlantic Pine Forest described from continental Europe (Munaut, 1966), although predating major pine woodland phases that have been described from some Irish midland raised bogs (McNally and Doyle, 1984).

O'Connell (1990), in a palynological study of blanket bog in Connemara, distinguished between basal peat and clearly recognizable blanket peat with a pollen assemblage typical of modern blanket bog above that level. The transition to recognizable bog vegetation occurred at 6000 BP.

In other locations in the west of Ireland, however, it is clear that blanket bog expansion followed Neolithic settlement. For example, Caulfield (1978; 1983) detailed the spread of blanket peat over Neolithic and Bronze Age farm settlements in Belderg, north Mayo (the townland adjacent to Glenamoy), with preserved pine stumps in that area dated at 4220, 3835 and 3200 BP, revealing a differential development within a confined geographical area. In the adjacent Glenamoy valley (which includes the Bellanaboy site mentioned above), there is no evidence of pre-bog settlement, revealing yet further variations in the initiation and spread of blanket bog in a restricted geographical area. O'Connell (1990) through his palynological work confirmed that the widespread initiation of Atlantic blanket peat followed human interference post 4000 BP.

Based on this evidence a general picture emerges for the development of Atlantic blanket bog systems in the west of Ireland. Initially there were foci of peat development in depressions in coastal plains, from which blanket peat spread outwards. Similarly there would have been foci for development in the upland areas where the precipitation:evaporation ratios were favourable. Such development has been described for Norwegian mountain blanket bogs where peat initiation commenced both on hill tops and in depressions at lower elevation, with subsequent spread upslope and downslope to give a relatively uniform peat blanket (Solem, 1986; 1994). The intervention of man would have promoted blanket peat development in areas lying at a distance from the initial focal points.

Clearly, there were several factors influential in this development including:

1. climatic deterioration,
2. soil deterioration and podzolisation, and
3. human impact.

Peat characteristics

Atlantic blanket peat varies in depth from 1.5 m to 7 m with an average of 3.5m. The peat is highly humified (5 to 6 on Von Post's Scale). It is typically waterlogged (900% of dry weight) owing to the excess of precipitation over evaporation in the area and the impervious nature of peat (Burke, 1969). The peat has low thermal conductivity so that relatively high soil temperatures are typical in this mild climatic region. The absolute minimum at 10 cm over a 15-year period at Glenamoy in County Mayo, was 0.9°C, while the annual mean at 5.4m was 10.3 ± 0.4°C. The ombrotrophic peat has a low ash content, ranging from 1.2% to 2.4% of dry weight. Low amounts of nitrogen, phosphorus and dust-derived elements such as aluminium are typical, while there are relatively high amounts of marine-derived elements such as chlorine, magnesium and sodium.

Relatively high pH values are typical of this peat type. Sparling (1967) quoted an average pH of 4.2 based on 187 estimates, while O'Sullivan (1974) quoted pH 4.5 based on 48 measurements in a wet July period. The pH varies, however, depending both on the subhabitat and the climatic conditions in the period prior to measurement (Doyle, 1982). In exceptionally dry summer periods (which do occur sometimes in the west of Ireland) relatively low pH values have been recorded. For example, in the dry summer periods of 1971 and 1973 the pH of low hummocks measured 3.44 ± 0.13 (n = 31), while that in adjacent hollows measured 3.64 ± 0.15 (n = 44) (Doyle, 1982). Differential drying in subhabitats influences pH and the mobility of plant nutrients and plant toxins (such as aluminium) and may play an important role in determining the distribution of plant species on the bog surface (Doyle, 1982).

In many places on Atlantic blanket bog the peat surface is noticeably compacted, a result of the impacts of the frequent fires and constant grazing that were a feature of these areas in historic times, and indeed have been features of these areas since Neolithic times (Moore *et al.*, 1975; Doyle, 1982).

Blanket bog vegetation mosaics (blanket mires)

In interpreting blanket peatlands, it is important to recognize the ombrotrophic nature of these systems. At the same time it is necessary to understand that the predominantly ombrotrophic areas are associated with a set of habitats and vegetation types that constitute a typical mosaic or blanket mire. Such a mosaic reflects (i) the extensive nature of these peatlands, (ii) the morphology of the terrain, (iii) the differential development of peat as related to slope, and (iv) the development of pools, lakes and drainage systems that dissect the bogs.

The ombrotrophic parts on deep peat deposits support vegetation that is classified (using European phytosociological nomenclature, cf. White and Doyle, 1982) as:

Association:	Pleurozio purpureae-Ericetum tetralicis
Alliance:	Calluno-Sphagnion
Order:	Eriophoro vaginati-Sphagnetalia papillosi
Class:	Oxycocco-Sphagnetea

The association comprises the ombrotrophic vegetation of low altitude blanket bogs along the western seaboards of Ireland, Britain and Norway (cf. Doyle and Moore, 1980). The alliance contains the vegetation of deep ombrotrophic bogs of Western Europe; the order, the vegetation of deep ombrotrophic bogs of oceanic areas worldwide; and the class, the vegetation of deep ombrotrophic bogs together with that of wet heathlands worldwide.

Atlantic blanket bog vegetation has been classified under different titles by British workers. The most important synonyms include the Tricoporeto-Eriophoretum (McVean and Ratcliffe, 1962) and NVC-M17 *Scirpus cespitosus - Eriophorum vaginatum* blanket mire (Rodwell, 1991).

In early works, Irish Atlantic blanket bog vegetation was placed together with the vegetation of ombrotrophic surfaces of raised and mountain blanket bogs in the *Pleurozia purpurea - Erica tetralix* association (Braun-Blanquet and Tüxen, 1952), but subsequently separated out from these other types by Moore (1968), who distinguished three distinct associations for the ombrotrophic surfaces of the major Irish types:

Atlantic blanket bog:	Pleurozio purpureae-Ericetum tetralicis
Mountain blanket bog:	Vaccinio-Ericetum tetralicis
Raised bog:	Erico-Sphagnetum magellanici

The original name used by Braun-Blanquet and Tüxen (1952) was redefined by Moore (1968) and subsequently used to denote only the vegetation of Atlantic blanket bog.

The distinction of the three major Irish bog types is based on both floristic and other grounds (see Table 2.3).

Table 2.3. Distinguishing features among the major Irish ombrotrophic bog types.

Atlantic blanket bog	Mountain blanket bog	Raised bog
	Floristic distinctions	
Schoenus nigricans	*Vaccinium myrtillus*	*Andromeda polifolia*
Pleurozia purpurea	*Empetrum nigrum*	*Vaccinium oxycoccus*
Campylopus atrovirens	*Rubus chamaemorus*	
Pinguicula lusitanica	*Andromeda polifolia*	
Polygala serpyllifolia		
Pedicularis sylvatica		
Potentilla erecta		
Molinia caerulea		
	Altitude	
Below 200 m a.s.l.	Above 200 m a.s.l.	
	Peat pH	
4.2	3.4	3.4
	Maximum depth	
7 m	7 m	15 m
	Peat type	
Mainly cyperaceous	Mainly cyperaceous	Stratified: fen peat, *Eriophorum* peat, wood peat, *Sphagnum* peat

Vegetation characteristics of Atlantic blanket bogs

Atlantic blanket bog vegetation in Ireland is graminoid-dominated, with *Schoenus nigricans* and *Molinia caerulea* as the major components. *S. nigricans* thrives where peat is compacted and permanently waterlogged, while *Molinia* is prominent on relatively dry slopes. Unlike blanket bogs elsewhere in western Europe, *Eriophorum vaginatum* does not play a dominant role, indeed it appears to be replaced in importance in the Irish situation by *S. nigricans*. While *Scirpus cespitosus* is present it is generally confined to shallow and well-drained peats within blanket bog complexes. Atlantic species such as *Narthecium ossifragum* are obvious, but do not grow with the vigour that is typical of this species on some raised bogs in the western parts of continental Europe.

Ericoid dwarf shrubs are present but are generally depauperate by comparison to such shrubs growing in other bog types. Both *Calluna vulgaris* and *Erica tetralix* are present but owing to almost permanent waterlogging of the surface peat (Gimingham, 1960; Bannister, 1966) remain in a reduced state.

The prominent *Sphagnum* species include *S. papillosum, S. capillifolium* and *S. magellanicum*, while *S. fuscum* forms isolated hummocks and the Atlantic *S. imbricatum* is found in some places. In general, the cover of *Sphagnum* on these blanket bogs is sparse by comparison with their contribution to the vegetation on ombrotrophic surfaces of raised bogs.

An obvious feature of many areas of Atlantic blanket bog is the abundance of the algal aggregate *Zygogonium ericetorum*, which plays an important role in maintaining the surface peat in a permanently waterlogged condition (Doyle, 1982).

The characteristic species combination of Atlantic blanket bog vegetation in Ireland (i.e. the species that are useful in the phytosociological definition of the vegetation) include:

Association character species

Schoenus nigricans
Pleurozia purpurea
Campylopus atrovirens

Differential species
(from raised and mountain blanket bogs, species that are typical in Ireland of wet heathland situations)

Pedicularis sylvatica
Pinguicula lusitanica
Polygala serpyllifolia
Potentilla erecta

Alliance/order/class species

Calluna vulgaris
Erica tetralix
Myrica gale
Eriophorum vaginatum
Molinia caerulea
Narthecium ossifragum
Drosera spp.
Campylopus paradoxus
Sphagnum capillifolium
Sphagnum papillosum
Sphagnum subnitens
Sphagnum tenellum
Cladonia impexa
Cladonia uncialis

Notable absences

Some species that are generally considered typical of deep peatlands in Europe, namely *Vaccinium oxycoccus* and *Andromeda polifolia*, are absent from ombrotrophic surfaces of Irish Atlantic blanket bogs. While *V. oxycoccus* is found within the blanket bog area, it is confined to obvious surface depressions in which schwingmoor habitats have developed (Doyle and Foss, 1984). Bog rosemary (*A. polifolia*), which is typical of both raised and mountain blanket bogs in Ireland, is not to be found as a native of Atlantic blanket bogs (Foss *et al.*, 1984).

Associated minerotrophic vegetation types

Apart from the vegetation that is typical of ombrotrophic surfaces, the blanket mire systems support a variety of drainage features: together these support vegetation types that form a characteristic mosaic (Doyle, 1990b). The characteristic vegetation types include the following, designated by their general titles together with their phytosociological classification (using the nomenclature of the Braun-Blanquet system, cf. White and Doyle (1982), for a classification of Irish vegetation):

Vegetation of lakes and pools in Atlantic blanket bog areas
  Littorelletea uniflorae
    Isoeto-Lobelietum
Vegetation of shallow, relatively still waters in streams within the blanket bog areas
  Phragmitetea
    Eleocharito-Hippuridetum
    Caricetum paniculatae
Rheotrophic mire vegetation in drainage channels of Atlantic blanket bogs
  Scheuchzerio-Caricetea
    Sphagno-Juncetum effusi
    Carici nigrae-Juncetum articulati
Shallow depressions on Atlantic blanket bog surfaces
  Oxycocco-Sphagnetea
  Scheuchzerietalia palustris
    Sphagno tenelli-Rhynchosporetum albae
Vegetation of seepage areas at heads of drainage channels
  Oxycocco-Sphagnetea
  Scheuchzerietalia palustris
    E. angustifolium-Sphagnum recurvum v. *mucronatum* community
Vegetation of isolated schwingmoors in the Atlantic blanket bog areas
  Oxycocco-Sphagnetea
  Scheuchzerietalia palustris
    Scheuchzerietum
Heathland vegetation at drain edges and on rock outcrops within the blanket bog areas
  Calluno-Ulicetea
    Calluno-Ericetum cinereae

Vegetation dominated by *M. gale* - in seepage areas near drainage channels
Franguletea
Myricetum gale
Willow scrub in deep drainage channels within the blanket bog areas
Franguletea
Osmundo-Salicetum atrocinereae

Peat islands

An insight into the ecology of the ombrotrophic parts of Atlantic blanket bogs in Ireland is gained from an examination of the vegetation of peat-based islands that are a feature of larger pools and lakes within the blanket bog areas (Doyle, *et al.*, 1987). These islands are, owing to their distance from shore, isolated from two of the major ecological impacts on open ombrotrophic surface - fire and grazing. As a result there is a marked contrast between the vegetation of the islands and that of open bog areas. For example, *Schoenus nigricans* is no longer found, there is increased vigour of *Calluna vulgaris* and *Molinia caerulea*, and there is a marked increase in species numbers (42 spp m^{-2} as opposed to 21 spp m^{-2} on the open surface). For example, a number of species that are epiphytic on old *Calluna* stems are confined to island locations. The regular occurrence of *Juniperus communis*, and of centrally placed *Racomitrium lanuginosum* hummocks topped with *Empetrum nigrum*, are characteristic of islands. Such moss hummocks with *Empetrum* were once a notable feature of some areas of blanket bog such as Sheskin Lodge (Osvald, 1949), but are now almost unknown in open bog situations. In addition, the peat substrates of islands and of mainland sites are different - island peat is loose and relatively well-drained, while that of the mainland is compacted, algal-covered and waterlogged. Since there is no equivalent vegetation on the mainland shores of the pools and lakes containing islands, it is evident that the observed differences in flora and vegetation are not simply a reflection of enhanced drainage, but are indicative of the influence of long-term fire and grazing, which must therefore be considered as major overriding influences on the ecology of such systems.

Chapter three:

Tropical Bog Forests of South East Asia

Dr John O. Rieley[1], Dr Susan E. Page[2] and Dr Peter A. Shepherd[3]
[1]University of Nottingham, UK; [2]University of Leicester, UK; [3]East Midlands Environmental Consultants, UK.

Introduction

Peat is distributed worldwide from polar to tropical regions in both hemispheres wherever suitable climatic and edaphic conditions prevail (Table 3.1). Lowland tropical peats originate from rain forest trees and, consequently, their physical and chemical properties and vegetation communities differ greatly from their temperate counterparts. Approximately 12% of the global peatland area occurs in humid tropical zones especially South East Asia, mainland Asia, the Caribbean, Central America (central and southern), and South America (Immirzi and Maltby, 1992); most of this peat is sub-coastal although a small amount is located in mountainous districts of Central Africa, Amazonia and South East Asia.

Table 3.1. Estimated maximum pre-disturbance peatland area in different continents (ha x 10^6).

Continent	Maximum area	% of total
Africa	4.86	1.1
Asia	43.61	10.7
Central America	2.61	0.6
North America	171.13	41.9
South America	6.17	1.5
Europe & former USSR	180.25	44.1
Pacific	0.28	0.1
TOTAL	**408.91**	**100.0**
TROPICAL	**49.71**	**12.2**

(modified from Immirzi and Maltby, 1992)

Few detailed investigations have been carried out on the genesis, development, ecology and environmental importance of tropical peats or the impact of man upon them. Tropical peatlands are being developed rapidly often with disastrous consequences for both the resource and the human communities that become dependent upon them. World attention is now focused upon peatlands globally because they are a major store of carbon which has implications for greenhouse gas fluxes and climate change. Although the proportion of carbon dioxide entering the atmosphere from peatland degradation and destruction may be small compared with other fossil sources, it is increasing, especially in the tropics (Armentano and Verhoeven, 1988).

Distribution, location and human habitation

In South East Asia peatlands occur near the coasts of Sumatra, Kalimantan, Irian Jaya, Papua New Guinea, Brunei, Peninsular Malaysia, Sabah, Sarawak, south-east Thailand and the southern islands of the Philippines (Immirzi and Maltby, 1992) (Table 3.2). Indonesia possesses the largest area of peat in the tropics (up to 27 Mha), although the estimates for the extent of this resource vary considerably (Table 3.2). Of the 2.5 Mha of peat in Malaysia 25% occurs in Peninsular Malaysia and 75% in Sarawak (Anon, 1987). There are extensive swamps in the lowlands of Papua New Guinea, some of which are peat-forming; the higher estimate of 2.9 Mha (Table 3.2), however, includes 2.2 Mha of peat in the highlands (Wayi and Freyne, 1992). In Thailand much of the peatland has been converted to agriculture; that which remains undrained occurs mostly around the Gulf of Thailand (Urapeepatanapong and Pitayakajornwute, 1992; Vijarnson, 1992).

Table 3.2. Estimate of undisturbed peatland area in South East Asia.

Country	Area (ha x 10^3)
Brunei	10
Indonesia	17000-27000
Malaysia	2250-2730
Papua New Guinea	500-2890
Philippines	104-240
Thailand	68
TOTAL	**19932-32938**

(based on data from Immirzi and Maltby, 1992)

Chronological relationships of tropical peatlands

Radiocarbon dating of peat samples from Sarawak (Wilford, 1960; Anderson, 1964), Central Kalimantan (Sieffermann *et al.*, 1988), Sumatra and West

Kalimantan (Diemont and Supardi, 1987) show significant variations in the dates of origin from 800 to 4575 BP for coastal and basin peats. The fastest rate of accumulation during this period was 20 cm per 100 years in the early stages of development. Recent dating has confirmed the greater age of high peat in Central Kalimantan which is over 9000 BP, the rate of accumulation of which varied between 14 and 50 cm per 100 years but was fastest between 9600 and 8400 BP.

Ecological and environmental characteristics of tropical bog forests

Climate characteristics

In South East Asia, lowland peatlands occur in regions of wet, humid, equatorial climate with annual rainfall of 1500 mm to more than 5000 mm; relative humidity usually exceeds 90%. Daily temperatures are high (30-34°C) and the diurnal range is greater than seasonal variation. Persistent cloud limits the diurnal temperature range to <11°C and very high temperatures (i.e. in excess of 37°C) are infrequent. Some extensive areas of peatland, however, experience a slightly seasonal climate where evapotranspiration exceeds precipitation during the dry season. Peat swamps and their associated forests exert a major influence on the meso- and micro-climate of large areas of landscape.

Substrate

Tropical peat swamps have an organic layer greater than 50 cm thick (Andriesse, 1988), with organic content higher than 75% and pH usually less than 4. Depth of peat varies from shallow (50-100 cm) to deep (greater than 3 m) (Radjagukguk, 1992); depths of up to 16 m have been recorded in Sumatra and 17 m in Sarawak (Anderson, 1983). Their surface is often markedly convex and not subject to riverine flooding although they may be inundated with rainwater in the wet season; thus tropical peat swamps are ombrogenous (Anderson, 1964). Lowland tropical peat is relatively heterogeneous, consisting of slightly or partially decomposed debris of the former forest. Well preserved tree trunks, branches and coarse roots are found within a matrix of dark brown amorphous organic material. The overall properties of tropical peat are a result of many factors including wood content, degree of decomposition, mineral content, stratification and compaction, which influence bulk density, hydraulic conductivity and water holding capacity (Driessen, 1977; Anderson, 1983; Esterle *et al.*, 1992). These peats are composed mainly of fibric materials and their mineral content is extremely low (loss on ignition usually >90% (Table 3.3); sometimes as high as 99%) (Anderson, 1983); higher ash contents result from sediment deposition during floods. There are few comparative data on the chemical composition of undisturbed tropical peats, although some limited information is available on 'exchangeable', 'soluble' and 'total' nutrients in Malaysian and Indonesian peats (Andriesse, 1974; Driessen, 1977; Anderson, 1983; Ahmad-Shah and Soepadmo, 1992; Ahmad-Shah *et al.*, 1992; Kyuma *et al.*, 1992).

Peat types of South East Asia

Three major categories of lowland tropical peatlands have been described based upon their topographical location, mode of formation and maximum age of the peat deposits (Sieffermann *et al.*, 1988, 1992; Rieley *et al.*,1992). These are (i) coastal peatlands of the maritime fringe and river deltas; (ii) basin and valley peatlands of river valleys; and (iii) high peats of low altitude watersheds. Peatlands in the first category are situated inland of mangrove swamps at, or only slightly above, sea level (1-2 m a.s.l.); the peat is shallow (0.5-2 m) and thought to be still accumulating, the surface of the bog plain is almost flat. Basin and valley peatlands occur further inland at slightly higher altitudes (5-20 m a.s.l.) than coastal peatlands with which they are usually contiguous; their water table is permanently high and the peat may be up to 20 m deep (Whitten *et al.*, 1987). High peats, which have only been described from Central Kalimantan (Sieffermann *et al.*, 1988; Rieley *et al.*, 1992; Sieffermann *et al.*, 1992) overlie tropical podzols and alluvial clays; their depth can exceed 10 m (Rieley and Page unpublished data) at the present day but it is thought that these peatlands are degrading and they may have been deeper in the past.

Table 3.3. Range of total mineral contents of surface peat samples collected along a 25-km transect through peat swamp forest in Sarawak.

Characteristic	Content
pH	3.26 - 3.58
ash (%)	0.66 - 4.28
C (%)	39.69 - 48.48
N (%)	1.48 - 2.13
P (ppm)	203.0 - 520.0
K (ppm)	194.0 - 488.0
Na (ppm)	172.0 - 340.0
Ca (ppm)	170.0 - 722.0
Mg (ppm)	544.0 - 1197.0

(based upon data from Anderson, 1983).

Vegetation

Peat swamp forest is less diverse than dry land rain forest (lowland dipterocarp), although it is never the less an important reservoir of biodiversity (Anderson, 1963; Silvius *et al.*, 1984). The species composition and vegetation types of peat swamp forest are not uniform across the South East Asian region. Many of the plant species are restricted to this habitat and some endemic trees have been described from Malaysia and Thailand (Ng and Low, 1982). There is little information on upland peat swamps in South East Asia except that they support montane forest, grass, sedge and *Pandanus* spp.

The natural vegetation of lowland peat swamp varies from a mixed swamp forest near the margins (with up to 240 spp ha^{-1}) to pole forest in the interior (lower tree diversity, dominated by one or a few species, such as *Shorea albida* in parts of Sumatra and Borneo, with an average of only 30 to 55 other spp ha^{-1})(Anderson, 1963; Silvius *et al.*, 1984). Peat swamp forest vegetation is a catena of forest types which replace each other from the swamp perimeter to the centre in response to changes in nutrients, depth of peat and waterlogging. The principal feature that characterises the catena is a replacement of higher forest (36-42 m) by a denser, lower 'pole forest' in which the diameter of the tree trunks is rarely greater than 30 cm and the canopy no more than 20 m high (Anderson, 1963; 1964). The centre of some peat swamps may be very dry (Anderson, 1983) and dominated by an open heathy forest (padang) in which most of the trees are less than 15 m in height. The trees of the low pole and padang forests show an increasing tendency to xeromorphism and chlorosis of foliage reflecting both the decreasing fertility of the peat substrate and adaptation of the trees to the stressful environmental conditions.

Hydrology

Pristine tropical peat swamps have a water table close to, or above, the surface throughout the year (Driessen, 1977). The water table fluctuates according to ambient rainfall. It is higher in the wet season when pools form on the forest floor and large areas may be flooded; the surface is seldom inundated during the dry season when a combination of lateral seepage, surface evaporation, evapotranspiration and low rainfall combine to lower the water table to depths of 50 cm or more below the peat surface (Rieley *et al*,, 1996; Takahashi and Yonetani, 1996). Surface run-off along anastomosing water tracks transfers water from the centre to the perimeters of these peatlands; this is particularly evident during the wet season. The balance between water inputs, outputs and storage in tropical peat swamps is little understood and hydrological budgets have not yet been published for them.

Carbon fixation, flux and storage

It has been accepted for a long time that peatlands are major carbon sinks (Moore and Bellamy, 1974; Bohn, 1978); it is only recently, however, that carbon fluxes between the atmosphere and peatlands have been systematically investigated and quantified. There is a lack of information on the carbon store in and fluxes from tropical peatlands.

Estimates of the total carbon content of Indonesia's peat swamp forests, based upon information on bulk density of different peat types, total areas, mean depths, and percentage carbon contents, range from a minimum of 16 x 10^9 tonnes (Sorensen, 1993) to a maximum of 39 x 10^9 tonnes (Immirzi and Maltby, 1992). It is unclear whether or not tropical peatlands are active carbon sinks because detailed investigations of their carbon flux have not been carried out. Immirzi and Maltby (1992), however, suggested a pre-disturbance sequestration rate of between 41.5 and 93.4 x 10^6 tonnes carbon per year for all tropical peatlands, whilst Sorensen's (1993) data suggest a lower value of between 18 and 54 x 10^6 tonnes carbon per year. These estimates, however, are subject to major errors. In order to obtain more accurate data on carbon sequestration and storage by tropical peatlands it is essential to know the exact area of different types of peat swamp forest and to obtain more, accurate information on peat depths, rates of accumulation, bulk density, carbon content and carbon flux.

Discussion

In their natural state, tropical peatlands contribute to biodiversity and perform important environmental and landscape functions. In the face of increased development and human settlement, there is an urgent need to demonstrate the direct and indirect environmental and socio-economic values of this ecosystem. Peat and associated freshwater swamp forests are important water catchment and control systems (Boelter, 1964); they absorb and store rain and river flood water during wet periods. As a result, they reduce river flood water peaks and provide water during dry periods (Andriesse, 1988). Coastal peat swamps also act as a buffer between salt and fresh water hydrological systems. The forests are a source of several locally important products, e.g. building materials, rattans, medicines, latex, resins, tannins, dyes, fruit and fish; owing to their waterlogged nature and inaccessibility swamp forests are not used for shifting agriculture. Peat swamp forests are increasingly recognised as a valuable timber resource within the South-east Asia region, especially as more of the dry land forests are cut-over or converted to agriculture. Commercially important tree species include *Shorea* spp. (Meranti), *Gonystylus bancanus* (Ramin), *Calophyllum* spp. (Bintangor) and *Dryobalanops rappa* (Kapur). In Indonesia many peat swamp forests have the potential to be managed sustainably, as extraction techniques employed are low-technology, utilising mainly manual labour, which cause minimal damage to the ecosystem.

Peat swamp forest is of only marginal use for growing crops, yet conversion to agriculture together with non-sustainable logging, are major and increasing threats. In Malaysia, under current practice, it is predicted that all of Sarawak's peat swamp forest will be logged before the year 2000 (Anon, 1987); in Peninsular Malaysia, large areas of shallow peat have been converted to agriculture of plantation oil palm and rubber or perennial crops (e.g. pineapple). In Indonesia, there is less intensive cultivation of plantation crops but, increasingly, the forest on peat of less than 3 m depth is being cleared, especially in East and South Kalimantan and East Sumatra, often in association with transmigration settlement of people from the overpopulated islands of Indonesia. Peat over 3 m thick cannot easily be brought into agricultural use, as it is extremely acid and infertile and, following conversion, may be subject to flooding. Drainage and cultivation of peat soils leads to rapid oxidation and shrinkage of the peat which may ultimately reveal potential acid sulphate substrates of very high acidity, toxicity, and therefore of very low agricultural value.

It is essential to carry out audits of the biodiversity and natural functions of tropical peat swamp forest in as many regions as possible in order to prepare guidelines for their environmentally sustainable management within the context of integrated landscape planning policies.

Acknowledgements

We gratefully acknowledge the British Council, the South East Asia Rain Forest Research Programme of the Royal Society, the Conservation Foundation and Earthwatch for their financial support.

Part Two

The Peat Archive

An Introduction

Dr Richard Hingley
Historic Scotland, UK.

This session of the conference examines the immense potential of wetlands as a source of information about the past. Some of the chapters and the topics discussed overlap to an extent with the *Mires as Cultural Landscapes* session of the Convention (see Part five). Both sessions highlighted the considerable importance of wetlands as a historical archive.

The chapters address a wide range of topics regarding information that can be obtained through the study of peat. Dr Buckland and Ms Whitehouse address the evidence that plant and animal remains can provide about the past whilst Dr Charman discusses the value of palaeoecology. Dr Blackford outlines the evidence provided by studies of volcanic dust (tephra), while Dr Jones discusses the potential of the wetlands as an archive of evidence for the pollution of the environment from the Industrial Revolution through to the present day. Dr Mackay provides a case study of the history and causes of blanket mire erosion from the Iron Age to today.

Two of the chapters focus on aspects of wetland conservation. Dr Coles argues for an integrated approach to wetland conservation, including both the archaeological/palaeoecological dimension and the nature conservation interest. Dr Charman highlights the requirement for a protection and conservation policy for palaeoecological resources within British wetlands.

Dr Charman's presentation led to a discussion of how best to preserve palaeoecological wetland resources. Archaeological structures in wetlands can be protected through their scheduling under the Ancient Monuments and

Archaeological Areas Act 1979. However, positive management is often required because scheduling does not protect fragile organic remains from desiccation (see Barber and Crone, 1993 for a case study). It was argued by some of those present that further consideration needs to be given to the possibility of using nature conservation designation to protect and conserve palaeoecological resources.

As part of the provision of enhanced protection for sensitive wetland information in Scotland, Scottish Wildlife Trust, Scottish Natural Heritage, the National Museums of Scotland, the Royal Commission on the Ancient and Historical Monuments of Scotland and Historic Scotland are co-operating to compile databases on archaeological and palaeoecological evidence from raised bogs. It is hoped that this information will be built into a planning and education resource for the Scottish wetland archive. The databases should assist with the process of integrating these aspects into a broader nature conservation strategy for peatland conservation.

Many of those present at the Convention appeared to agree that the future aim should be a fully integrated approach to wetland conservation.

Chapter four:

Ancient Past and Uncertain Future

Professor Bryony Coles
University of Exeter, UK.

Introduction

Sir Harry Godwin, pioneer of pollen analysis and the study of vegetation change, first introduced the idea of peat as an archive, or source of information about the past. In one of his two admirable autobiographical surveys, appropriately entitled *The Archives of the Peat Bogs*, he describes the development from the 1930s onwards of his own understanding of the wealth of evidence available. Parallel with this was a growing awareness of the threats to peatlands. Godwin was a strong advocate of the need for conservation, writing that 'On both ecological and archival grounds the case for conservation ... must be undeniable. Nor is there much time to be lost.' (Godwin, 1981 p.216). The title of Godwin's other autobiographical work, *Fenland: Its Ancient Past and Uncertain Future* (1978) sets a similar tone of appreciation of the record contained in wetlands and concern for its preservation. It is the intention of the present paper to combine the same two strands and to demonstrate how, some years on from Godwin's work, the archive can be seen to be increasingly valuable and at the same time increasingly threatened.

The fact that peat holds information about the human past is well-known to a number of conservation bodies (e.g. O'Connell, 1992; Scottish Natural Heritage, 1995), and the Scottish Wildlife Trust has emphasised the value of the information by including the Peat Archive as a section in the present publication and in the conference which preceded it. An examination of the significant characteristics of the archaeological evidence from peatlands may reinforce this understanding, and perhaps make some contribution to the long-term success of conservation measures. Examples will be taken from a range of freshwater wetlands including, but not restricted to, raised bogs.

Ancient past: the wetland archive

Archaeologists are concerned with objects that have survived from the past. From dry land contexts, the evidence is primarily inorganic, for example from a ploughed field a scatter of small flint chips which typologically can be assigned to the early post-glacial period. From a contemporary wetland context, similar flint microliths might be found, but given anaerobic, waterlogged conditions organic evidence may also be preserved. The flints can be seen to be components of a larger object, stuck with birch bark gum into a wooden shaft to make a projectile point, and the point itself lodged along with several others in the hindquarters of a large elderly bull auroch, whose skeleton survived in the peat of a former pool near Prejlerup in northwest Zealand (Coles and Coles, 1989 p. 95). The quality of the wetland evidence leads to a better understanding of the sparse dry land record, in this instance allowing archaeologists to interpret the scattered microliths from the plough-soil as the last traces of one or more Mesolithic projectile points. Their context, however, is lost and it is unlikely that the number of weapons or the nature of the quarry could be identified.

Archaeologists are also concerned with structures. In dry land contexts, the evidence consists largely of post-holes, bedding trenches, ditches and earthworks. The relationships of these to each other and their associations with artefacts allow for relative dating, and reconstruction of the sizes and sequences of buildings. But it is often difficult to identify individual structures with certainty, or to determine how many buildings were in contemporary occupation. Cunliffe's recent excavations at Danebury illustrate well, how a thorough research project can recover the broad outlines of settlement history whilst lacking the evidence necessary for a detailed and precisely dated reconstruction. On the well-drained Hampshire chalklands, only a small fraction of the original record has survived (Cunliffe, 1984 pp.173-189).

A wetland context may preserve a larger fraction of the evidence for past structures and settlements than is commonly found on dry land. Where people built their houses along the shores of former lakes, posts and floors and hearths may survive; the posts admittedly truncated but sufficient to allow for detailed reconstruction on the basis of dendrochronological analyses. On the northern shore of Lake Neuchâtel in Switzerland, many prehistoric settlements are known, one of which at Hauterive-Champréveyres was recently excavated ahead of motorway construction. The plan of the site shows a mass of posts (see Fig. 4.1), many of which have been dated by dendrochronology - that is, the last year of growth of the parent tree has been determined. From this evidence, together with the evidence of hearth and artefact distribution, it has been possible to interpret the site history in some detail (see Fig. 4.2). Building began in 3810BC, with one large house set up early that year and another five houses before the year was out. In 3807 and 3804BC smaller shelters went up, and in 3801BC a solid oak fence or palisade partially enclosed the village. More small shelters followed. From the first year of building until 3793BC, repairs were a constant feature of village life; the lack of activity thereafter suggests that by 3790 BC at the latest the village had been abandoned (Egloff, 1989). It will be evident from the plan that some posts remain unaccounted for; mostly these are species other than oak, and they can not yet be dated by dendrochronology although rapid developments are being made in this field (e.g. Billamboz *et al.*, 1992).

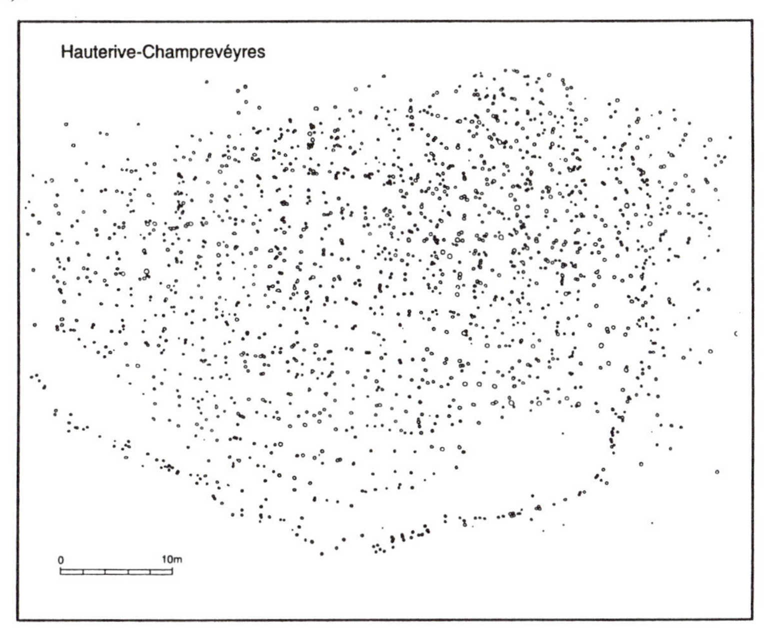

Fig. 4.1. Hauterive-Champréveyres. The posts of a Neolithic village, built on the shores of Lake Neuchâtel in Switzerland. Based on Egloff, 1989.

The wood preserved in wetland contexts, together with the associated palaeoenvironmental evidence, can be used for study of the landscape contemporary with a settlement, and something of the ways in which people used and altered their surroundings can be discerned. Good examples of such studies can be seen in the work of Casparie on Dutch raised bogs (Casparie, 1993), or Schlichtherle and his colleagues on the peat moors and lake shores of southern Germany (Schlichtherle, 1990 and accompanying papers).

The above examples have underlined the value of the wetland record by contrasting it with the impoverished dry land record for a similar aspect of the past. But sometimes there is no dry land record surviving, and what we know is known only from wetland contexts. Before pottery came into use, people must nevertheless have had containers and carriers of some sort, made presumably of organic materials. The wetland site of Friesack in northern Germany provides evidence dated to the early post-glacial for just such articles. Here, a sheet of birch bark was found that had been folded up at the ends to make a shallow trough, together with the remains of several willow-bast string nets that could have been used as carrier bags or for hunting or fishing (Gramsch, 1992). Friesack is but one of many sites where the preservation of organic evidence has significantly enlarged our knowledge of the past.

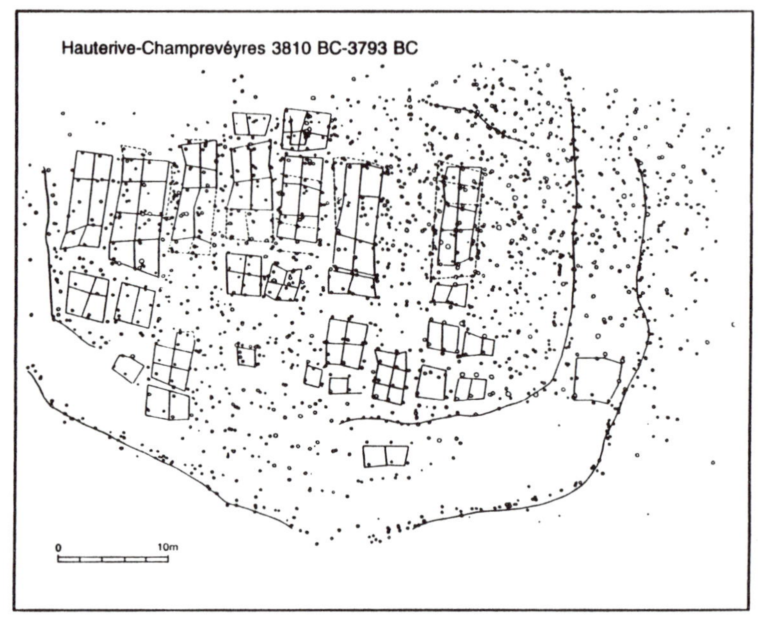

Fig. 4.2. Hauterive-Champréveyres. Identification of structures and dating of the site on the basis of dendrochronological analyses. Based on Egloff, 1989.

The value of the palaeoenvironmental record should also be emphasised, and in particular the frequent direct association, in wetland contexts, of evidence for past environmental conditions with evidence for past human activity. Fig. 4.3 illustrates this most important aspect of the archive by reference to the Sweet Track from the Somerset Levels, where virtually every element of the wooden structure can be used to provide information about the people who built it and the environment which they lived in and exploited. A wooden plank, for example, can be studied in terms of the carpentry skills required to produce it using only stone and wooden tools. Knowing the age and the girth of the parent tree from tree-ring analysis, 400 years old and a metre or so in diameter, felling techniques can be estimated and appreciated. Identification of the wood species as oak is relevant to the understanding of both felling and plank production. At the same time, the plank provides evidence for reconstruction of forest species and character, indicating tree cover that included mature oaks with straight, branch-free trunks for at least the first three metres, that being the length of the knot-free plank.

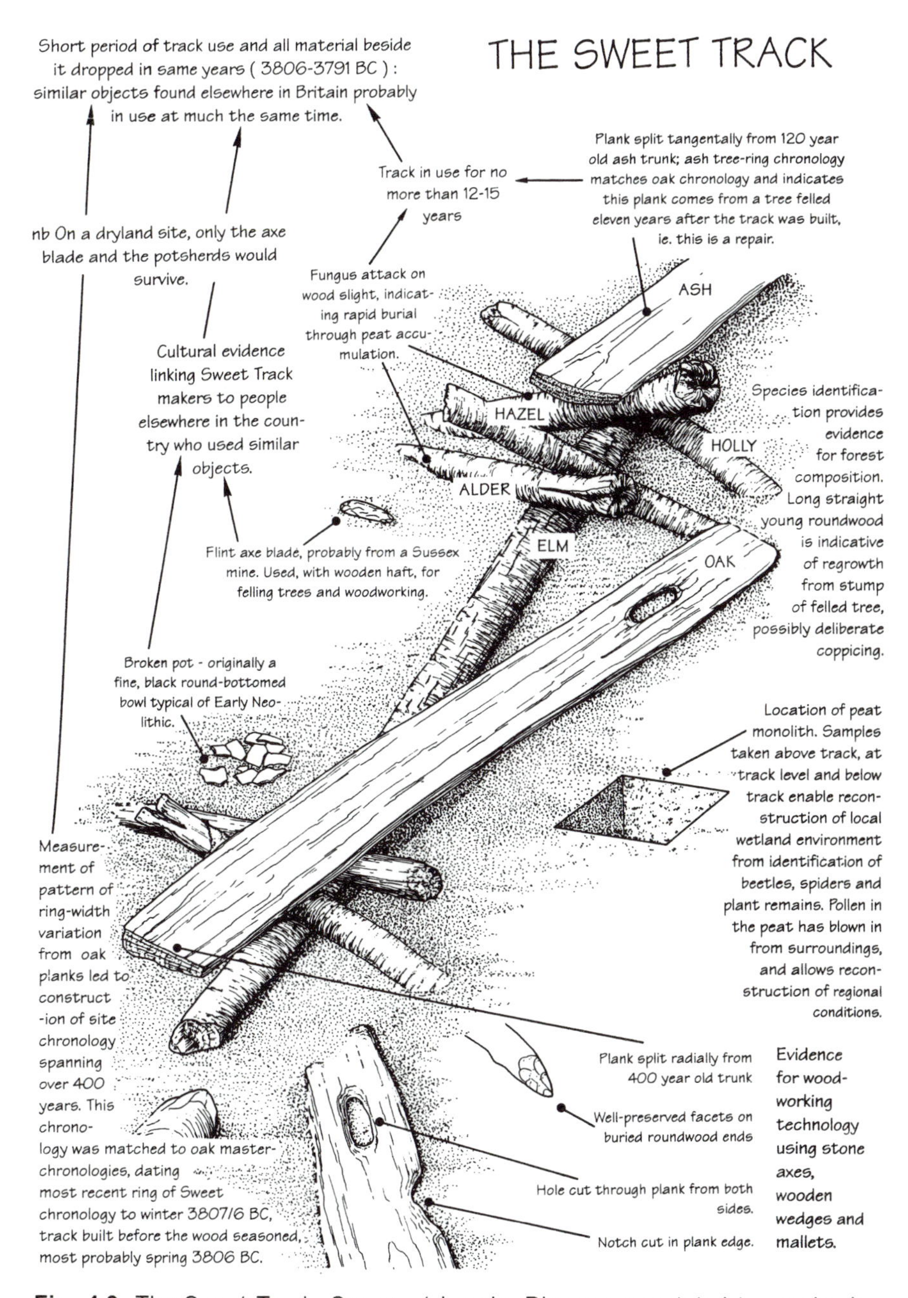

Fig. 4.3. The Sweet Track, Somerset Levels. Diagram annotated to emphasise the inter-relationships of different categories of evidence, in particular the close association of environmental and artefactual evidence.

The peat surrounding the trackway preserves pollen grains and insect exoskeletons both of which provide a record of past environments, as does the peat matrix itself. Other chapters in this volume discuss the significance of such records (e.g. Whitehouse *et al.*, Chapter 5, this volume; Charman, Chapter 6, this volume) again the key point here is the direct association with evidence for human activity, giving the possibility of determining whether or not people have affected the course of environmental change. It has been argued elsewhere that from an early stage they must have affected the ecology of wetlands, and that an understanding of their influence can contribute to the success of present-day schemes for the restoration and management of wetland habitats (Coles, 1995b; Louwe Kooijmans, 1996).

Uncertain future: threats and preservation

The above is a small sample of the wealth of evidence for the past which has been retrieved from wetland contexts (a recent broader survey is available in Coles and Coles, 1996). All of the evidence, it should be noted, has been discovered through the disturbance of the protective, waterlogged context. Friesack, for example, was found early this century when drainage channels were cut to improve rough land for agriculture, and it was excavated in the 1970s ahead of renewed drainage to lower the water table still further. The Sweet Track was discovered through drainage in advance of peat-cutting, and excavated where peat-cutting would otherwise have destroyed it. Hauterive-Champréveyres, as noted above, was excavated ahead of motorway construction.

Drainage, extraction and development remain three powerful threats to the survival of the archive, although paradoxically they still provide the most frequent occasions for discovery; as yet, techniques are poorly developed for sensing waterlogged organic structures and artefacts buried in peat. However, this should not deter us from recognising that the majority of wetlands preserve evidence for the period of time during which they have evolved, whether it be the full post-glacial record of the Solway mosses or the shorter span of a relict river channel cut off from the main current only a few centuries ago. The character of the record will vary from context to context, and in places it will be predominantly palaeoenvironmental, but all undisturbed wetlands should be accorded the highest priority for preservation of their archival content. Effective action in this respect, it can be argued, is best achieved in concert with the protection of wetland habitats and wildlife, so long as certain differences between the living heritage and the archive are recognised.

The wetland archive, as we have just seen, is buried and hidden until disturbed, in contrast to the flora and fauna of a wetland which, however elusive, can usually be identified and perhaps quantified following patient observation and monitoring. One should not, therefore, expect archaeologists and palaeoenvironmentalists to be as specific about the character and value of a particular wetland as a conservation officer might be, particularly if there has been little or no disturbance to the surface. This does not mean absence of an informed assessment of potential, however, as can be seen in the various publications of the English Heritage wetland surveys (e.g. Van de Noort and Davies, 1993; Hall and Coles, 1994; Waller, 1994; Hall and Wells, 1995; Middleton *et al.*, 1995).

Another, and perhaps more fundamental difference, is that a living wetland ecosystem can recover from damage whereas the archive cannot. Recovery of the living wetland may not entail a full return to previous conditions, but in many cases it is expected that a viable wetland ecosystem can be re-established, or there would be no attempts at wetland rehabilitation or restoration. The archive cannot be rehabilitated: what is damaged is lost forever. If, following drainage, the adze-marks on an ancient plank are eaten away by invading insects and worms, no amount of re-wetting will restore that evidence. A hurdle trackway largely destroyed by peat-cutting (see Fig. 4.4) will not expand from the few remnants left behind if peat cutting ceases and is followed by conservation management. The archive, it must be understood, is a finite, non-renewable resource, and it should be valued as such.

Schemes for the rehabilitation or restoration of wetlands are likely to affect the archive during the transitional period from previous land use to stable conservation management. In a sense, conservation managers could be termed developers during this phase, and their responsibility for due care of and respect for the archive should be the same as that of any other developer. Their activities can affect the archive in various ways. For example, digging peat to dam the drains on a raised bog may cause physical damage to a buried structure, or it could reveal a bog body; all digging should therefore be undertaken in the presence of an observer trained in the recognition of archaeological evidence. Restoration of former peat extraction land may involve the creation of bunds, lagoons, reed beds and the like, and encouragement of new types of vegetation cover, all of which could affect the buried archive either by physical damage or by alterations to hydrology and chemistry.

Restoration should, therefore, be undertaken with an awareness of the archaeological potential of an area, and efforts made to minimise damage. In the long run, conservation management must surely be to the benefit of conservation of the archive as well as conservation of the living heritage. The transitional period may be something of a threat to the archive, and its non-renewable nature should always be acknowledged and allowed for in management plans.

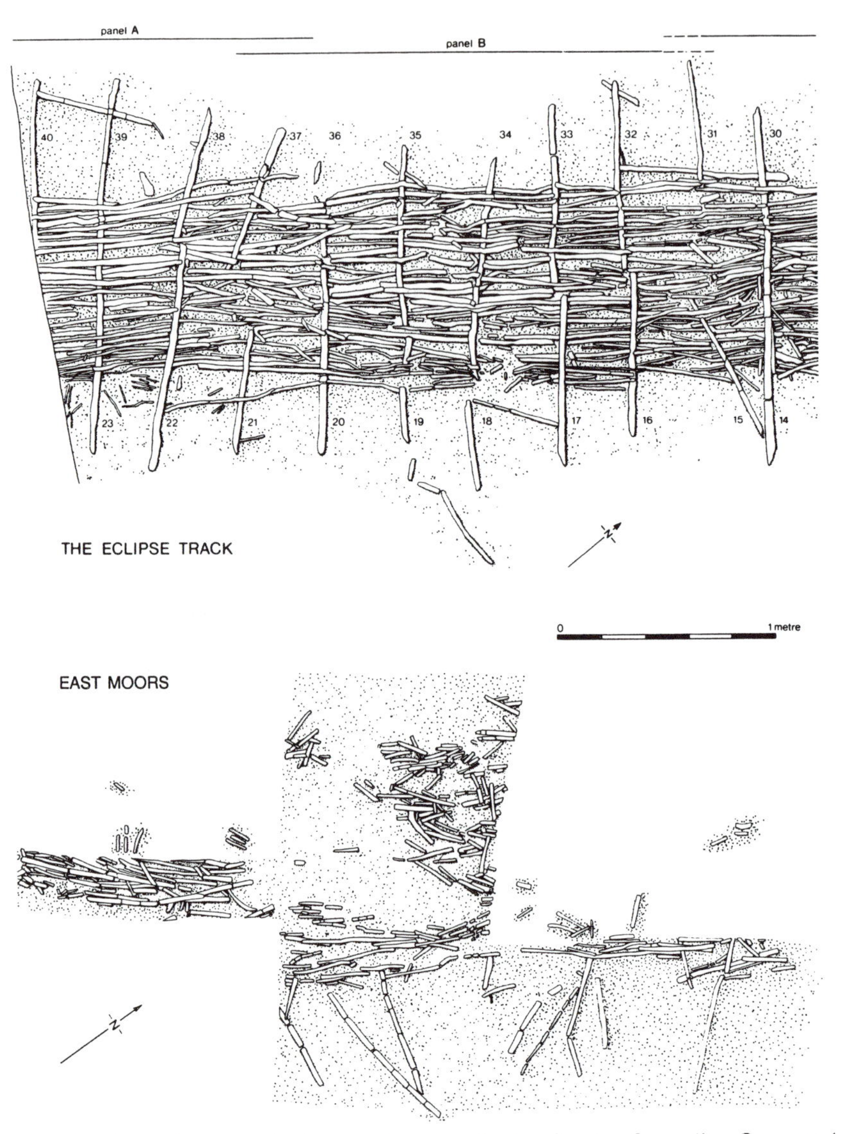

Fig. 4.4. East Moors and Eclipse, two hurdle trackways from the Somerset Levels. The East Moors trackway was heavily damaged by peat cutting prior to excavation, in contrast to the Eclipse track. There is no way in which the damaged sections of the East Moors track can be restored to their original state, which must have been something like that of the Eclipse trackway.

In recompense for the caution which care for the archive imposes on conservationists, the archive can offer enhancement of the living heritage in two respects, if not more. Firstly, there is the enormous, and, as yet, little understood and little exploited contribution to be made through a better understanding of wetland development and the influences upon it, including human. Secondly, protection of wetland archaeological sites has been shown to enhance local wetland wildlife, and in this respect it can be regarded as a form of restoration or, in some cases, wetland creation. Good examples are provided by work carried out in Cambridgeshire at Flag Fen and at Stonea, and, on a slightly larger scale, by the archaeologically-driven conservation work around the Lac de Chalain in eastern France where protection of eroding and desiccating shoreline deposits has led to a rise in the water table, a return of wetland flora and fauna, and a concomitant enhancement of the tourist value of the lake (Coles, 1995a; Pétrequin, 1996).

In the interval between the presentation of this paper to the Scottish Wildlife Trust's Peatlands Convention in July 1995 and the preparation of the written version in October 1995, in England alone two significant wetland discoveries have been made. In Cheshire, on the line of a small by-pass, a former mere has been identified which contains a long palaeoenvironmental record and likely evidence for human activity from the early Mesolithic onwards (Tindall, pers. comm.). In Sussex, near Eastbourne, excavation of an amenity lake has revealed what seems to be a complex wooden structure of later prehistoric date (Woodcock, pers. comm.). Neither discovery was predicted, both were consequent to the disturbance of a former wetland, and both will enhance our understanding of the past. One, it will be noted, came about as a result of road-development and the other as a result of what one might term development for wetland creation. These discoveries underline the archival potential and the vulnerability of all wetlands, and they serve as a reminder of the constant need for vigilance and awareness in the protection of the archaeological and palaeoenvironmental record.

Acknowledgements

Much of the research on which this chapter is based was funded by English Heritage and the British Academy, and support has also been provided by the University of Exeter. Sue Rouillard has assisted with the illustrations and Tina Tuohy with text preparation. County archaeologists Adrian Tindall (Cheshire) and Andrew Woodcock (E. Sussex) have kindly provided information concerning recent wetland discoveries.

Chapter five:

Peatlands, Past, Present and Future; Some Comments from the Fossil Record

Ms Nicki J. Whitehouse, Ms Gretel Boswijk and Dr Paul C. Buckland
University of Sheffield, UK.

Introduction

Peatlands constitute a unique series of ecosystems, which support distinctive animal and plant communities; they are also one of the most threatened ecosystems in the world (Williams, 1990). Within their deposits are preserved the remains of their past biodiversity and by detailed studies of these remains, it has been possible to document their formation, from their inception to their present, often degraded, form. These deposits provide 'proxy data' of past climatic fluctuations, in, for example, changes in stratigraphy and degree of humification, and varying animal and plant assemblages. These also provide valuable indications of local hydrological conditions, changes in water regimes and evidence for direct human impact upon the bog. Peatlands are the only terrestrial ecosystems which lay down a continuous three-dimensional record of their autochthonous history, as well as that of surrounding animal and plant communities.

The importance of the peat archive has rarely been recognised by ecologists and conservationists and proposals for bog regeneration and conservation have often taken little heed of the information contained within the sediments, nor have they provided mechanisms for its survival. This chapter discusses the nature of the archive, and why it is important for these deposits to be preserved. Particular reference will be made to the fossil record of Thorne and Hatfield Moors, in the Humberhead Levels, South Yorkshire (see Fig. 5.1), which, despite their status as National Nature Reserves, continue to be milled on a large scale for horticultural peat. Whilst recognising the importance of other lines of evidence, this discussion will emphasise the Coleopteran (beetle) and tree ring data, particularly from the pine-wooded mire preserved at the base of the peat.

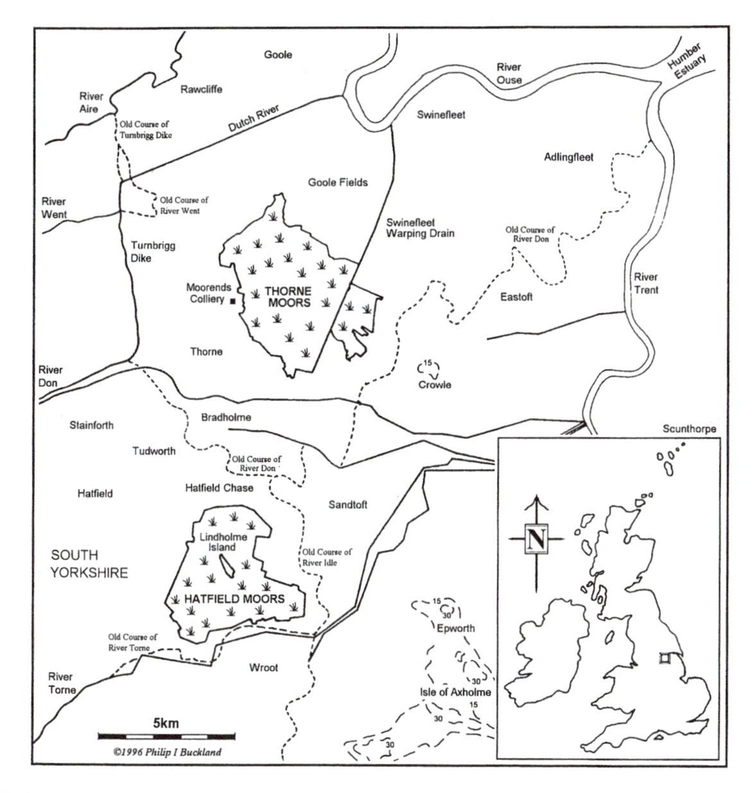

Fig. 5.1 Location map of Thorne and Hatfield Moors, S. Yorkshire.

The peat archive

It is perhaps unfortunate that, in mainland Britain, the information preserved in wetlands is the responsibility of two agencies; archaeology is the preserve of English Heritage, Cadw Welsh Historic Monuments and Historic Scotland, whilst natural history, including Quaternary geology, is the brief of English Nature, the Countryside Council for Wales and Scottish Natural Heritage. Whilst the former bodies may show concern where actual waterlogged remains are under threat, the peat archive is only occasionally considered. Regeneration of a bog, which may seem the prime objective to the natural historian, achieved by cutting and grading the landscape to a lower level, more in keeping with the prevailing water table, may be wholly destructive of any contained archaeology. More importantly, the record of the development of the mire biota, once destroyed, can never be re-generated. This particular aspect of peatland conservation has, until recently, been largely ignored by the conservation bodies although recent signs show that at least English Heritage recognises the importance of wetland conservation (Coles, 1995a). However, we are still a long way from an integrated management approach which takes into consideration the importance of the Holocene fossil record and which extends beyond the margins of the bogs (cf. Buckland, 1993; Buckland and Dinnin, 1993). This is amply demonstrated by the limited reference to the value of the palaeoecological record of mire development in the recent government sponsored *Restoration of Damaged Peatlands* publication, where the section on palaeoecology and archaeology amounts to about a page (Wheeler and Shaw, 1995b p. 30). Active cutting of peat is advocated in order to obtain a *tabula rasa* to create a 'restored wetland landscape' (N.B. it is impossible to recreate a plant and animal assemblage several thousand years in the making). Such an approach has much to commend it to the mineral extraction industries, yet the palaeoecological record would suggest that the initiation and development of true raised mire may take several hundred years, whilst poor fen can be rapidly achieved, although without much of its former invertebrate species diversity. In contrast with Wheeler and Shaw's (1995b) over-optimistic ecological approach, Joosten's (1995c) salutary comment from the palaeoecological record is that there is no evidence that bogs as self-regulating systems can be restored after severe anthropogenic damage within the human time frame.

Raised mires contain the most complete and reliable record of climatic and environmental change during the Holocene. The growth of this type of bog is related to atmospheric inputs, with a direct relationship between the growth of the water dome within the mire and precipitation levels (i.e. the balance of precipitation minus evaporation) (Ingram, 1982; Bragg, 1988). This is coupled with the rate of decomposition of accumulating peat (Clymo, 1991). Research by Barber (1985; 1994) suggests that, in the absence of drainage changes or human influences, and given that the main inputs into the bog are atmospheric, there is likely to be a good correlation between climate, the bog water-table and the nature of its vegetation; thus, the peat archive can be viewed as a proxy-record of changing climate. By contrast, minerotrophic and rheotrophic mires are subject to other inputs into their system, resulting from natural and cultural changes in their catchment. These present a composite picture of the local environment, which may mask the climatic signal (Barber, 1993). Blackford (1993), however, notes that blanket peats, which are a mixture of ombrotrophic and minerotrophic, are a largely untapped source of data which may complement other proxy records.

The organic remains preserved within the bogs are both allochthonous and autochthonous, consisting of pollen and plant and animal macrofossils, as well as records of past pollution and other atmospheric inputs, such as dust, charcoal and volcanic ash (tephra). The waterlogged and anaerobic nature of peat offers good preservation of these organic and mineral remains and species-level identification of plants and animals is often possible. Many groups, however, which may be highly sensitive to internal and external changes which affect the locality, such as the mites (Acarina) have yet to be more fully investigated. As on modern bogs, the invertebrates are most numerous both in terms of individuals and taxonomic diversity but it is usually the plant fossils, particularly the pollen, which have received most study. Work upon plant macrofossils has indicated that highly humified peats, developed over long periods of time, produce less well preserved or even partially decomposed remains, although invertebrate remains may continue to be identifiable even when most plant remains are not. Peat may develop rapidly providing a more continuous, higher resolution, record compared with other sediments. Barber (1993) suggested that the average accumulation rate is c. 1 mm per year, although this remains a matter of some debate (Foster and Wright, 1990; Clymo, 1991; Korhola, 1992) and has occasioned some discussion in the archaeological context of the bog bodies from Lindow Moss in Cheshire (Barber, 1995; Buckland, 1995).

The peat archive has the advantage of being suitable for a variety of dating techniques, such as C^{14}, ^{210}Pb, and other isotope analyses. Amongst a number of new techniques, tephrochronology offers the opportunity to present isochrones between different locations (Dugmore *et al.*, 1995) and check the apparent synchronicity of changes both in vegetation (Blackford *et al.,* 1992) and humification interpreted as being driven by climate. The presence of 'bog oaks' and 'firs' is a well documented occurrence throughout Britain and Europe (e.g. Anderson, 1967) and tree ring analysis of bog oak has been of great importance in extending the absolute chronologies for Britain and Ireland used for archaeological dating and for the calibration of the radiocarbon curve (Baillie, 1992).

The importance of the peat archive: a case study of the Humberhead Levels

An important aspect of the peatland archive is the record of changes which have affected the mire in the past, such as the temporary expansions of woodland across the surface of the bog, perhaps providing an analogue to what is happening to many sites at the present day (Chambers, Chapter 18, this volume). Peat often contains information about early woodlands growing on the bog through the presence of tree macrofossils (particularly Scots pine) and their associated fossil insect fauna. Research into this area can provide valuable data in understanding processes of mire initiation and development, as well as the history of the invertebrate fauna. This has important implications regarding any attempt at regeneration of cut-over bogs, such as those at Thorne and Hatfield Moors (Humberhead Levels, South Yorkshire). These sites are internationally renowned, long registered as Sites of Special Scientific Interest and recently declared as National Nature Reserves, at least in part. They lie at the southern limit for a number of species and the northern for others (Heaver and Eversham, 1991) and are notable as having developed in an area of low rainfall; indeed, the moors receive the minimum amount of rainfall considered necessary for ombrotrophic peat growth (c. 600 mm per annum) (Money, 1995). They are therefore likely to be particularly sensitive sites for research into climate and environmental change.

The sub-fossil insect evidence

Numerically and taxonomically, the fossil insect fauna preserved within the peat provide the most detailed record of change. Past biodiversity is reflected in these assemblages, charting the environmental conditions, the succession of insect host plants and the differing water regimes throughout the bogs' development (Roper, 1996). The number of Holocene sites studied for their insect remains is small and archaeological sites tend to outnumber natural deposits (Buckland and Coope, 1991). The work has highlighted a range of Coleoptera species which have either become very restricted in their distribution, or, which have become extinct within the British Isles (Buckland and Dinnin, 1993). The total known number of pre-Linnean British extinctions recorded as fossils currently numbers 35, and each study of suitable natural deposits, largely preserved in peatlands, adds more to the list. Many of the extirpated species and early assemblages recovered are associated with mature, relatively undisturbed woodland. Osborne (1965) was the first to point out the Old Forest (*Urwaldtier*) element in early to mid-Holocene fossil beetle assemblages, and Thorne Moors provides a good example of such an assemblage.

Thorne Moors is the most recent Holocene site which displays many elements of an *Urwald* fauna. Insect assemblages from the base of the peat, taken from the southern area of the Moor, dated to 3080 ± 90 BP included seven species no longer found in Britain (Buckland and Kenward, 1973; Buckland, 1979). One of the most frequent fossils was the Cucujid *Prostomis mandibularis*, which is now virtually restricted to isolated populations within the Central European forests (Palm, 1959). The recent peat milling over large areas of the northern part of the Moors, has removed peat to within a few centimetres of the underlying Lake Humber Clay-silts, and has thrown up many thousands of well-preserved pines, birch and some other deciduous trees. In this area, peat growth appears to have begun nearly one thousand years earlier than in the south, and limited sampling has shown that about a quarter of the fossil insect assemblages comprised saproxylics, including a large number of rare and locally extinct Coleoptera (Roper, 1996; Whitehouse, 1996).

Many of the extirpated species are associated with pine woodland, such as the Ostomid, *Temnochila coerula*, previously unrecorded in Britain (Whitehouse, 1996). It is described in the European literature as a rare '*Urwaldrelikt*', living under the bark of conifers (Reitter, 1911) and it has a decidedly central-southern European distribution (Fig. 5.2). Three members of the saproxylic weevil genus *Eremotes* no longer found in Britain were also recovered: *E. sculpturatus*, a new species to the British fossil list, *E. elongatus*, and *E. punctulatus*, the former recovered previously from deposits at Church Stretton (Osborne, 1972) and the latter from Stileway, Somerset Levels (Girling,

1985). Associated with these species the British *E. ater* was recovered, a species today largely associated with northern pine forests except where re-introduced as a consequence of forestry activity (Buckland, 1979). Three other non-British pinicolous species were identified. A complete individual of the jewel beetle, *Buprestis rustica,* was recovered from its pupal chamber in a fossil pine by P. Skidmore during fieldwork on the northern part of the Moors. The weevil *Pissodes gyllenhali* is associated with shed pine litter and bark (Freude *et al.*, 1983) and the Cucujid *Cryptolestes corticinus* lives under the bark of firs and pine (Freude *et al.*, 1983). Both these species are new to the British fossil list (Whitehouse, 1996). The burrows of the longhorn beetle, *Arhopalus rusticus,* now largely restricted to Caledonian pine forests, were evident in many of the fossil pines. The abundance of insects associated with dead and decaying timber suggests a proliferation of dead, dying and mature trees around the site. The loss of Scots pine habitat is reflected in these invertebrate extinctions, many of which had, as a result of forest clearance by the Bronze Age, become restricted to marginal habitats, such as developing peat bogs.

Palaeoentomology has been used successfully in the reconstruction of climatic change, such as some of the climatic fluctuations during late Pleistocene glacial and interglacial episodes; such results have proven more sensitive that those obtained by other methods, such as pollen analysis (Osborne, 1980; Atkinson *et al.,* 1987). Unfortunately, given the difficulty of differentiating between severe habitat loss and the role of climatic change in influencing species' distribution, Holocene climatic change has been far less easy to infer from the fossil insect record. Nevertheless, the modern distributions of some of the species recovered from Thorne and Hatfield Moors imply climatic change since the Bronze Age. The distribution of *Eremotes punctulatus, E. elongatus* and *Temnochila coerula,* which range across Europe (including Italy and Greece in their current distribution) may suggest warmer summers, although their occurrence in southern Scandinavia may suggest a preference for a more continental climatic regime than that currently prevailing. However, the bark beetle, *Scolytus ratzeburgi,* was particularly frequent in fossil birches within the same deposit as *T. coerula* (Roper, 1996). Considered together, they suggest a more complex interpretation, reflecting the differing histories of woodland and forest in the West and Baltic regions (Fig 5.2). The way woods have been managed, as well as cleared, has had considerable impact upon their associated biota and may explain some of the apparently contradictory disjunct distributions. Ice core evidence suggests that the most significant downturn in Holocene temperatures took place during the medieval period (Mayewski, pers. comm.) and many species, like the water beetle *Gyrinus colymbus* (Girling, 1984) may have finally succumbed at this time. As a result of peat-cutting, wetland drainage and forestry, the paucity of suitable medieval and post-medieval deposits, makes this model difficult to validate. The survival of the record through to the recent past must be seen as an urgent priority in wetland conservation.

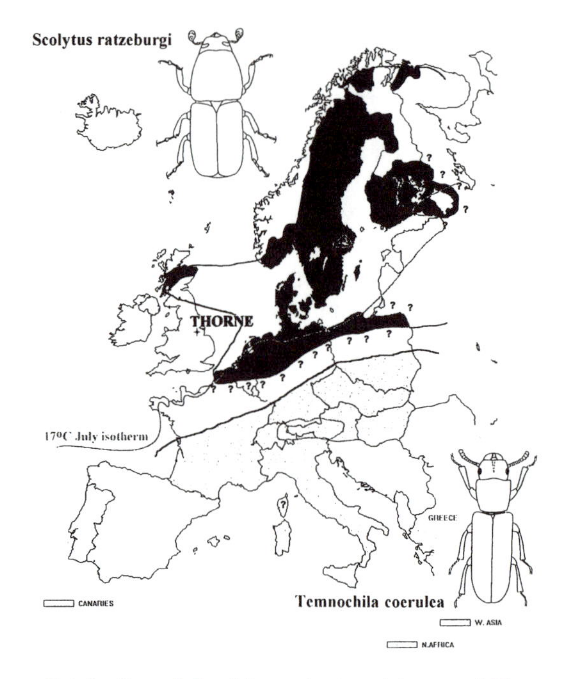

Fig. 5.2. Modern distribution of *Scolytus ratzeburgi* Jan. and *Temnochila coerulea* Westw. with 17°C July mean isotherm. These species were recovered as fossils within the same context, in deposits of c. 4000-3000 BP on Thorne Moors. The distribution illustrates some of the current problems in the interpretation of such data. For example, do these distributions reflect variation in oceanicity versus continentality in the late Holocene climate? Note: actual species distribution is much more discontinuous. Question marks indicate a lack of information, rather than known absence of species. Drawing: Philip I Buckland.

The continuity of the peat archive enables the transition from forest to 'culture-steppe' (*sensu* Hammond, 1974) to be observed through changes in the fossil record. British forests had reached their maximum diversity by the mid-Holocene, c. 5000 BP, and both the pollen record and the fossil insect data suggest that, by the Bronze Age, a significant reduction of forested areas had occurred; by 1000 BP it is doubtful whether anything more than patches of secondary forests remained (Greig, 1982).

The fossil trees

Pine woodland is seen to be part of the natural vegetational succession, giving way eventually to *Sphagnum* mire communities (McNally, 1990; Hulme, 1994). Tree remains are generally found within a thin layer towards the base of the peat, giving a visual impression of dramatic woodland decline, coinciding with peat accumulation and triggered by widespread climatic change (McNally, 1990). Yet the results from studies of fossil pines have shown that the temporal relationship between similarly stratified remains can be quite broad, even within a localised region. Scots pine tree ring chronologies from a site on Garry Bog (County Antrim, Ireland) spanned 800 years (Brown, 1991), whilst, overall, there is a wide range of radiocarbon dates from sub-fossil pines in Britain and Ireland. (e.g. Birks, 1975; McNally and Doyle, 1984; Dubois and Ferguson, 1985; Lageard, 1992) .

Tree ring width can be a valuable index of both edaphic and climatic change (Fritts and Swetnam, 1989) and dendrochronological study may provide precise calendar dates to archaeological sites. In recent years tree ring analysis has been applied to the study of mire pine from several different sites in Britain and Ireland (Bridge *et al.*, 1990; McNally, 1990; Brown, 1991; Gear and Huntley, 1991; Lageard, 1992). Given that trees which have grown in the same area over the same timespan have similar growth patterns, it is possible to identify trees which were contemporary. In Britain, oak has been the main species used for dendrochronology, mainly because it was the principle building timber since prehistoric times and long absolute-dated oak chronologies have been established. The Belfast Long chronology extends over 7000 years (Baillie and Brown, 1988). Although most pine chronologies have been dated by radiocarbon, work carried out by Brown (1991) in Ireland successfully dated pines against the Belfast Long oak chronology.

The data produced through tree-ring analysis provides an opportunity to study the dynamics of a palaeo-woodland. The population structure, duration and continuity can be studied and related to changes in the immediate environment. For example, germination and mortality trends can show whether the woodland was multi-aged and had a gradual decline, or if there were distinct growth flushes. Variations in growth rates and longevity may relate to the influence of immediate growth conditions, such as shifts in the mire hydrology. Chronology building and dating, establishes the temporal relationship of trees, through which trends in colonisation and retreat can be observed. For example, a recolonisation period resulting when mire surfaces may have dried sufficiently to allow the germination of seedlings, may be short-lived and, therefore, overlooked by conventional conventional pollen studies (Gear and Huntley, 1991). By correlating data between sites it may be possible to observe common trends which may be related to the impact of climate.

The beetle assemblage from Thorne Moors presents an image of a mature, multi-aged, pine forest. Current research focusing on both oak and Scots pine macro-fossils from the Moors considers the demise of the woodland and the initiation of mire growth. At Thorne Moors, most of the trees suitable for dendrochronology, oak as well as pine, appear to have been preserved rapidly in a wet bog environment. Whole pine trunks, some complete with bark, others with charred edges, have been found *in situ*. Tree ring analysis of the oaks has resulted in a dated chronology (3777-3017 BC) which overlaps with the C^{14}

dates obtained from the base of the peat across the northern area of the Moors (Boswijk, 1996). A series of separate Scots pine chronologies is being established. From these, a pattern of varying growth rates and longevity is beginning to emerge for the mire woodland. What has yet to be clarified is whether the pine chronologies are part of a long single-phase woodland, or represent distinct episodes of die-back and recolonisation. Evidence of fire covers over 5 km^2 at Thorne and there is similar evidence from Hatfield. Both insect and plant macrofossil evidence combine to show that fire appears to have run through a moribund forest, as, in most cases, trees appear to have failed to recover after fire damage (Boswijk, 1996). It is, however, still uncertain whether a single event or several is involved and the role of man, if any, is also unknown.

The importance of fire habitats

Fire, as an element within ecosystems, has been largely ignored, yet peat deposits frequently suggest its former importance. Charcoal fragments are common in peat deposits which are being processed for insect remains, suggesting that fire has erupted on peat bogs not infrequently in the past, particularly during periods of relative drought. This is evident also from the recent fire which affected Thorne Moors during summer 1995. Much of the fire affected the cut-over areas, where the dry peat was particularly flammable. Some of the fire also spread to a section of the vegetated area. This clearly provides an ideal opportunity to study the effects of fire upon the mire surface. Elements in the insect fauna are also influenced by the frequency of fire. Kaufmann (1948) suggests that the longhorn beetle, *Arhopalus rusticus,* a frequent fossil at Thorne, is attracted to fire-damaged pines. In Britain *Scolytus ratzeburgi* appears to be confined to the uplands of Scotland, with a preference for mature birch stands. However, in the deposits at Thorne, the species has been found in both mature and young birch. Its disappearance from the Lowlands is not easily explained but its distribution may be affected by the frequency of burns, since it has sometimes been found in fire-damaged woodland, attacking the damaged trees. This may be one of the species' requirements, resulting in its restriction to areas of the north with coniferous and birch woodlands, which are more flammable than other types of forest. Two other non-British species recovered from the Moors are also associated with fire, *Eremotes elongatus,* known from charred Scots pine, and *Pelta grossum*, (Buckland, 1979), associated with *Urwald* damaged by forest fires (Palm, 1951), a habitat which has become increasingly rare. Charcoal, and charred stumps - particularly of pine - have been recorded from peat bogs in Ireland, Scotland and England (Lageard, 1992; Bradshaw, 1993; Whitehouse, 1996). These, combined with the beetle remains, raise questions regarding the role of natural fire within temperate coniferous woodlands and mire systems - is there a natural fire cycle, influential in maintaining a forested environment, and what role, if any, did natural or anthropogenic fire play in the development of the mire ecosystem?

Conclusion

In such a short contribution it is difficult to summarise all aspects of the importance of the peat archive to future conservation strategies. The trajectories of mire and other ecosystem developments are preserved in waterlogged sediments and much of the record is one of cautionary tales of extirpation and major contractions in distributional range. We still do not know when much of this took place, whether it was a gradual process in the face of progressive forest clearance and sediment influx to wetlands, or the impact of a precipitate downturn in temperatures which ushered in the 'Little Ice Age' during the late medieval period. The changes are not restricted to the Coleoptera and the disappearances of the Rannoch rush, *Scheuchzeria palustris* and the moss, *Sphagnum imbricatum* from the lowlands, including the Humberhead Levels, have been discussed elsewhere (Buckland *et al.*, 1994; Stoneman *et al.*, 1993). Past three-dimensional landscape change cannot be plotted from a single column of peat and there is an urgent need for the careful consideration of the spatial elements of wetland preservation. This would enable future study, not only in terms of more sophisticated palynology, but also of other groups, in particular of invertebrates, where knowledge of the Holocene record remains rudimentary. Much of the dendrochronological evidence and macro-invertebrate record continues to be lost, not only to peat cutting, but also to the more insidious processes of pumping down of water tables.

The fossil record indicates that there can be no quick fixes in wetland restoration (*contra* Wheeler and Shaw, 1995b) and that, for many apparently mobile groups, like the Coleoptera, habitat continuity is essential (cf. Buckland and Johnson, 1983). Thorne and Hatfield Moors remained essentially unchanged for three thousand years, until modern large scale peat cutting effectively destroyed them. A conservation strategy which only considers the evidence of a few plant species, rather than the whole peatland ecosystem, is essentially cosmetic. Surviving biota must be compared with past biota in order to understand the future. What follows, in conservation terms, requires, not a quick fix, but a strategy of planned intervention over several generations. It is debatable whether, by the end of the millennium, there will be enough of the lowland wetland record remaining to apply its lessons to the future.

Acknowledgements

This chapter owes much to members of the Thorne and Hatfield Conservation Forum, in particular, Brian Eversham and Peter Skidmore. In addition, Cathy Groves, Jennifer Hillam and Eva Panagiotakopulu commented upon previous drafts of the chapter. The authors are grateful to Paul Mayewski for discussion of the latest evidence from the Greenland Summit ice core. The University of Sheffield provided funds to allow the authors to take part in the Peatland

Symposium. Thanks also to English Nature and Levingtons, who allowed access to the study area.

Chapter six:

Peatland Palaeoecology and Conservation

Dr Daniel J. Charman
University of Plymouth, UK.

Introduction

The theme of a large component of this conference volume is that of peatland archives within which a number of sub-themes have been recognised. It is apparent from these and other writings (Buckland *et al.*, 1994; Buckland and Dinnin, 1994; Chambers, 1994; Charman, 1994a) that peatland archives which reflect the past ecology of peatlands and their environment are of immense scientific value in a diverse range of subjects from palaeobotany through to reconstruction of climate change and the human history of the landscape. These values can be broadly categorised as; historic value in the development of Quaternary science, vegetation history reconstruction, palaeoclimatic reconstruction and archaeological value (Charman, 1994a). Furthermore, it has been argued (Buckland *et al.*, 1994; Charman, 1994b), that there has been insufficient recognition of these values in the design and implementation of conservation policies in the UK and that there is an urgent need to tackle this problem if valuable sites are to be conserved. However, while there is clearly a good argument for the conservation of the peatland archive resource on its own merits, it is also crucial to appreciate the links between peatland palaeoecology and conservation of the present ecosystem and its biological resource. This chapter outlines some of the important lessons to be learned from palaeoecological research for the present and future protection of the biological resource on peatlands. The continuation and future value of this research depends on the conservation of an adequate number and distribution of suitable sites on which to work. The final part of the chapter outlines the criteria for the assessment of palaeoecological value on peatland sites and sets out the priorities for a strategy for evaluation and conservation of the national palaeoecological resource.

Naturalness and the basis for conservation: allowing for natural processes and fluctuations

Palaeoenvironmental studies offer the opportunity to study natural processes on a timescale longer than is possible by real time studies. Palaeoecological research can give us an understanding of long term vegetation dynamics on peatlands and what we should expect in terms of community stability. Furthermore it helps define and assess naturalness of the peatland system and plant communities. Many valuable conservation habitats are often assumed to be in their natural state and indeed it may be suggested that they owe their particular characteristics to long term stability and lack of human interference. This may often prove to be an unfounded assumption as Segerstrom *et al.* (1994) have shown for a forested swamp peatland in Sweden. These habitats are valued particularly for their rich biodiversity and rare species and it has been suggested that they are refugia from fire and other disturbance. However, pollen and macrofossil analysis shows that the forest was actually under cultivation 500 years ago and that the plant communities are highly dynamic rather than being very stable. Segerstrom *et al.* (1994) suggest that this calls for a re-evaluation of conservation strategy for such areas and the provision of a network of sites which allow the operation of natural species dynamics. There are other examples where current ecological conditions have been assumed to be a consequence of long term stability but are subsequently shown to be of relatively recent origin. The oligotrophic vegetation of floating bogs in south western Ontario is one such case (Warner, 1993), where a dramatic change in vegetation was consequent to increased runoff following European settlement of the area. Human disturbance has also played an important role in the development of the British peatland landscape. The best known example is that of the initiation of blanket peats which are frequently associated with human activity at various times from the Mesolithic onwards (Moore, 1993). During the thousands of years of subsequent growth and development, the plant communities on these peatlands have varied considerably. One of the striking features is that *Betula* and *Pinus sylvestris* have frequently grown on mire surfaces in many parts of the country, although this has sometimes been for only a very short time (Gear and Huntley, 1991).

It is important that such information is used sensibly and constructively in conservation policy. 'Naturalness' is much valued as a basis for conservation (e.g. Ratcliffe, 1977) and a simplistic argument is that ecosystems that can be shown to be in part a product of human intervention are somehow less valuable in conservation terms. A more constructive viewpoint is to adapt management of the ecosystem in the light of such information. In the case of the Swedish swamp forests, Segerstrom *et al.* (1994) suggest that conservation policy needs to concentrate on the provision of more widespread varied habitats in a spatial configuration which allows species to migrate and respond to external influences rather than on individual site protection. In the case of the blanket mire habitats in upland Britain, the fact that the peatlands have not developed in complete isolation from human influence is not an argument to destroy them or regard them as devalued ecosystems. Palaeoecological research has shown that they are ecosystems of great antiquity. They are also almost unique in the international context. However, it may be that conservation policy should allow a more diverse range of peatland vegetation to develop, including the natural recolonisation of some areas of peatland by native species of trees and shrubs. This issue is tackled more extensively by Chambers (Chapter 18, this volume).

Landscape scale processes and conservation

Up until fairly recently, sites for nature conservation were viewed almost in isolation from the rest of the landscape. In Britain, the Sites of Special Scientific Interest (SSSI system) has encouraged this attitude as statutory bodies have had to justify the inclusion of every part of a site in terms of its scientific interest. Recent ideas have been changing this philosophy, especially for peatland sites and the idea of buffer zones is now broadly accepted although still difficult to apply in practice. However, although the theoretical basis for the inclusion of buffer zones is relatively well understood, there are few good practical demonstrations of the interlinked nature of peatland landscapes in the UK. Most British ecosystems are already severely fragmented, but the Flow Country in northern Scotland still provides a large area of intact blanket mire landscape. Palaeoenvironmental research on mire development at the Cross Lochs in Sutherland (Charman, 1994c; 1995) has traced the development of several individual 'mesotopes' (*sensu* Ivanov, 1981) which would be considered separate sites in the survey and evaluation for nature conservation (Lindsay *et al.*, 1988). Table 6.1 indicates the developmental sequences at each of three mesotope sites. The early sequence of development at all three sites is remarkably similar, with individual deposits probably developing in isolation from one another with extensive dry ground between each embryo system. Sedge fen with sparse birch cover was able to colonise infilled basins and subsequently from c. 6500-6000 radiocarbon years before present (BP) ombrotrophic peat forming communities colonised the sites. The development of poor fen communities on the two present fen sites is quite recent (c.2850 BP) and is coincident with a transition to *Sphagnum* dominated vegetation at the watershed mire site, indicating increased surface wetness. Two factors may have caused this change. Climate may have been important but it was probably also related to the closing of peat cover over the area and a consequent decrease in the potential for discharge of excess water between separate mire mesotopes. Although the water discharged as surface and sub-surface runoff is oligotrophic, its concentration at the head of some slopes has resulted in the development of the poor fen communities and the surface patterning which is such a unique feature of some of the fens in this region. These are clearly landscape scale processes which are operating and they demonstrate the critical link between apparently separate parts of a blanket mire system. The patterned fens would not have developed in the absence of the watershed blanket mire area and any disruption to the hydrology of that area would certainly have an impact on the ecology and continued growth of the fens themselves. These long term landscape processes are only observable by investigating the palaeoenvironmental record.

Table 6.1. Developmental history of three mesotopes in the Cross Lochs blanket mire. Cross Lochs fen A and B are both patterned fens situated on the fringes of the more extensive ombrotrophic watershed blanket mire mesotope. Codes (e.g. CLO5) refer to macrofossil assemblage zones.

Cross Lochs ombrotrophic	Cross Lochs Fen A	Cross Lochs Fen 'B'	General developmental sequence for patterned fens
	CLA6c: ? 600-0 BP. Minerotrophic *Sphagnum*-sedge poor fen	CLB6: Minerotrophic *Sphagnum*-sedge poor fen.	Recent change: acidification &*Sphagnum* growth.
CLO5 & CLO6: 2700-0 BP. Wetter *Sphagnum* community	CLA6b: 2850-?600 BP. Poor sedge fen.	CLB5 Poor sedge fen. Some Ericaceae.	Poor fen communities. Gradual transition. Continued burning.
CLO3 & CLO4: 6050-2700 BP. Ombrotrophic *Eriophorum*—Ericaceae. Burning begins.	CLA4, CLA5, CLA6a: 6400-2850 BP. Ombrotrophic *Eriophorum*--Ericaceae. Pine 4250-3900 BP. Consistent, large scale burning begins.	CLB4: Ombrotrophic *Eriophorum*--Ericaceae. Burning begins.	Ombrotrophic mire communities. Regular burning established.
		CLB3: Mixed ericaceae & fen sedges with birch.	Transition to ombrotrophic status: ericaceous communities
CLO2: 10400-6050 BP. Bryophyte-sedge fen with occ. birch	CLA2 & CLA3: 9150-6400 BP. *Sphagnum*- sedge fen with birch	CLB2: *Sphagnum*-sedge fen. Some birch.	Basin infill. Sedge fen with birch on drier sites.
CLO1: Limnic. Late glacial or early Holocene	CLA1: 9400-9150 BP. Bryophyte-sedge fen	CLB1: Limnic. Early Holocene (Post c.9000 BP)	Mire initiation: early peat forming communities; bryophyte flushes, shallow water.

(Based on Charman, 1995).

Response to climate change

The use of peat deposits to reconstruct past climatic change is well established (e.g. Barber, 1981; Blackford, 1993). The best records of climate change are derived from ombrotrophic systems and particularly raised mires (Barber, 1981; 1982; Barber *et al.*, 1994b). Macrofossil analysis has been the most frequently used technique in this work, focusing particularly on the fluctuations in *Sphagnum* species and species groups, which are closely linked to mire surface wetness. Since raised mires obtain all their moisture from precipitation, the record of mire surface wetness is a proxy measure for the precipitation-evapotranspiration balance and therefore of climatic change (Barber, 1981; 1994). Blanket mires have also been used effectively in reconstructing climate change for specific periods of time (Blackford and Chambers, 1991) on the basis of the same surface wetness - climate linkage, but using measures of humification as the proxy for this.

Reconstructions of mire surface wetness are an important aspect of palaeoclimatic research for assessing the past natural variability of climate change. However, they also allow the assessment of the impact of past climatic change on peatlands and therefore may give some insights into future responses to natural and anthropogenic perturbations in climate and hydrology. Barber (1981) examined the correspondence between peat stratigraphy and the climatic indices for the last 900 years of Lamb (1977). Wet shifts in the peat stratigraphy were associated with reduced temperatures and increases in summer rainfall, particularly during the late 14th and 15th centuries and the late 18th century. For example, Lamb (1977) suggests that mean annual temperatures increased from c.9.5°C to 10.5°C from 1100 to 1200AD and subsequently fell again during the 14th and 15th centuries. Stratigraphic changes associated with such temperature changes are complex but at one site a relatively unhumified *Sphagnum imbricatum* peat was replaced with a 'strongly growing' hummock of *Calluna-Eriophorum* with *Sphagnum imbricatum* and was then superseded by a wet pool with *Sphagnum* Sect. *subsecunda* and Sect. *cuspidata* and *S. papillosum* over this time (Barber, 1981). Absolute predictions of species change are difficult to make on the basis of the complex pattern found in the macrofossil record, but it is clear that - even if modest predictions of climatic change over the next 100 years are fulfilled - existing extensive wet *Sphagnum* carpets on ombrotrophic mires are likely to dry out significantly. This will severely threaten some of the most valued conservation microhabitats on peatlands in the UK.

Peat stratigraphy can therefore be used to assess potential change in plant communities in response to future climatic perturbation. The response of the system can be predicted in general terms but actual estimates of water table changes are more difficult. Recent research suggests that quantitative estimates from testate amoebae, a group of protozoans, may be possible (Warner and Charman, 1994). This offers the possibility of more precise predictions of mire hydrological changes in response to future climatic change.

Recovery from anthropogenic disturbance

Palaeoecological studies have played an important part in assessing long term impacts of past human activity. One of the foci of much recent interest in conservation management of peatlands has been in the regeneration of cut-over ombrotrophic mires. Specific arguments and techniques are discussed elsewhere in this volume, but the long term perspective can only be provided by palaeoenvironmental work. It is particularly important in assessing the potential for natural regeneration and for establishing a baseline against which to assess the success of interventionist policies of conservation management. Sequences of vegetation change can be determined in former peat cuttings and the rate of renewed peat accumulation can be calculated. Joosten (1985) provides a detailed reconstruction of environmental change from a variety of preserved micro- and macrofossils in old domestic peat diggings abandoned in 1857. Up to 1939, *Sphagnum papillosum* was the main peat forming species and peat accumulation occurred at an average rate of 240 g m^{-2} $year^{-1}$. Following this *Molinia caerulea* became a significant component of the vegetation and peat accumulation increased to an average of almost 490 g m^{-2} $year^{-1}$, although decay is presumably incomplete in these upper layers. Present vegetation after 130 years consists almost entirely of ombrotrophic mire species (Joosten, 1985, p279). Larger scale commercial peat cuttings have also been investigated. The re-flooded workings of the Dutch canal system on Thorne Waste, South Yorkshire were abandoned in 1920 and investigated by Smart *et al.* (1986; 1989). Their stratigraphic data show that between 2 and 20 cm of recent material had been deposited since abandonment and that a variety of vegetation sequences were followed. Colonisation of the cut *Sphagnum* peat surfaces was primarily by ericaceous species but also by *S. fimbriatum*, *S. recurvum* and *Drepanocladus* spp. However, even after 65 years of revegetation, plant communities are characterised by poor-fen species, although in most cases ombrotrophic bog species are present. This study also examined stratigraphy of uncut baulks which demonstrated the effect of drainage and disturbance associated with cutting, with the loss of the original *Sphagnum* species and their replacement by bryophytes more typical of dry conditions. The initial phases of revegetation are similar in many respects to those detected by more recent long term vegetation monitoring (Meade, 1992), but the failure to establish ombrotrophic conditions after more than 60 years at Thorne Waste suggests that natural revegetation of extensive peat cuttings will be an extremely long term process. The more successful revegetation of the Peel site (Joosten, 1985) is over relatively restricted domestic cuttings.

The needs of palaeoenvironmental research

The examples briefly outlined above demonstrate the relevance of palaeoecological study to the conservation and management of peatlands.

However, these examples are selective and there are many other areas where the palaeoecological perspective is equally important, such as in historical assessment of pollution impacts (Headley *et al.*, 1992), extent and timing of past erosion (Tallis, 1985a) or the status of rare species (Stoneman *et al.*, 1993). In order to maintain and build on this work, the specific needs of palaeoenvironmental research must be considered in addition to the general values outlined earlier (Charman, 1994a). Since there are few palaeoecologists directly involved with statutory conservation agencies, it is important that these needs are effectively communicated to those responsible for the selection, designation and management of peatland conservation sites. These needs can be summarised under the following categories.

1. Distribution of sites

Good geographical coverage of sites is crucial in maintaining a high quality national palaeoenvironmental resource. For the reconstruction of national and international vegetation changes (Huntley and Birks, 1983; Birks, 1989), a network of well distributed sites is crucial. Likewise regional climatic reconstructions are only possible with well distributed, comparable site types (e.g. Stoneman, 1993). Within regions, a diversity of site types and sizes is important (see 4. below). The location of individual sites may further affect their value for palaeoenvironmental research, for example where the deposit is in close proximity to an archaeological site.

2. Length of record

The length of record is clearly important. While a long record is desirable in most cases and particularly if it is a continuous record of the Devensian Late glacial/Holocene periods (c.15,000 years), even fragmentary records may be valuable in particular regions where alternative sources of palaeoenvironmental data are lacking. Thus a remnant 7000-5000 BP record from a cut or eroded surface may still retain a palaeoenvironmental value. Short but high resolution records are also of particular interest.

3. Resolution of the record

The resolution of the record determines the precision with which environmental changes can be characterised and dated. Some studies benefit from very high temporal resolution so that a very rapidly accumulating deposit is potentially of higher value than a slowly accumulated peat. For example Rimsmoor, Dorset (Waton and Barber, 1987) has accumulated 18 m of peat in c. 8000 years at a maximum rate of 2.6 mm year^{-1}, potentially allowing 2-5 year resolution of environmental change.

4. Site typology

Different site types respond differently to external forces such as climatic change. In order to effectively discriminate autogenic (local) from allogenic (external) palaeoenvironmental changes, it is important to conserve the peat archive in a suitable range of sites within individual regions. Equally, different techniques are more readily applied to particular site types (e.g. raised mires are more suitable for *Sphagnum* macrofossil analysis).

5. Site size

Site size is an important consideration in palaeoenvironmental work as it affects the source area for pollen records (Jacobsen and Bradshaw, 1981). Every effort should be made to retain a series of different sizes of sites within an individual area.

Conservation strategies for palaeoecology

As stated at the beginning of this chapter, the calls for increased recognition of the archival value of peatlands in conservation policy and management have been growing over the last few years. There have been some positive moves towards this recognition in a number of areas. Firstly, the general concept is now an established feature of many peatland protection strategies (e.g. Stoneman, Chapter 45, this volume). Secondly, a number of specific sites have been identified as being important for the known geological value of their deposits through the Geological Conservation Review (GCR) (e.g. Campbell and Bowen, 1989; Gordon and Sutherland, 1993). These are welcome developments but may not necessarily ensure the establishment and permanent well being of the entire resource. One of the difficulties in assessing the archival value of peatlands is that it is difficult to provide a quick field based assessment of research potential. As a result, only intensively investigated sites such as those in the GCR are afforded clear protection. On many peatland SSSIs, even basic palaeoecological information such as the extent and depth of the deposit and site stratigraphy are unknown. In the light of this it is difficult to arrive at a balanced view of where important deposits are to be found, even within the existing SSSI network. It should also be remembered that SSSI status is not a guarantee that the archive will be preserved given the potential damaging impact of site management for the biological resource such as during re-contouring for 'regeneration' of cut-over mire.

A strategy is needed to address the remaining problems of conservation of peatland palaeoecology in Britain. There is not room to explore this in depth here, but as a minimum, this should include the evaluation of existing peatland SSSIs according to the values outlined in Charman (1994a) and the site characteristics introduced above. Much of this information would be available already in the published literature or from sources such as the recently established database of palaeoecological sites in Scotland (R. Tipping, pers. comm.). A further step forward would be the evaluation of additional sites outside the SSSI network. While this would be a major undertaking, there is a good deal of published information and selective fieldwork could provide further data if necessary. The ultimate goal should be a thorough evaluation of the peatland palaeoecology resource with a prioritised list of existing and required site protection measures.

Conclusion

The peatland palaeoenvironmental record is worth preserving in its own right but this chapter has shown there are also many lessons for conservation policy and practice to be learnt from it. These arguments are especially relevant as we move into an era of unprecedented climatic change (Houghton *et al.*, 1992). It is time to give this aspect of peatland conservation the attention it deserves by a thorough review of site quality and protection.

Chapter seven:

Volcanic Ash in Peat

Dr Jeff J. Blackford
University of Durham, UK.

Introduction

When volcanoes erupt in an explosive manner, water, gases and particulate matter are injected into the atmosphere. The most explosive types are 'Plinian' and 'Ultraplinian' eruptions, named after a description, by Pliny the Younger, of Vesuvius erupting in AD 79. Material ejected during the explosive phase can be travelling upwards at several hundred metres per second (Francis, 1993), which together with the effect of thermal convection can put material high into the stratosphere. The particulate matter is subsequently redeposited, with the largest fragments falling close to the point of eruption within seconds or minutes, but with the smaller particles staying airborne much longer, and travelling much greater distances. If dust penetrates the lower stratosphere, a cloud of the finest particles can circle the Earth within 2-6 weeks (Chester, 1993), subsequently spreading latitudinally. Once in the upper atmosphere, volcanic ejecta can alter the balance of incoming and outgoing radiation. This causes changes in the degree and location of atmospheric absorption and reflection of radiation, and therefore alters temperatures and circulation patterns (Rampino and Self, 1982; Sear *et al.*, 1987; Handler, 1989; McCormick *et al.*, 1995). Small particles that are erupted into the troposphere become entrained in airflows that are themselves modified by the eruption. Volcanic ash typically smaller than 120 µm may be transported thousands of kilometres before falling to the surface.

Various eruptions, mostly of Icelandic volcanoes, have produced fine ash in sufficient quantity, and projected them to a height great enough, to reach Scandinavia and northern Germany (Thorarinsson, 1981; van den Bogaard *et al.*, 1994). In 1989, Andrew Dugmore of the University of Edinburgh located and identified an ash layer from Caithness (Dugmore, 1989). Since then volcanic ash has been found in Shetland and Orkney, the Hebrides, mainland Scotland, Ireland and England. These finds come from peat deposits and lake sediments, and in some cases have been identified and linked to a specific eruption. The purpose of this chapter is to briefly outline how volcanic ash-layers (tephra layers) can be found in peat deposits, to discuss what they can be used for once located, and then to discuss why peat deposits are particularly valuable in this context.

Finding and identifying tephra layers in peat

In ombrotrophic peat such as raised mires or the watershedding parts of blanket peats, there is little minerogenic material. A fall of tephra will add substantially to the mineral component of any given layer of peat. If the tephra fall is of sufficient concentration, the layer may be visible to the naked eye. Further from the source of the material, however, this is not the case, and detection methods are needed to find less concentrated falls of smaller shards. In some cases, layers may be shown by X-radiography of core sections (Dugmore and Newton, 1992), and occasionally magnetic properties of the core may be useful (van den Bogaard *et al.*, 1994). Both of these methods, however, will fail to detect the finest ash layers, which at present can only be found by optical scanning of the mineral fraction of contiguous samples. This fraction of the peat is obtained by removing the organic material, either by combustion (Pilcher and Hall, 1992) or by acid digestion (Dugmore *et al.*, 1992). Combustion alters the geochemistry of the particles, but is the most rapid preparation for the initial location of tephra. The material left after either preparation method is mounted on a light microscope slide and volcanic dust is searched for manually. Tephra particles, which resemble shards of broken glass, can be identified by their morphology and optical properties (Heiken, 1974). Tephra shards are characterised by angular edges, random shapes, a smooth surface even under phase contrast, vesicular forms, column-like structures and sometimes tints of yellow or brown colour (Westgate and Gorton, 1981). Under polarised light, volcanic glass has no refractive capability and cannot be seen. A microscope fitted with a rotating stage and polariser can be used, therefore, to separate tephra from most other siliceous minerogenic fragments extracted from peat samples. Although the observer cannot be absolutely sure of the origin of an individual fragment until the geochemistry is known, when a sample contains many particles that fit this description, especially in ombrotrophic peat, there is very little else they can be. Plant-cell silica deposits, including symmetrical and consistently shaped phytoliths, can usually be distinguished by shape and by a very fine surface roughness.

By using a known quantity of sediment, it is possible to estimate tephra concentration, in terms of numbers of pieces per cm^3 (Blackford *et al.*, 1992) or alternatively the results can be quantified by calculating the percentage of mineral fragments that are tephra, or tephra-like (Bennett *et al.*, 1992).

Once located, ash layers can be identified by analysing the geochemistry of the individual particles; 'fingerprinting' (Dugmore, 1991). This is possible because each eruption, even of the same volcano, produces tephra with a different chemical composition (Self and Sparks, 1981; Dugmore, 1989; Dugmore *et al.*, 1992). The technique employed is EMPA (electron microprobe analysis), which has been shown to produce replicable and distinctive major-element abundance data for each individual particle (Sweatman and Long, 1969; Dugmore *et al.*, 1992) when applied under carefully controlled experimental conditions (Hunt and Hill, 1993).

Uses

Dating and correlation

Peat bogs have been well documented as sources of environmental information (see various chapters in this volume). The technique most regularly used to directly determine the age of the environmental changes revealed by bogs has been radiocarbon dating. This involves a significant error in correlation and dating, due to inherent problems of contamination, calibration and an inevitable laboratory counting error (Pilcher, 1993). Once calibrated, a typical radiocarbon date from peat can fall within a range of up to 500 calendar years (Blackford, 1993; Pilcher, 1993). This degree of precision is not sufficient to allow the record in peat bog cores from different areas to be exactly - or confidently - compared, although improvements in calibration have been made and higher-precision dates are possible. Tephrochronology, using volcanic ash layers as marker horizons, appears to be at least a partial solution to the problems of dating and correlation for a number of reasons:

1. *Precise correlation.* Because most of the erupted particulate matter >10 μm falls out of the atmosphere within days, volcanic ash can provide precise correlation wherever layers in different places can be shown to be from the same eruption.
2. *Precise dating.* The exact date of some historical-age eruptions is documented. Locating and identifying the tephra from these eruptions, then, can be used to precisely date the sediment in which the tephra is found. Precise dating may also be possible for prehistoric eruptions if the links between ice core records, tree-ring sequences and tephra layers can be firmly established.
3. *Radiocarbon accuracy and precision.* A single age-estimate by radiocarbon measurement is less accurate than an estimate based on a number of determinations of the same sediment. The organic sediments surrounding tephra layers can be dated several times, improving the precision, or wiggle-matched dates can be obtained (e.g. Hall *et al.*, 1994). Once a high-precision date is obtained for a tephra layer from one site, then that age-estimate can be used wherever the tephra is identified without further radiocarbon measurement.
4. *Future potential.* As research into tephra progresses, previously unrecorded layers are located. At present, their age and origin may be unknown, but if they

are successfully analysed geochemically, they may be used for dating and correlation in the future, when the same layer is recorded elsewhere. Volcanic ash layers of unknown age and origin can still be used for precise correlation.

5. *Numbers of layers present.* Core-correlation is possible between many fixed points. In the British Isles there may be 15 or more different layers (Tephra Conference and Workshop, Cheltenham, December 11th 1994). Closer to the source of the volcanic ash this figure is much increased; in Iceland, a single peat profile can contain over 100 tephras.

Tephra layers, then, can be used as a means of correlation, that is of showing that one or more points in different sedimentary sequences are of the same age. For the time period covered by most peat bogs, there is no better way of correlating records from different bogs because it can be assumed that the ash-deposits fell within days of each other in different places. Re-deposition of particles after the original ash fall is possible, and is a significant problem in lake sediments. Although the tephra layer may be dispersed slightly in peat, the horizon of original deposition can still be estimated by pinpointing the maximum concentration of particles (Blackford *et al.*, 1992).

An example of correlation potential in areas distant from the eruptive source can be shown from work by John Pilcher and co-workers at Belfast (Pilcher and Hall, 1992), whose preliminary results showed that at least 12 layers were present in three mires, with perhaps nine tephras present in all three (Fig. 7.1). The information contained in records from those mires can therefore be matched exactly at certain points. Furthermore, the presence of some of those layers is also to be expected in other regions, allowing correlation across the range of ash fall-out from each eruption. Tephra layers present in peat bogs may also be correlated with deposits in lake sediments, ocean sediment cores and ice cores. This would allow the correlation of environmental records, including proxy climate records, from a variety of sources. It may also prove possible to link the tree-ring chronologies with ash layers from mires, as major eruptions appear to cause a detectable tree-ring narrowness event in some records (Baillie and Munro, 1988; Baillie, 1991). Precise dating, as well as correlation, is possible in examples where the ash-layer is of known age (Einarsson, 1986). Within the European region, the ages of several classical-age Mediterranean eruptions, and most large-scale, post-11th century, Icelandic eruptions are known.

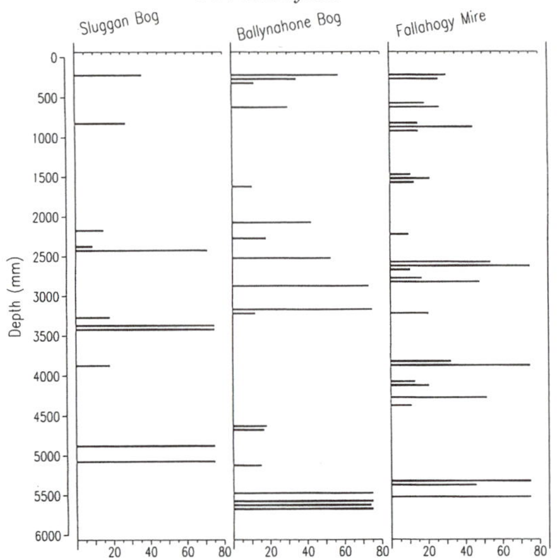

Fig. 7.1. Tephra peaks in peat cores from three northern Irish raised mires (after Pilcher and Hall, 1992). Units are the number of volcanic ash particles present on half a slide prepared by the combustion method. Samples containing less than 10 pieces have been omitted. These graphs show the number of tephra layers available, even at a distance of >1000 km from Iceland, the likely source.

For both dating and correlation, tephra layers can only provide part of the dating solution. This is because the tephra layers may not coincide with the horizons or features that need dating. Age-estimates for peat or sediment between tephra layers can be obtained only by extrapolation or by radiocarbon dating. What a sequence of tephras can do, however, is to provide a chronological framework for the whole sequence, with; (i) some precise dates, (ii) a series of estimated dates, and (iii) a number of exact points for correlation.

Assessing volcanic impacts

The impacts of volcanic eruptions have long been a matter of speculation and study. The effect of the historic eruption of Krakatau in 1883 was noted at the time, apparently linked with reduced insolation and surface temperatures (Rampino and Self, 1982). The eruption of Mt. Pinatubo in 1991, probably the biggest explosive eruption of the century, caused a lower-troposphere cooling of 0.5°C globally, and 0.7°C in the northern hemisphere (Dutton and Christy, 1992; McCormick *et al.*, 1995). This is equivalent to the entire global average rise attributed to anthropogenic greenhouse-gas production since the beginning of the industrial revolution. Previous eruptions, whose effects were not so closely monitored, have been even bigger (Rampino and Self, 1982). One way of assessing the impacts of unmonitored eruptions, including impacts at various distances away from the eruption, is to look at the fossil and sub-fossil (preserved rather than mineralised) remains of indicators of environmental conditions either side of volcanic ash layers (Edwards *et al.*, 1994). Studies in distant regions, for example 500-2000 km away, should show any regional-scale effects such as climatic perturbation, rather than being dominated by the direct local effects of the eruption.

An example is shown by the apparent coincidence between the explosive eruption of Hekla approximately 4000 years ago (Hekla-4) and the decline in the northern-most pine tree population in the UK at that time (Blackford *et al.*, 1992). This could have been caused by a short-term climatic change, an acid deposition event, or it could be a coincidence (Blackford *et al.*, 1992; Grattan and Charman, 1994; Hall *et al.*, 1994; Edwards *et al.*, 1996), and needs testing in various ways (Birks, 1994). Whatever the relationship, the tephra can be used as a marker horizon for further investigations of the history of pine, as Hekla-4 is also present in Ireland and Scandinavia - an example of the use of tephra for precise correlation.

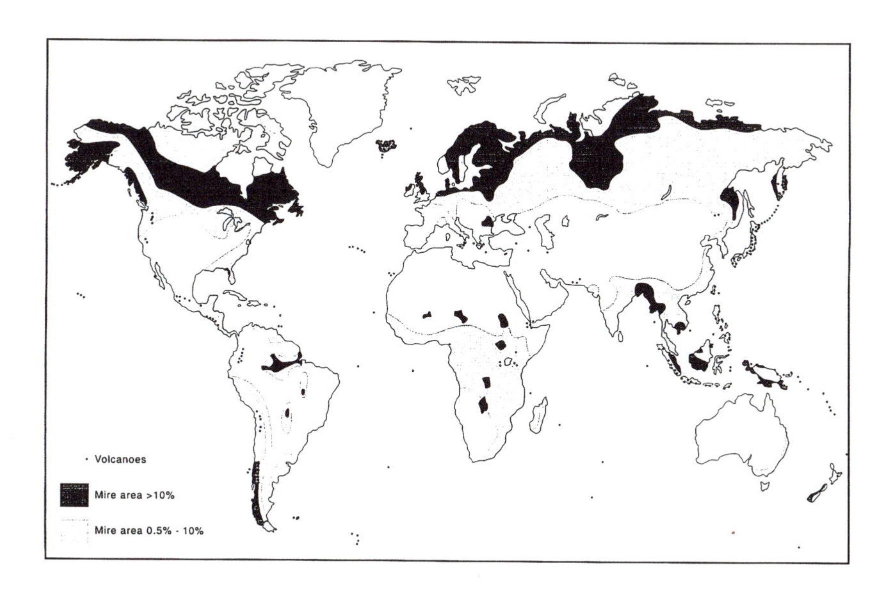

Fig. 7.2. Peat distribution on a global scale (after Gore, 1983; Lindsay *et al.*, 1988). Areas in light shading are those within which between 0.5% and 10% of the land surface is peat covered. Darker shaded areas contain >10% peat. Circles represent the position of active volcanoes (after Decker and Decker, 1989; Chester, 1993; Francis, 1993). This map shows the overlap in ranges of tephra sources and peat deposits.

On a global scale, there is considerable overlap between areas that have been covered by frequent ash-falls during the past millennia and the areas of peat accumulation (Fig. 7.2). As well as in the European region, studies of stratigraphic correlation, palaeoclimatology and volcanic impacts have been, or could be, implemented in regions including South and Central America, Central Africa, West and North West North America, Kamchatka, Japan and the Kuril Islands, South East Asia and New Zealand.

Conclusion: the importance of volcanic ash in peat

The record preserved in peat is of particular importance for the following reasons:

1. The tephra layers are well preserved, usually geochemically unaltered, and remain in place. They can be readily located in, and extracted from, raised mire and blanket bog cores.
2. They can be used to directly date and exactly correlate the climatic, ecological, agricultural and archaeological information present in the peat record.
3. The presence of volcanic ash layers in peat deposits provides a means of testing the effects of prehistoric volcanic eruptions on local, regional and global scales.
4. The global distribution of peat and that of volcanoes means that there are many areas where tephra layers are present in mires.

Peat bogs contain a range of environmental indicators, showing past conditions both at the site of the peat itself and in the surrounding region. The types of indicators used include pollen, seeds, leaves, wood, charcoal fragments, fungal spores, algal remains, amoebae, insects, animal remains and the isotopic ratios of oxygen and carbon. Individually or collectively these variables can be used to study hydrological conditions, temperature, agriculture, forest history and ecological processes. At any given time it is probable that at least one of the indicators present in the peat record is close to an environmental threshold. Within the range of testate amoebae present, for example, at least one species may be living at or near the limits of its habitat range. Individual species will therefore be sensitive to even short-term climatic perturbations. Changing proportions of different indicators, each with a different tolerance, can also be used in this context. It is also possible to use the indicators available in a combined manner, each providing a slightly different signal, to reconstruct former conditions and changes in environmental parameters. The palaeoecological and palaeoclimatic records from mires are greatly enhanced by the improved dating possibilities of tephrochronology, and for the first time such records can be exactly correlated.

Birks (1994) suggested that to investigate the impacts of prehistoric eruptions on vegetation, as shown in the palaeoecological record, requires appropriate project design, sampling, implementation and statistical techniques. Suitable sample sites are needed that are close to, or at, the current and former range boundaries of responsive species. Peat bogs currently provide the required sites, with a unique spatial coverage and a wide range of potential environmental indicators. If enough mires are preserved and their waterlogged, preserving conditions maintained, then the presence of tephra layers in peat bogs can provide important information about the climatic and other environmental effects of ancient volcanic events.

Acknowledgements

I am grateful to the staff of the Scottish Wildlife Trust for organising a stimulating conference and inviting me to contribute. Dr J. Pilcher allowed the derived diagram (Fig. 7.1) to be used. Figures 7.1 and 7.2 are used with the permission of Edward Arnold and Elsevier Scientific Publishing. David Hume produced Fig. 7.1. Andrew Dugmore, Paul Buckland, Jane Boygle, Bob McCulloch, Peter Hill and Olavi Taikina-Aho have in the past advised on tephra matters.

Chapter eight:

Peat Erosion: Some Implications for Conservation and Management of Blanket Mires and Heather Moorland in the UK

Dr Anson W. Mackay
University College London, UK.

Introduction

The Peatlands Convention, held in Edinburgh in 1995 by the Scottish Wildlife Trust focused on the need for a detailed inventory and conservation strategy for peatlands throughout the United Kingdom. In this chapter, I focus on the nature of erosion in peatlands (i.e. blanket mires and heather moorland) and the implications that erosional processes may have for the conservation and management strategies of these ecosystems.

The nature of peat erosion

Peat erosion is widespread throughout the upland peatlands of Great Britain: for example, in northern Scotland and the Outer Hebrides (Osvald, 1949; Walker and Walker, 1961; Lindsay *et al.* 1988); south-west Scotland (Stevenson *et al.*, 1990); north-west England (Mackay and Tallis, 1996); northern England (Bower, 1959; Radley, 1962; Tallis, 1965; 1985a; 1985b; 1987) and south-west England (Taylor, 1983), Wales (Bostock, 1980; Robinson and Newson, 1986; Francis, 1990). Four types of erosion predominate and are associated with a particular topographical situation. These are:

1. *Summit-type erosion* occurs in peat on higher, flatter ground (Osvald, 1949; Radley, 1962; Tallis, 1964a) and, when left unchecked, results in exposure of the underlying parent substrate, leaving only residual, and often widely separated, peat mounds (which are themselves often largely unvegetated);
2. *Type I gully erosion* is restricted to deep peat on flat areas, and forms an intricate pattern of gullies that occur very close to one another (Bower, 1959);

3. *Type II gully erosion* is more common on the sloping peat areas, and forms a system of sparsely branched channels aligned nearly parallel to each other (Bower, 1961);
4. *Marginal erosion* occurs at the edges of the main peat mass.

Controversy still prevails, however, as to the actual causes of peat erosion. Various theories have been proposed: climate change (Conway, 1954; Bower, 1961); biotic factors such as burning (Pearsall, 1950; Radley, 1965; Stevenson *et al.*, 1990), grazing (Shimwell, 1974), atmospheric pollution (Tallis, 1964b; Ferguson and Lee, 1983) and forest clearance (Mackay, 1993; Tallis, 1995); and that it simply occurs as a natural endpoint to blanket peat development (Lewis, 1906a; Johnson, 1957). Recent evidence points to different types of erosion being caused by different processes. For example, some recent studies have emphasised that gully type erosion is a long-standing feature of many blanket mire systems, perhaps pre-dating intensive upland land-use and pollution (Mackay, 1993; Tallis and Livett, 1994; Tallis, 1995). It is suggested that gully erosion was perhaps initiated by the active cutting back of upland streams, which in turn was initiated by forest clearance from upland slopes during the late Iron Age and Romano-British periods. Other studies however, have shown that erosion in some areas is of a more recent origin. For example, Mackay and Tallis (1996) suggest that summit-type erosion on the Bowland Fells in north-west England was initiated by fire during the early 1920s, and came about as a result of a combination of unusual circumstances: a period of below-average rainfall in the region in the early 1900s, resulting in lowered water tables in the peat; exceptional summer drought in 1921; and a decline in management standards because of shortage of gamekeepers after the First World War. A combination of conditions, as outlined above, may also have been responsible in causing erosion at other sites, e.g. on the Abbeystead Estate, Bowland (Moseley and Walker, 1952); North York Moors (Maltby *et al.*, 1990); Alport Moor, southern Pennines (Tallis and Livett, 1994).

Knowledge of the timing of the onset of erosion is therefore clearly important in assessing the possible roles of putative causal agents, such as moorland management, atmospheric pollution and recent climate change. However, dating of the inception of erosion may usually only be achieved by indirect methods, as definite documentary evidence is rarely available. Techniques employed include: estimations of contemporary rates of erosion (thereby allowing the calculation of the approximate length of time needed to remove a volume of peat equivalent to that lost in the study area in question (Tallis, 1981a; b; Labadz *et al.*, 1991); palaeolimnological techniques, for example accelerated sediment accumulation rates or increases in organic matter found in sediment profiles (Stevenson *et al.*, 1990)); and palaeoecological techniques such as pollen and macrofossil analysis of nearby peat profiles from uneroded areas (Tallis, 1985a; b; 1987; Mackay, 1993; Tallis and Livett, 1994; Mackay and Tallis, 1996).

Why protect these habitats from erosion?

Heather moorlands in the UK and Ireland are the best examples of their kind in western Europe, and our blanket mires are of even greater international significance (Lindsay *et al.*, 1988). Ecologically, both are internationally important habitats for invertebrates and many birds of prey (Thompson *et al.*, 1995). For example, the Bowland Fells are one of the few remaining areas in England where both dwarf-shrub heath and blanket mire are extensive together, and as such they have become an especially important breeding ground for red grouse (*Lagopus lagopus* ssp. *scoticus*), golden plover (*Pluvialis apricaria*), hen harrier (*Circus cyaneus*), merlin (*Falco columbarius*) and peregrine (*Falco peregrinus*) (Mackay and Tallis, 1996). The last three are afforded special protection under the Wildlife and Countryside Act (1981), by virtue of their rarity and vulnerability. Of these, the hen harrier is the most notable, with the Bowland Fells representing its only regular breeding locality in England. Economically these habitats are important as grazing land for sheep and for rearing red grouse (e.g. White and Wadsworth, 1994), as reservoir catchment areas (Labadz *et al.*, 1991), and as recreation and tourist attractions (e.g. Tallis and Yalden, 1983; Lancashire County Council, 1990). Increased erosion leads to a change in the hydrological balance of these habitats, causing them to dry out further, thus promoting the spread of graminoid species at the expense of a mosaic of vegetation (which may lead to a reduction in invertebrate biodiversity (Usher and Thompson, 1993)). It is also, almost certainly, a contributory factor in the increase in the incidence of wildfires (see below).

Blanket mires and heather moorland are also important scientific archives for historical changes in vegetation in the UK over most of the Holocene, as well as being rich in archaeological remains (Barber, 1993; Buckland, 1993). Increasingly, over the last few years, there has been a resurge in palaeoecological contributions to aspects of global climate change (e.g. Gadjewski, 1993), vegetation science (e.g. Birks, 1993), mainstream ecology (e.g. Davis, 1994) and archaeology (Buckland, 1993). A continuous loss of this record therefore, by erosion, would be a serious loss to many academic disciplines. Furthermore, the incidence of erosion may have indirect, but important, implications for global climate change, again through the general lowering of water tables and increased oxidation of the peat column (Heathwaite, 1993). Recent studies have suggested that these hydrological changes may produce an increase in the fluxes of important greenhouse gases such as CO_2 and N_2O (Moore and Knowles, 1989; Freeman, Lock and Reynolds, 1993).

Prevention of erosion of peatland ecosystems

It is important to stress here that some forms of erosion are long standing features of many peatland ecosystems and may even be an integral part of mire development. Thus, there may be little we can do to reverse the effects of certain types of erosion on the surrounding landscape. However, other forms of erosion are more recent, and can be linked to anthropogenic activities in combination with other factors, such as accidental burns during periods of prolonged dry climate (e.g. Mackay and Tallis, 1996).

To prevent erosion occurring, it is necessary to maintain an intact surface vegetation cover. Possibly the main threats to an intact surface vegetation come from catastrophic burns (either from poorly managed muir-burns, or from those that arise accidentally) and the prevention of recolonisation, and the extension of existing bare areas of peat. Other factors have also been implicated in destroying the vegetation cover of peatlands, e.g. atmospheric pollution (Ferguson and Lee, 1983), although pollution may have been more responsible for the prevention of recolonisation of already existing bare peat surfaces.

Accidental fires

The importance of accidental moorland fires in the large-scale degradation of upland peat landscapes has really only recently been realised. Poorly managed muir-burning can also result in uncontrolled fires (Phillips, 1981), although severe fires that destroy large areas of heather moorland and blanket mire more usually arise accidentally during the summer months in years of severe drought (Tallis, 1981c). Catastrophic fires, such as these have been documented across England (Moseley and Walker, 1952; Radley, 1965; Phillips, 1981; Tallis, 1981c; Maltby *et al.*, 1990), can destroy the matrix of peat turning it to ash (Maltby *et al.*, 1990). Subsequent recolonisation by vascular plants is often slow, and is usually preceded by the stabilisation of the peat surface by lichen and bryophytes (Tallis, 1981b; Maltby *et al.*, 1990). Peat that does not undergo recolonisation becomes exposed to large temperature fluctuations which can promote the further breaking and fragmentation of the peat surface, which, in turn, leads to greater peat removal by increased water run-off and higher wind velocities above the peat surface (Fullen, 1983).

The occurrence of wildfires is linked to prolonged drought and to lowered water tables in the peat mass; conditions which may be enhanced in the UK under predicted scenarios of global warming (Heathwaite, 1993). Consequently, fire degradation of our moorlands is also likely to increase. This has important implications for the conservation of our uplands. If these fires are primarily accidental, as Tallis (1981c) suggests, then a drive for increased awareness among the public (similar to that carried out for areas of woodland and forest) is undoubtedly needed to protect these vulnerable ecosystems from future damage. A detailed study by Lancashire County Council (1990) into tourism within the Forest of Bowland found that the levels of visitors increased dramatically with exceptionally good weather conditions over the summer season, thus increasing the risk of fire through thoughtlessness.

Sheep population densities

Bare peat is very susceptible to erosional processes. Probably the most significant factor which acts to prevent recolonisation is high sheep stocking densities. Management of heather moorland and blanket mire usually represents a compromise between management for red grouse and that for sheep (although the Highlands of Scotland are also managed for red deer). Between 1947 and 1980 up to 20% of heather moorland in England and Wales has been lost due to the rapid increase in sheep stocks resulting from EC farm subsidies (Hill Livestock Compensation Allowances) given out in the 1970s and 1980s (Thompson *et al.,* 1995). Grazing of moorland vegetation by sheep not only induces the decline of *Calluna* and the appearance of grasses such as *Agrostis* and *Festuca* (Nicholson and Robertson, 1958) and sedges such as *Eriophorum vaginatum* (Rawes and Hobbs, 1979) but it has also been implicated in the prevention of recolonisation of bare peat surfaces by higher plants and bryophytes (Tallis and Yalden, 1983; Turtle, 1984; Grant *et al.*, 1985) and even in causing localised areas of erosion during periods of foddering.

Revegetative trials in the Peak District have shown that recolonisation of bare peat is possible by excluding sheep from affected areas (Tallis and Yalden, 1983). More recently, Anderson and Radford (1994) have demonstrated that sheep need not be completely excluded for recolonisation to occur, but that the density of the sheep population should be reduced to below 0.5 ewes ha^{-1}. In the longer term, however, more substantial shifts in policy are needed to reduce sheep numbers on moorland areas. Major gains will only be achieved if subsidies for farmers, based on numbers of sheep, are reduced, and farmers are instead encouraged to meet new conservation objectives and EC legislation requirements (Thompson *et al.*, 1995). To this end, the Government has just produced a White Paper on rural England (17-10-95) which continues to press the EU to reduce farm subsidies, and instead redirect them towards more 'environmentally beneficial and sustainable farming'.

Conclusions

Erosion poses a serious threat to the integrity of heather moorland and blanket mires within the UK. Paradoxically, within these internationally important habitats, we still know very little about the actual extent, and therefore possible future effects, of peat erosion in the UK. There have been a few notable studies carried out to address this problem, including one for the southern Pennines (Phillips *et al.*, 1981) and another, more recently, for Scotland (Grieve *et al.*, 1994). Phillips *et al.* (1981) estimated that up to 75% of blanket mires in the southern Pennines were affected by erosion. Grieve *et al.* (1994) found that - of the total area covered by a soil erosion survey of Scotland - up to 6% was affected by peat erosion. These studies emphasise that the problem of peat erosion is extremely widespread (probably more so than most people would have initially thought) and therefore deserves immediate attention. Lindsay (1995) reminds us that effective conservation can only occur if we have a good knowledge of exactly what we are trying to conserve, i.e. a detailed inventory. The time must now be right to detail the condition of our peatlands, noting the types and extent of erosion present, and use this information to target environmental policies to those habitats most at threat.

Acknowledgements

AWM would like to thank Dr D.J. Adger for comments on the manuscript, and to ENSIS Ltd. for providing funds to attend The Peatlands Convention 1995, in Edinburgh.

Chapter nine:

Pollution Records in Peat: An Appraisal

Dr Jennifer M. Jones
Liverpool John Moores University, UK.

Introduction

Evidence for contemporary trends in the deposition of organic and inorganic pollutants is readily accessible. Far less is known about past deposition rates of these elements. Retrospective studies of pollutant flux rates could provide insight into the magnitude of human impact on geochemical mobilisation. Certain sedimentary environments provide potentially useful contexts for the study of temporal trends in atmospheric deposition of pollutants. Those which have attracted most interest include snow and ice cores (Boutron and Görlach, 1990), lake and marine sediments (Galloway *et al.*, 1982; Rippey *et al.*, 1982) and peat cores (Livett *et al.*, 1979; Jones and Hao, 1993).

Pollutant deposition to peatland surfaces

At any one point in time and space, the atmosphere comprises a heterogeneous mix of dusts and gases derived from various sources. The relative contributions of these sources will differ at local, regional and global scales. Ultimately, all particles are removed from the atmospheric system by one of several mechanisms: sedimentation, impaction and diffusion (collectively termed dry deposition) and rainout and washout (collectively termed wet deposition). Particles derived from fossil fuel combustion are significant sources of heavy metals including aluminium (Al), cadmium (Cd), chromium (Cr), copper (Cu), iron (Fe), nickel (Ni), lead (Pb) and zinc (Zn). Some of these metals are potentially toxic to plants, animals and - ultimately - humans.

As a receptor of pollutants (both gaseous and particulate), a peatland presents a heterogeneous surface: several vegetation communities may be present and there may also be small-scale variability in topography e.g. peat hummocks. The main receptor surfaces on a peatland will comprise bryophytes (e.g.

Sphagnum), higher plants (e.g. *Calluna vulgaris*), open pools, and bare peat areas. All these receptor surfaces have different roughness factors.

The growth form and relatively high surface area of bryophytes ensure that they are particularly efficient at trapping particles. They have no vascular system, therefore there is good exchange between the plant tissue and the atmosphere. *Sphagnum*, in particular, has a high cation exchange capacity. These properties have resulted in several authors using moss bags filled with *Sphagnum* as low-cost deposit gauges to monitor spatial variability in contemporary pollutant deposition (Gailey and Lloyd, 1993).

Several authors have alluded to the presence of soot deposits in peatlands. Godwin (1981) recounts his experience of a trip to the Pennine blanket peat areas with the Swedish palaeoecologist, Osvald: '*I recall that Osvald, upon his first excursion to the south Pennine Eriophorum moors was staggered to find that his pale-grey flannel trousers (quite suitable for a Swedish trip) were soon generously striped with black soot collected from the stems and leaves of the vegetation: it was a sharp reminder of the influence of the industrial north.*'

Similarly, Conway (1949) working in the same area reported the presence of a soot layer in the surface peat of Ringinglow Bog. More recently, several groups have attempted to apply semi-quantitative soot-counting techniques to the determination of trends in particle deposition in peat cores.

Contrasting microtopographical structures on the bog surface also influence pollutant deposition. For example, *Eriophorum* hummocks act as obstacles for the impaction of particles carried in sub-horizontal laminar air flow. Particle deposition here differs markedly from that on to an open bog pool surface. Therefore, before individual element behaviour or physicochemical characteristics of the peat environment are even considered, it is apparent that there is considerable potential for complex deposition processes to occur.

Peat as a pollution monitor

If peat is to be used as a pollution monitor, several criteria need to be satisfied. Firstly, the peat should be ombrotrophic (i.e. rainwater-dependent). This characteristic ensures that the atmosphere is the primary source of particles and their associated heavy metals. Consequently, raised mires or blanket peats are the most appropriate environments for such studies. Secondly, there should be confidence that the core represents an intact time sequence. Therefore, care should be taken to choose sites where there is no evidence that significant peat erosion or peat winning has occurred. Pollen- or ^{210}Pb-based chronologies aid the determination of more meaningful pollutant fluxes. Thirdly, and most importantly, the approach relies on the assumption that the elements do not undergo any changes following deposition on the bog surface. This is by far the most contentious of the assumptions.

An appraisal of pollution monitoring using peat cores

One of the earliest attempts to relate the heavy metal deposition record preserved in peat to human activities was made by Lee and Tallis (1973). They showed that chemical analysis of dated horizons in blanket peat revealed good correlation between major industrial events and the deposition of lead. Subsequently, Livett *et al.* (1979) explored the potential of the approach further but still based their work on the premise that 'certain heavy metals remain virtually immobile once they are incorporated into peat'. It is only relatively recently that this assumption has been contested.

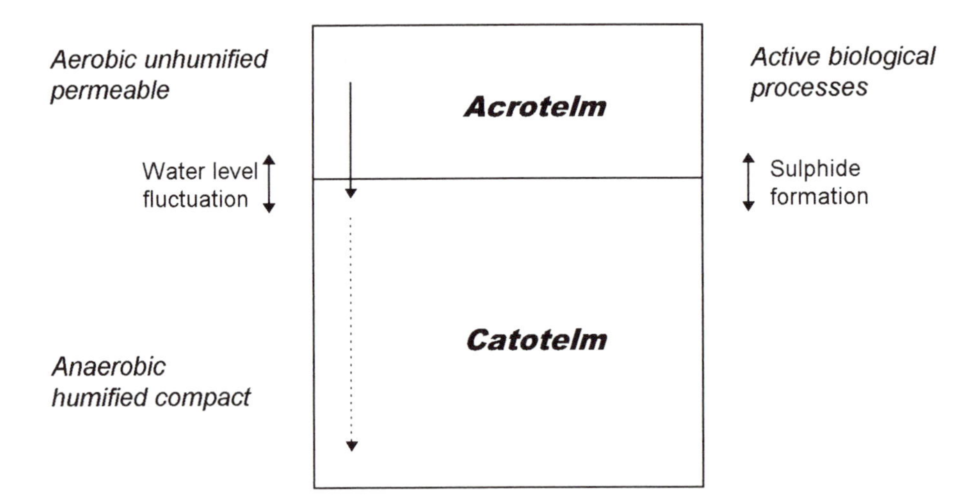

Fig. 9.1. Schematic diagram to demonstrate physicochemical differences between the acrotelm and catotelm.

The peat environment is both botanically and hydrochemically diverse. A peat body can be sub-divided into the upper acrotelmic peat and the lower catotelm (Fig. 9.1). The acrotelm can vary in depth from 10 cm to 60 cm. It is a zone where aerobic microorganisms are active due to the diffusion of oxygen from the atmosphere or rainwater, where the peat may be relatively unhumified and, consequently, more permeable (Hobbs, 1986). Conversely, the catotelm comprises more compact (and thus less permeable) peat where biochemical transformations are mediated predominantly by anaerobic bacteria. The acrotelm-catotelm boundary is determined by the level of the lowest water table within the peat. In biogeochemical terms, then, the peat environment is one of varying pH and oxygen status, in turn interacting with differences in plant type. Consequently, any element deposited to such peats may be influenced by the effects of several processes including plant uptake, dissolution, evaporation, diffusion, capillarity, adsorption, complexation, precipitation and co-precipitation as a response to variation in pH and oxidation conditions. It is all the more difficult then to accept that heavy metals do not undergo change once deposited.

During the 1980s numerous authors in the UK, Europe and North America evaluated the persistence of metal records in peat. In a comprehensive assessment of the biogeochemistry of peat, Clymo (1983) reported that 'some elements can form relatively insoluble sulphides and probably become less mobile in consequence when they pass into the anaerobic zone where sulphide is produced'. He also noted that 'much relocation of inorganic substances may occur' in the acrotelm. However, he concluded that 'there is still much to be learned about the chemistry of peat'. Certainly, depth-element profiles show considerable variability in the patterns of metal concentration. Table 9.1 summarises the outcomes of numerous investigations of heavy metal analyses of ombrotrophic peats. It is clear that there is little consensus about the persistence of heavy metals in peat thus casting doubt on the potential of peat cores for pollution monitoring.

Table 9.1. Summary of reported reliability of selected heavy metals for pollution history reconstruction.

Location	Al	Cd	Cu	Pb	Ni	Zn	Authors
Pennines, UK	*	*	+	+	*	-	Livett *et al.* (1979)
Pennines, UK	*	*	+	-	*	-	Jones (1987)
Galloway, UK	-	*	+	+	*	+/-	Clymo (1990)
Scandinavia	+/-	*	*	-	*	-	Damman (1978)
USA	_	*	+	+	*	+/-	Norton (1987b) Norton and Kahl (1987)
Canada	-	*	+	+	-	+	Shotyk *et al.* (1990, 1992)

+ (reliable); - (unreliable); +/- (ambiguous); * (element not analysed)

In 1987, Jones carried out sequential chemical extraction of Cu, Pb and Zn in ombrotrophic blanket peat. This technique attempts to extract metal species in sequence from highly mobile to relatively intractable forms. The author concluded that a considerable proportion of Pb and Zn deposition records are in a form predisposed to transformation and mobilisation. Thus these two metals are poor surrogates of pollutant deposition. In contrast, the Cu record was dominated by more chemically intractable forms and thus, comparatively, may be less mobile (Jones, 1987). This supports the view of Williams (1988) who contends that Pb and Zn are mobile in peat until they form relatively insoluble sulphides at the acrotelm-catotelm interface. Urban *et al.* (1990) also suggest post-depositional mobility of Pb in peat by proposing a mechanism where leaching is aided by high concentrations of dissolved organic carbon. Shotyk *et al.* (1992) corroborate this with their view that aqueous species of Pb and Zn will predominate in the acid bog environment. Jones and Hao (1993) applied factor analysis to heavy metal concentration data for Pennine peats. Their work indicated commonality between Al, Cu, Pb and Zn. If Pb and Zn are mobile it could be inferred that Al and Cu are also mobile. However, numerous authors corroborate the persistence of copper in the peat record (Livett *et al.*, 1979). The behaviour of Al in peat is more problematic. Aluminium may derive from both lithospheric and anthropogenic sources and this may account for the variation in results observed. Some authors believe that Al is mobile in peat while the evidence of others is less precise. Results of the analysis of Scottish peats show evidence of Al mobility; in contrast, those from Whixall Moss imply some persistence of the Al deposition record (Jones, 1985).

Conclusion

Despite over twenty years of research, the value of peat-based pollution history reconstruction remains questionable. Given the range of physicochemical processes operating in bogs, persistence seems unlikely. However, there is substantial evidence to support the view that Cu, at least, does maintain a coherent deposition record. The use of sequential chemical extraction together with multivariate statistics may further aid discrimination of complex data. The identification of persistent elements would not only allow pollution history reconstruction it could also provide a time-marker to substantiate other dating methods.

Acknowledgements

The author acknowledges the contribution of Jicheng Hao (formerly of Liverpool John Moores University) to the evidence for heavy metal speciation in ombrotrophic peat.

Part Three

Peatland Biodiversity

An Introduction

Dr Rob Stoneman
Scottish Wildlife Trust, UK.

For many who attended the Peatlands Convention, the reasons for conserving bogs are clear and simple. Bogs are part of our wild heritage and having access to wild country - of whatever type - is something we 'like' to have and something we feel we 'ought' to have. Joosten (Chapter 44, this volume) refers to this as an emotional and ethical need. Unfortunately, our love of all things boggy holds little sway against, for example, the considerable financial profit realised by peat extraction or the siting of a new bypass. It is important, then, to define the beauty and wonder of our peatlands in a little more detail. This Part examines one element of the delight - biodiversity.

Of course, we can only dip into the myriad of biodiversity the world's peatlands harbour. Here we look at *Sphagnum*, waders and dragonflies as a taster of the diversity of peatland life. What is revealed is the particularly unusual nature of this life.

Professor Clymo shows how *Sphagnum* creates the unusual conditions of waterlogging, high acidity and low nutrients - conditions in which Sphagna thrive.

Yet, in amongst the carpet of multi-coloured, spongy *Sphagnum* mosses, lives a wide variety of other plant-life including carnivorous plants: sundews in Europe, pitcher plants in N. America (*Sarracenia* spp.) and S.E. Asia (*Nepenthes* spp.). The unusual nature of the environment is emphasised by these odd shaped, insect-eating plants supplementing meagre soil (peat)-nutrients with their prey.

The special nature of peatland is further explored by Mr Brooks through its dragonfly fauna. Fully, two-thirds of Britain's 38 dragonfly species breed in acid bogs whilst 11 of those are virtually restricted to this habitat. Peat bogs may be highly acid with unproductive waters, yet the greatest dragonfly biodiversity is found in peatland habitats.

Whether faunal diversity is high or low, it is nearly always specialist. Dr Whitfield provides an example of this. For the twitcher, a trip to a bog is unlikely to produce an extensive bird list. Yet, Dr Whitfield's chapter shows the immense importance of conserving peatlands for some of Britain's rarest breeding birds - namely golden plover, dunlin and greenshank. Indeed, the blanket bog of Lewis may hold the highest breeding densities of golden plover in the world.

His chapter also highlights the cost of damaging these habitats by revealing the drop in wader numbers as afforestation takes its toll across the once pristine Flow Country.

The study of peatland biodiversity provides both a reason for their conservation and also gives a greater understanding of how we can conserve this remarkable habitat.

Professor Clymo's exposition on *Sphagnum* demonstrates this. He gives, for example, the theoretical background for the use of *S. fimbriatum* as a colonising species across bare peat surfaces (e.g. Sliva *et al.*, Chapter 32, this volume) as a pre-cursor to recreating raised bog conditions (Wheeler and Shaw, 1995b). Clymo shows that the species can cope with higher pH and calcium than other species allowing it to colonise peat where oligotrophic conditions cannot be assured. Once colonisation occurs, the dynamic between plant growth, precipitation and microbial activity serves to maintain high water-levels and change chemical conditions for oligotrophic *Sphagnum* species to grow.

Biodiversity offers a virtuous circle. We can enjoy it for its variety and wonder whilst its understanding leads us forward to more effective strategies for the conservation of that very enjoyment.

Chapter ten:

The Roles of *Sphagnum* in Peatlands

Professor Richard S. Clymo
Queen Mary and Westfield College, UK.

Introduction

The genus *Sphagnum* is morphologically isolated as the only one in its family, order and class of bryophytes. It contains perhaps 150 to 200 species, of which about a score are common. The best known of its special features is the two types of leaf cell (hyaline and chlorophyllose) arranged in a regular elongate diamond pattern. A similar but less differentiated pattern is seen in the fossil *Protosphagnum* from the Russian Permian about 200 million years old (Neuberg, 1960), and remains of plants indistinguishable from present day species are seen in some brown coals from Yunnan Province, south-west China, about one million years old (Lu and Zhang, 1986).

Most, but not all, species of *Sphagnum* grow in (and are restricted to) habitats that have acid, solute-poor, water. Most such systems accumulate peat - the dead, and partly decayed, remains of the plants that once grew at their surface. Northern hemisphere peatlands of this kind cover about 350 Mha (a square of side about 1900 km if all were to be herded together), have an average depth of about 2.3 m, have an average dry bulk density of about 0.11 g cm^{-3}, and an average carbon percentage of about 52% in their dry mass (Gorham, 1991). The total carbon in such systems is about 250 Pg (250 Gt) - about the same as the total mass of carbon in atmospheric carbon dioxide. We may guess that about half the peatland carbon is in *Sphagnum,* alive and (mostly) dead: *Sphagnum* is far and away the most successful bryophyte and, by the criterion of the amount of carbon incorporated, is possibly the most successful plant genus of any kind.

What makes *Sphagnum* so successful? There seem to be three main reasons:

1. it thrives in (most species can only tolerate) water with an unusually low concentration of dissolved solutes;
2. it makes the water around itself unusually acid;
3. it decays unusually slowly.

The first two of these factors result in *Sphagnum* being most abundant in situations where the live plants depend entirely on precipitation for their water. In these acid, solute-poor, conditions few vascular plant species can survive, and most of those that can are specialised. Slow decay leads to the accumulation of peat and to the most conspicuous example of natural ecological engineering - the huge peatlands of Minnesota, James Bay and the west Siberian plain.

It is well known that *Sphagnum* holds water like a sponge. Most of this water is held in capillary films between imbricate (overlapping) leaves, not (as is commonly supposed) in the hyaline cells (though these cells do contribute). Differences in water relations among species are the main cause of the restriction of particular species to particular microhabitats (e.g. pool, hollow, lawn or hummock) - but these differences do not account for the general success of *Sphagnum*.

Finally, *Sphagnum* has formidable powers of regeneration given suitable conditions and time.

In this chapter, I consider first the general economy of a growing peatland surface, then the three main causes of *Sphagnum's* success, then the water relations of *Sphagnum* and finally the regenerative powers of *Sphagnum*. Further detail on most of these may be found in Clymo and Hayward (1982), though there has been a lot of work on *Sphagnum* since that review was made.

The general economy of a *Sphagnum*-dominated peatland surface

Imagine a rainwater-dependent peatland approached on foot on a fine day. From a distance it may seem to be mainly covered by dwarf shrubs (mostly members of the *Ericaceae*) and linear-leaved sedges (such as those in the genera *Eriophorum, Trichophorum* and *Rhynchospora*). Closer inspection reveals *Menyanthes* in pools, *Rubus chamaemorus* on hummocks, and insectivorous plants of genera such as *Drosera, Utricularia* and *Sarracenia*. Closer still one can see that all these rooted vascular plants are set in a nearly continuous carpet of *Sphagnum*. These moss plants grow at the apex producing a continuous series of new branch primordia. The internodes and branches extend as the branches produce the characteristic leaves with green chlorophyllose and hyaline cells. The branches are of two types: spreading, and pendent around the stem. The pendent branches act as a wick allowing water to reach the apex externally (Hayward and Clymo, 1982). The branches and leaves form a porous but optically dense canopy which absorbs all but 1% of the incident light in its top 2-3 cm (Clymo and Hayward, 1982). Below this euphotic zone, light is so limited that respiration exceeds photosynthesis, the leaves (and most of the branch and stem cells) die, the green colour disappears, and the plants become light brown or straw-coloured. Nevertheless all the leaf and branch structures remain and the whole canopy, in which dry matter occupies barely 1-2% of the volume, is very porous. Water and gases can permeate freely, molecular oxygen in the air is abundant, and aerobic decay by fungi and bacteria is the dominant process in this moss litter layer. Some plants, such as *Rubus chamaemorus* and *Menyanthes,* decay rapidly and disappear almost completely leaving those that decay slowly, such as *Sphagnum* to dominate the moss litter (even if they did not do so to begin with) (Clymo, 1984a).

Let us follow what becomes of this newly dead moss litter. The green surface grows onward and upward gradually burying our layer. The load of new matter above our layer increases and the 10-20 fold greater load of associated capillary water increases too. At the same time the loss of matter by decay weakens the *Sphagnum* plant structures. Eventually the larger structures collapse, the dry bulk density increases 3-5 fold, and the space between remaining structures decreases by the same factor. The resistance to flow of water in channels is inversely proportional to a power, approximately four, of the width of the channel. As a result of the structural collapse, the resistance to the flow within the dead moss increases approximately 3^4-5^4 (i.e. 81-625 fold). Water, from above can no longer move easily downwards but is diverted sideways. As long as precipitation exceeds evaporation the peat, as it now is, below the collapse remains saturated. In periods of excessive precipitation the water table rises into the porous layer but, the higher it rises, the easier it is for water to flow away sideways, so the whole system is beautifully self-limiting. During a dry period the water table drops into the collapsed layer, but the first rains rapidly refill the pores and the water rises again to near the point of collapse. These mechanisms keep the water table within 3-5 cm of its mean position for 60-80% of the time (Clymo, 1992).

Just below the water table, aerobic decay continues but the rate of diffusion of oxygen in water is only 1/10,000 of that in air, so oxygen is used up faster than it can be replaced: the peat becomes anoxic. Bacteria able to exist anaerobically replace the aerobic fungi and bacteria. There is no obvious intrinsic reason why anaerobic metabolism should be slower than aerobic metabolism - anoxic sludge digesters in sewage works operate at a high rate - but in practice the decay rate of anoxic *Sphagnum* peat is about 100-1000 fold slower than that of oxic *Sphagnum* peat. Thus we get the spectacular accumulations of organic matter in peat bogs.

The whole process is dynamic: it depends on continued plant growth, continued precipitation, and continued fungal and microbial activity. It is, to a great extent, caused by *Sphagnum*, but it also creates the conditions in which most *Sphagnum* grows.

Relationship with solutes

The leaves of *Sphagnum* have only one layer of cells and no fatty cuticle (rooted vascular plants have much thicker, multi-layered, leaves and cuticles). There is no obstruction to direct absorption of solutes and, contrary to long-held beliefs, *Sphagnum* has an effective internal transport system for carbon- and phosphorus- containing compounds at least (Rydin and Clymo, 1989). In experiments *Sphagnum* is able to incorporate nitrate and phosphate rapidly from solutions of unusually low concentration (Clymo and Hayward, 1982), and has an inducible nitrate reductase (Woodin and Lee, 1987). Adding ammonium nitrate at 2 or 4 g m^{-2} $year^{-1}$ (= 0.7 or 1.4 mol m^{-2} $year^{-1}$ = 20 or 40 kg m^{-2} $year^{-1}$) stimulated the productivity of *S. balticum* but had no effect on the productivity of *S. magellanicum* growing in similarly wet lawns (Aerts *et al.*, 1992). But sodium di-hydrogen phosphate added at 0.2 or 0.4 g m^{-2} $year^{-1}$ (= 0.052 or 0.10 mol m^{-2} $year^{-1}$ = 2 or 4 kg m^{-2} $year^{-1}$) had no effect on the productivity of *S. balticum* but increased the productivity of *S. magellanicum*. These rates of addition are barely ten times those that fall in precipitation nowadays and are tiny compared with agricultural rates. Supplying nitrogen and phosphorus, at annual rates that were no more than one to five times the amount already in *Sphagnum* plants, reduced growth and proved lethal to some species (Clymo, 1987). This egregious behaviour has advantages and disadvantages: on the one hand, *Sphagnum* is able to flourish in situations where most plant species cannot; on the other hand, most *Sphagnum* species cannot spread into solute-rich situations.

There are similar relations with the concentration of calcium and hydrogen ions. In nature, most species of *Sphagnum* are restricted to waters that are acidic and have only low concentrations of calcium. In experiments (Clymo, 1973), however, the same species seem able to grow well in either high calcium concentrations or high pH (low hydrogen ion concentration); it is the combination of high calcium concentration and high pH that is lethal, and this is the combination almost always found in nature. A few species (examples are *S. squarrosum* and *S. fimbriatum*) are able to grow in waters that have moderately high calcium concentrations with, at least initially, high pH. These are the species that are able to establish in fens and thus act as John the Baptist to the main bog species of *Sphagnum*.

How does *Sphagnum* make the water around it acid?

Up to 30% of the dry mass of the cell walls of hummock species of *Sphagnum* is composed of uronic acid residues linked in long chain polymers (Clymo, 1963). A uronic acid is similar to a sugar but the terminal carbon has a charged acid COO^-H^+ attached instead of the neutral $-CH_2OH$ found in sugars. Such large polymeric molecules are not soluble in water: if they were, the plants would rapidly disintegrate. The uronic acids are made by the plant in the H^+ form but the bond between COO^- and H^+ is relatively weak so the H^+ may be competitively replaced by any other (positively charged) cation. The process is very rapid - it is essentially complete within 5-20 minutes. Rainwater, away from sea coasts, contains about 1 mmol l^{-1} of cations, mostly sodium, potassium, ammonium, calcium and magnesium. These compete with the H^+ on the uronic acid polymers and displace it into the water around the plants. This is the process of cation exchange, well-known in water softening, which in the case of *Sphagnum* makes the water acid. If the plants were not growing then, as with a water softener, there would come a time when all the exchangeable H^+ had been exchanged and the plants would no longer acidify the rain (and they would have been becoming steadily less effective from the start). But *Sphagnum* does grow and continuously produces new exchange sites loaded with H^+, so it continues to acidify rain that flows over it. In average conditions of precipitation and cation concentration and for *Sphagnum* growing in carpets, the plants can maintain a pH of about 4.0-4.2 around them (Clymo, 1967). On a hummock during periods with little precipitation and fast *Sphagnum* growth, a pH of 3.0 is possible both in theory and in practice (Clymo, 1984b).

There has been some argument (e.g. Shotyk, 1988) about the role of cation exchange in producing acidity in peat bogs because one can show that in many cases organic acid anions contribute a significant part of the total concentration of soluble anions in peat bog waters. But cation exchange with *Sphagnum* undoubtedly occurs, and on a scale sufficient to be able to account for the observed pH. The real problem is to explain where the soluble organic acid anions have come from. There is, in the cell walls of *Sphagnum,* a fraction of the polymers which is slowly soluble. This fraction contains many residues of a very unusual uronic acid: D-lyxo 5 hexo-sulopyranuronic acid also known as 5-keto-D-mannuronic acid or 5-KMA (Painter, 1991a; b; 1995). This is able to form cross-links between chains and to result in large soluble negatively charged (anionic) polymers, named sphagnans by Painter, that give the water a yellow or brown colour. Indeed it seems plausible that the colour of peat bog water is due to these molecules and not, as is often supposed without any evidence, to phenolic substances. It may therefore be that the rapid primary process, complete within half an hour, producing acidity around *Sphagnum* is cation exchange but that there is a secondary process involving the much slower leaching of residues of 5-KMA that were originally in the H form but which have become associated with other cations before both rejoin the H^+ in the water around the plants.

The sphagnans are important in other ways too. They may be responsible for the tanning of peat-bog bodies (such as Tollund Man and Lindow Man) and they seem to have bacteriostatic properties: brown bog water was taken on voyages by the Vikings because it remained fresh for longer than colourless water did. Painter (1991a; b;1995) suggests that the bacteriostatic property is a result of the ability of sphagnans to sequester metal cations thus depriving bacteria of essential metals. He also suggests that the same mechanism may, at least in part, account for the slow decay of *Sphagnum* peat.

Why does *Sphagnum* decay so slowly?

One must consider decay in oxic and anoxic conditions separately. In the surface oxic layers the rate of decay is related to temperature and water content but in a similar way for most species of peatland plant (Bliss *et al.*, 1981). There are, however, large and approximately linearly correlated differences in decay rate and concentration of nitrogen (Clymo and Hayward, 1982). Material, such as the leaves of *Calluna vulgaris,* with a nitrogen concentration of 1.5% on a dry mass basis, loses 50% of its dry mass in a year. But *Sphagnum,* with 0.5% nitrogen, loses only 15% in a year, and some species of *Sphagnum* lose even less. Such low rates of loss cause *Sphagnum* to become proportionately over represented in the material which is submerged by the rising water table and which becomes anoxic. This resistance to decay is supplemented by unpalatability to animals - vertebrate and invertebrate. There are few, if any, authenticated examples of an animal ingesting *Sphagnum* except inadvertently.

Slow decay of *Sphagnum* in anoxic conditions is perhaps the main reason why deep deposits of peat accumulate. But the reason(s) for slow decay are still obscure. The temperature is not particularly low: a metre or two down, the temperature is nearly constant at the annual mean which is 8-10°C in some 10 m deep peat deposits. Nor is the concentration of nitrogen particularly low: it seems that much of the nitrogen is sequestered so that as carbon is lost so the C/N quotient increases (Malmer and Wallén, 1993). Painter's suggestion that sphagnans sequester metals and so inhibit bacterial growth is a plausible hypothesis and should be tested.

The water relations of *Sphagnum*

Suppose that rain fell on a carpet of *Sphagnum* on a low hummock until an hour ago, since when water has drained away. In this state, the mass of water in the top few centimetres is about 30-40 times the oven-dry mass of the plants (Clymo and Hayward, 1982). Of this total, the water associated with the hyaline cells of the leaves and the porous cells of stem and branches is about six times the dry mass, and water in fine capillaries in the cell walls is about two times the dry mass. This means that water held in capillary spaces between leaves is about (22)-27-(32) times the dry mass. The percentages of water in fine capillaries: porous cells: between leaves is thus approximately 6: 17: 77. In wet conditions therefore most of the water is in relatively large spaces between leaves where it moves easily up and down outside the *Sphagnum* plants. During longer dry periods the water in these spaces is the first to be lost and the water in the porous cells moves less easily. In prolonged dry periods even water in the porous cells disappears and the plants become papery in appearance. But *Sphagnum* plants are more resistant to desiccation than is usually supposed and most that have become papery for several weeks will recover when rain comes. The ability to withstand desiccation is uncorrelated with the normal habitat: *S. auriculatum* which usually grows in hollows or pools is among the most resistant species; *S. cuspidatum* of hollows and pools is of intermediate resistance; *S. papillosum* of lawns is among the least resistant; while *S. capillifolium* of hummocks is of intermediate resistance (Clymo, 1973).

There are differences among species, however, that are closely related to the microhabitat in which each species usually grows. When the water table is only one centimetre below the surface the rate of evaporation of water from the surface is similar in all species, but when the water table is 10 cm below the surface then the evaporation from *S. capillifolium* of hummocks is greater than from *S. papillosum* of lawns, and this in turn is greater than from *S. cuspidatum* of hollows (Clymo, 1973). The cause of these differences is the numerous small spaces between leaves of *S. capillifolium* compared with the fewer, larger spaces in *S. papillosum* and the very few spaces in *S. cuspidatum*, which has few pendent branches and those it has do not form a continuous capillary path to the apex of the plant (Hayward and Clymo, 1982). As the water table drops so the water tensions increase and the larger capillary spaces empty. *S. capillifolium*, with its abundant small spaces, maintains a continuous external capillary path while most of the paths in the larger *S. papillosum* are broken. It is interesting to note that, in experiments, both *S. cuspidatum* and *S. capillifolium* grew best in hollows, although *S. capillifolium* was the least successful of the two. On hummocks, however, even though it was not at its best, *S. capillifolium* grew better than other species, because of its superior water-conducting ability (Clymo and Reddaway, 1972).

Regeneration

Sphagnum establishment in a new site is most likely to be from spores, which germinate to form a filament, then a plate, and from that develop the conspicuous three-dimensional growths with stem, branches and leaves. The apex may become so large that it 'divides' (there is discussion about the details of this process). The apex exerts dominance over the stem and branches below it, but if the apex weakens or dies then one or more new stems may grow from patches of cells in the axils of branches below, and these new apices and stems may take over from the original one (Clymo and Duckett, 1986). We do not know if any, or the majority, of the apices we see on the surface of a 6000 year old peatland are the same genet as the plants that originally established the peatland. The molecular techniques exist to decide this question, and the *Sphagnum* plants we see are haploid, but the necessary work would be tedious and expensive.

If one takes one centimetre thick slices of peat from a 30 cm diameter core of *Sphagnum* and the peat it has formed and keep them in sealed transparent bags in the light then there is abundant regeneration of *Sphagnum* down to 30 cm depth. This is from a depth where the peat may be 50 years old or more. The *Sphagnum* regenerates from spores (which may have washed down) and from leaves and, especially, from the patches of cells in the axils of branches (Clymo and Duckett, 1986). Whether or not this ability to regenerate is ever exercised in nature we do not know.

Conclusion

Sphagnum is not only the most successful of the bryophytes; it is a notably successful plant by any standards.

Chapter eleven:

Waders (Charadrii) on Scotland's Blanket Bogs: Recent Changes in Numbers of Breeding Birds

Dr D. Philip Whitfield
Scottish Natural Heritage, UK.

Introduction

The outstanding importance of the blanket bogs of northern Scotland for birds is well-known (Stroud *et al.*, 1987; Ratcliffe and Thompson 1988; Avery and Leslie, 1990; Ratcliffe, 1991; Pritchard *et al.*, 1992). Three species of waders (Charadrii) are especially characteristic of the northern flows: *Pluvialis apricaria* (L.) (golden plover), *Calidris alpina* (L.) (dunlin), and *Tringa nebularia* (Gunnerus) (greenshank). These waders were at the forefront of the debate between conservationists and developers over the impact of widespread commercial afforestation of the 'Flow Country' of Caithness and Sutherland during the late 1970s and 1980s (Stroud *et al.*, 1987; Lindsay *et al.*, 1988; Avery and Leslie, 1990). Many peatland sites were surveyed for birds during this period to: (i) derive bird population estimates, (ii) assess impacts of afforestation, and (iii) identify remaining areas with high breeding densities (Stroud *et al.*, 1987; Avery and Haines-Young, 1990; Avery and Leslie, 1990).

With the removal of tax incentives which assisted afforestation and UK Government Ministerial clearance for notification of up to half of the remaining peatlands as Sites of Special Scientific Interest (SSSI), the controversy abated (Avery and Leslie, 1990). There has been no published information, however, on how bird numbers have changed since such widespread protective measures came into effect.

In this chapter the results of recent surveys of peatland waders in Caithness and Sutherland are presented and compared with results from surveys in 1979-87. Comparisons are also made with estimates of current breeding numbers and changes in numbers for peatland waders on the Isles of Lewis and Harris in the Outer Hebrides.

Methods

Field methods and sites

Areas of 80 km^2 and 258 km^2 of blanket bog in Caithness and Sutherland, composed of 47 survey sites, were surveyed for birds during 1993 and 1994 respectively (Arnott *et al.*, 1994; Bates *et al.*, 1994a). The 'Constant Search Effort' (CSE) survey method of Brown and Shepherd (1993) was followed. The CSE method involves two site visits and was developed in areas of low breeding density. This is reflected in the distance criteria employed in making analytical decisions as to whether observations of birds seen on the same or different survey visits represent the same or a different 'breeding pair'. For *C. alpina*, individuals are deemed to represent different pairs on the same visit if the distance between them is at least 200 m, and pairs are considered to be separate from one another only if they are at least 500 m apart on the different visits (Brown and Shepherd, 1993). To produce accurate estimates these distance criteria should reflect the distances separating breeding pairs and distances birds move. A further 12 sites (covering 89 km^2) that had been surveyed at least once between 1979 and 1987 (Stroud *et al.*, 1987) were re-surveyed in 1993 or 1994 using the same method, the 'Line Transect' (LT) method of Stroud *et al.* (1987). These 12 sites were representative of the range of blanket bogs in Caithness and Sutherland (Lindsay *et al.*, 1988) and were selected without regard to their importance for birds.

Areas of 265 km^2 and 232 km^2 of peatlands on Lewis and Harris, representing 25 sites, were surveyed using the CSE method during 1994 and 1995, respectively (Bates *et al.*, 1994b; Shepherd *et al.*, 1995). Seven sites (covering 42 km^2) which were surveyed during 1987 (Stroud *et al.*, 1988) were re-surveyed in 1994 using the same (LT) method. A further four sites surveyed in 1987, using the LT method, were partially or wholly re-surveyed in 1994 or 1995 by the CSE method. All but one 'CSE' site and all re-surveyed 'LT' sites were on Lewis.

Analysis

The CSE analysis methods that determined which field records of birds were considered as 'breeding pairs' (details in Brown and Shepherd, 1993) were used for all species on CSE sites, except *C. alpina*. For *C. alpina* individuals were deemed to represent different pairs on the same survey visit only if they were at least 50 m apart and 'pairs' were considered to be separate from one-another only if at least 50 m apart on the different survey visits (200 m and 500 m, respectively, were the original CSE 'distance criteria' used by Brown and Shepherd, 1993). The 50 m criteria are more appropriate than the original CSE criteria over a range of *C. alpina* densities and unlike the original CSE criteria do not seriously underestimate *C. alpina* breeding numbers at high density (D.P. Whitfield, unpublished data).

The distance criteria used in LT analysis (as reported by Stroud *et al*., 1987) are unclear, except for *C. alpina* where 50 m is used, as birds have to be in 'the same area' on different visits to qualify as the same pair. In this study, 'the same area' was taken to be 500 m for all species except *C. alpina*. To negate any effects of interpretation differences, because the original LT analysis method was unclear, all pre-1993 surveys were reanalysed using the set distance criteria.

Statistical analyses of changes in populations were made on the basis of comparisons between estimated numbers of breeding pairs. For *C. alpina* in Caithness and Sutherland the unit of comparison was the number of birds seen on a single survey visit within the same 10 days in June, as this obviated the effect of apparent differences in original field surveyors' interpretation of calling/alarm calling in *C. alpina* and was broadly equivalent to the estimates of breeding pairs when all surveys were considered. When a site was resurveyed the estimated number of breeding pairs was judged to have changed if it was ±20% different from the previous surveyed number. The level of 20% was an arbitrary choice that was used for illustrative purposes only and to discount possible variation in breeding number estimates due to 'sampling error' (as judged by re-survey of the same site in the same or consecutive years). If, during the original survey period (1979-87), a site had been surveyed more than once then the latest survey was chosen for comparison with a 1993-94 survey.

On sites that were surveyed in the same year using both methods there was no significant difference between LT and CSE estimates of breeding numbers (using the above analysis methods) (D.P. Whitfield, unpublished data).

Results

Breeding densities

Densities of all three wader species in 1993-95 were greater on sites in Lewis and Harris than in Caithness and Sutherland (Figs. 11.1-11.3). These differences were significant for *P. apricaria* and *C. alpina* (around 2x and 8x greater densities, respectively, on Lewis and Harris: Kolmogorov-Smirnov Tests, $P<0.001$ for both species).

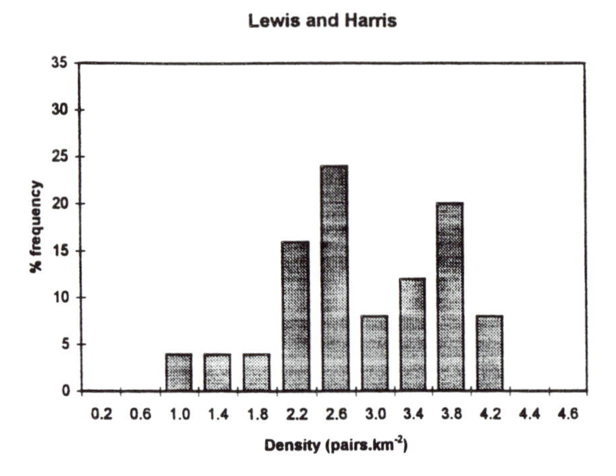

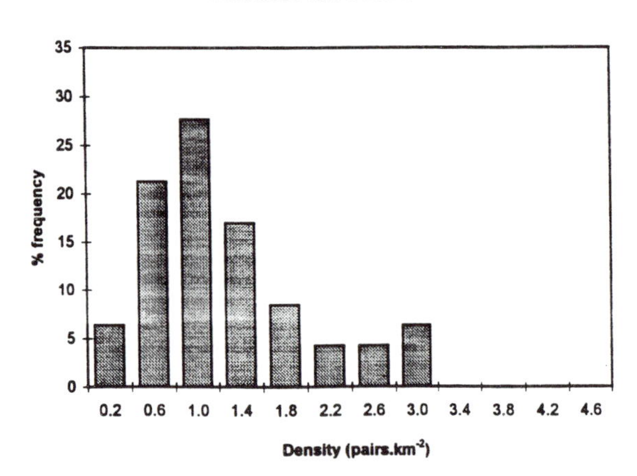

Fig. 11.1. Breeding densities of *Pluvialis apricaria* on peatlands of Lewis and Harris (median = 2.72 pairs km^{-2}, 25 sites) and Caithness and Sutherland (median = 1.43 pairs km^{-2}, 47 sites), 1993-1995.

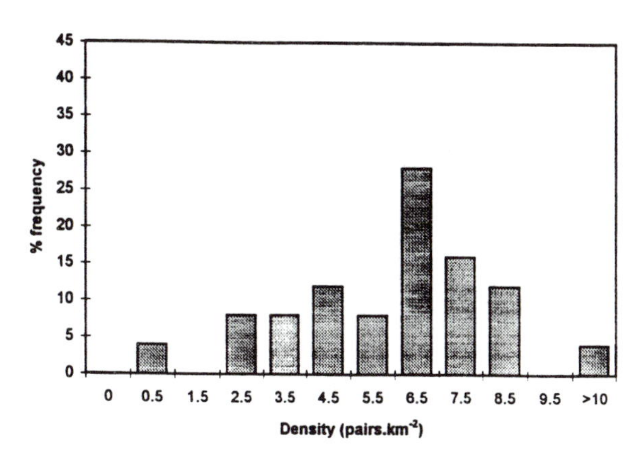

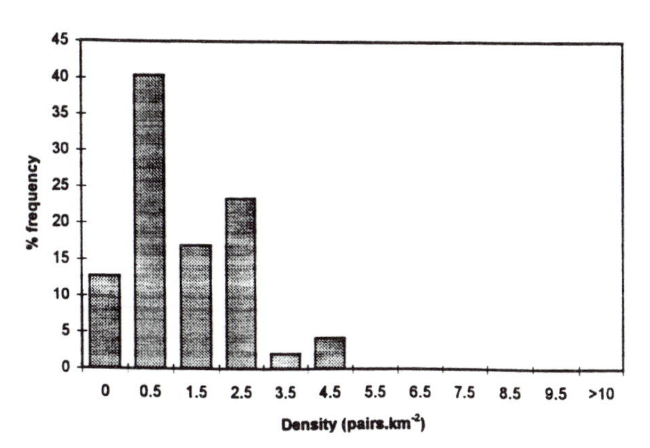

Fig. 11.2. Breeding densities of *Calidris alpina* on peatlands of Lewis and Harris (median = 6.52 pairs km^{-2}, 25 sites) and Caithness and Sutherland (median = 0.74 pairs km^{-2}, 47 sites), 1993-1995.

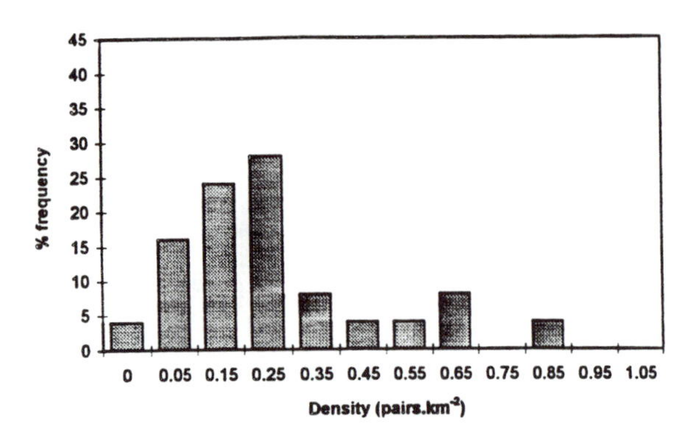

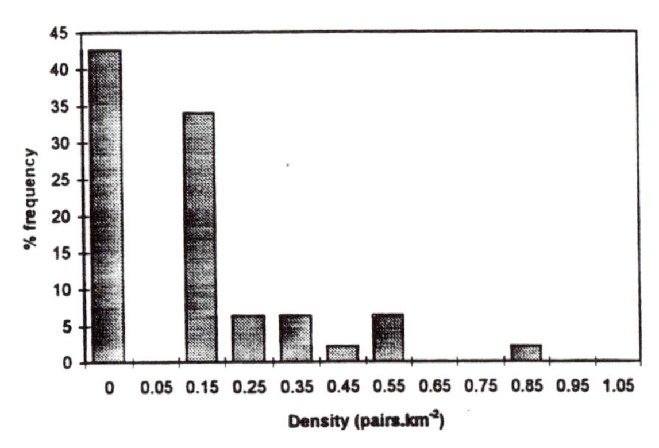

Fig. 11.3. Breeding densities of *Tringa nebularia* on peatlands of Lewis and Harris (median = 0.22 pairs km^{-2}, 25 sites) and Caithness and Sutherland (median = 0.14 pairs km^{-2}, 47 sites), 1993-1995.

Changes in breeding numbers: Lewis

Repeat LT surveys in 1994 of seven sites first surveyed in 1987 on Lewis showed that *P. apricaria* numbers had increased on most sites, *C. alpina* numbers were usually stable or had increased, and *T. nebularia* numbers showed a very slight tendency for decline (Table 11.1). Changes were significant only for *P. apricaria* (Sign Test, $P = 0.032$). There was no correlation between change in density and the original density (in 1987) for the three species.

With the observed annual rates of change (Table 11.1) and with a hypothetical starting population of 500 pairs of each species, then after 10 years there would be 1145 *P. apricaria*, 570 *C. alpina* and 450 *T. nebularia*. The slight decline for *T. nebularia* should be regarded cautiously because the size of the survey sites in relation to the spatial and temporal dispersion of *T. nebularia* will lead to variation in survey population estimates even with a stable population.

Inclusion of four sites with CSE survey in 1994/1995 and LT survey in 1987 confirmed the marked increase in *P. apricaria*, the slight increase in *C. alpina*, and suggested overall stability in *T. nebularia* numbers (Table 11.1).

Table 11.1. Changes in estimated numbers of breeding waders on seven peatland sites on Lewis and Harris between 1987 and 1994/1995. Figures in brackets include four sites with LT survey in 1987 and CSE survey in 1994/1995 (see Methods for details). % change per year calculated using only repeat LT survey sites.

	Number of sites:		
	P. apricaria	*C. alpina*	*T. nebularia*
Decreased	0 (0)	1 (1)	3 (3)
Stable	1 (2)	3 (6)	2 (5)
Increased	6 (9)	3 (4)	2 (3)
% change per year	+12.9	+1.4	-1.0

Changes in breeding numbers: Caithness and Sutherland

Table 11.2. Changes in estimated numbers of breeding waders on 12 peatland sites in Caithness and Sutherland between 1979-1987 and 1993/1994.

	Number of sites:		
	P. apricaria	*C. alpina*	*T. nebularia*
Decreased	7	5	5
Stable	3	6	5
Increased	2	1	2
% change per year	- 0.7	- 2.4	- 3.3

On most of the 12 re-surveyed sites, recorded numbers of all three species were either stable or had declined (Table 11.2). The decline was significant for *C. alpina* (Sign Test, $P = 0.022$), but not for *T. nebularia* ($P < 0.10$). There was no correlation between annual change in density and the original density (in 1979-1987) for the three species. With the observed annual rates of change and with a hypothetical starting population of 500 pairs of each species, then after 10 years there would be 465 *P. apricaria*, 380 *C. alpina*, and 335 *T. nebularia*.

Discussion

The *P. apricaria* population of Lewis and Harris appears to have doubled since 1987. The increase seems widespread and not just restricted to the highest density sites. The increase is remarkable not just because of its scale but also because elsewhere in the UK *P. apricaria* is declining (Parr, 1992; Gibbons *et al.*, 1993). The cause of the increase is unknown. The peatland habitat on Lewis has changed little over the relevant period (SNH, unpublished data). Although fewer sheep are now put out on to the bog (J. Cumming, pers. comm.), *P. apricaria* increases were recorded on peatland where sheep are still grazed. Recent research has shown that *P. apricaria* breeding on peatlands frequently feed on pastures (in-bye, re-seeds) as much as 10 km from their nest site during the incubation period (O'Connell *et al.*, unpublished data). Indeed, EU grant-aided conversion of crofts' hay meadows to sheep pasture during the 1980s may be implicated in the *P. apricaria* increase, and *P. apricaria* densities on Lewis appear to be greatest on bogs that are within ready 'commuting' distance of crofts (D.P. Whitfield, unpublished data; see also Ratcliffe, 1976).

Intensive observations in 1995 on one of the repeat survey sites suggested that *P. apricaria* densities may be higher than the survey estimates, probably because when spacing of pairs is significantly lower than the 500 m/1000 m distance criteria used in CSE analysis (as occurs frequently on Lewis), there will be an expectation of survey underestimating true breeding density (D.P. Whitfield, unpublished data). Even with the conservative survey estimates, however, local *P. apricaria* densities on Lewis are now comparable with the better-known high density sites in N England and S Scotland (Ratcliffe, 1976), but there are probably few if any other places in the world where such densities occur over such a large area (several tens of kilometres) (Cramp and Simmons, 1983).

Densities of *C. alpina* on Lewis and Harris are also high, especially for peatland habitat, and are far greater than found in the Flow Country. In the UK, *C. alpina* densities on Lewis are surpassed only by those recorded in the 1980s on the machair of the Outer Hebrides (Fuller *et al.*, 1986). Recent re-survey shows that *C. alpina* numbers on the machair have declined significantly, however, and may now be only 40% of the 1983 figures (RSPB, SNH, unpublished data). This, together with the decline of *C. alpina* in the Flow Country (this study), points to the peatlands of Lewis and Harris now being the best area in the UK for breeding *C. alpina*, with up to one-third of the national breeding population (D.P. Whitfield, unpublished data).

The combination of high densities, stable or increasing breeding numbers, and widespread, uninterrupted breeding habitat make Harris and, especially, Lewis the most important area for peatland waders in the UK. Although much of the blanket bog on Lewis is heavily eroded (Stroud *et al.*, 1988), this appears to have little detrimental effect on the breeding waders and, indeed, it may actually be beneficial.

There was evidence of declines in the populations of all three waders in the Flow Country, quantifying the subjective impressions of several observers (F. Symonds, pers. comm.). Stroud *et al.* (1987) estimated that by 1987 912 pairs of *P. apricaria*, 791 pairs of *C. alpina* and 130 pairs of *T. nebularia* were lost from the peatlands of Caithness and Sutherland as a direct result of commercial afforestation, leaving 3980, 3830 and 630 pairs, respectively, of the three species on the unplanted ground. Since 1987, according to the present study, 223 pairs of *P. apricaria*, 735 pairs of *C. alpina* and 166 pairs of *T. nebularia* have been lost from the Flow Country (using Stroud *et al.*'s (1987) population estimates as a baseline). Thus, for *C. alpina* and *T. nebularia,* similar numbers of breeding birds have been lost since new tree-planting ceased as were lost over the same length of time when afforestation was at its peak. The contemporary losses cannot be attributed directly to afforestation, but several indirect, 'knock-on' effects are expected on unplanted ground following commercial afforestation in the same area (Stroud *et al.*, 1987; Thompson, Stroud and Pienkowski, 1988; Avery and Leslie, 1990; Stroud *et al.*, 1990; Parr, 1993). Few catchments in the Flow Country remain unplanted (Stroud *et al.*, 1987), and it may be significant that on Lewis, where peatland birds are at such high density and are maintaining or increasing their numbers, there has been very little afforestation. The recent losses of waders in the Flow Country are serious and should prompt major concern and immediate research into the causes.

Acknowledgements

I am indebted to Mark Bates, Kevin Shepherd, Joan Cumming, Fraser Symonds, Nigel Buxton, Alison Rothwell, Nick Littlewood, Sue Holt, Des Thompson, all the field surveyors, and all my colleagues in SNH Area offices and RASD, whose assistance made this work possible.

Chapter twelve:

Peatland Dragonflies (Odonata) in Britain: A Review of their Distribution, Status and Ecology

Mr Stephen J. Brooks
British Dragonfly Society and Natural History Museum, UK.

Introduction

In his book on the dragonflies of Suffolk, Howard Mendel (Mendel, 1992) stated that 'nutrient poor, acid bog in southern England is potentially the richest of all dragonfly habitats'. This is undoubtedly true; indeed, in terms of dragonfly diversity, Britain's premier site is Thursley Common, a wet heathland with shallow peat deposits in Surrey, which supports 25 of the 38 British breeding species. In fact peatland bogs provide an important habitat for dragonflies throughout Britain. Why should this be? At first sight peat bogs, with their highly acidic, unproductive waters, appear to be unpromising habitats for aquatic insects, especially predators like dragonflies.

First, many species of dragonflies are able to tolerate acid water; they are often among the last animals to survive in lakes that have been badly affected by acid rain. As well as supporting populations of a few generalist species, peat bogs are essential for 11 species that are virtually restricted to this biotope in Britain. These species are not necessarily acidophilic (Brooks, 1994; Foster, 1995) but rather they are able to survive in acid water and the peat bog biotope provides the habitat structure that the species can exploit. The bog habitat may even be sub-optimal or put the species under stress but the population thrives in the absence of predators, due to reduced competition from other odonates that are sensitive to low pH, and because the *Sphagnum* offers a substratum that suits some species. Peatlands that have a mosaic of different dragonfly habitats support the highest diversity of species since tiny bog pools, seepages and larger ponds each have their specialist species.

Another reason that dragonflies do well in peatlands is that development can be fast in bog pools. The heat accumulation by *Sphagnum* mosses can accelerate egg development (Soeffing, 1986) and water in the shallow bog pools warms rapidly, thus promoting larval growth. Conversely the air temperatures over peat bogs are usually cooler than the surrounding countryside. This makes peat bogs suitable breeding habitats for species in which the adults are cold-adapted (Sternberg, 1990). Such species are often rare and localised in Britain.

Table 12.1. Status, distribution and habitat of British Odonata typical of peatlands.

Species	Status and distribution	Habitat
Pyrrhosoma nymphula	Common & widespread	Heath & moorland pools
Ceriagrion tenellum	Local; S. England & Wales	Small bog pools & seepages
Coenagrion hastulatum	Rare; Scottish highlands	Large pools & lochs in cold regions
C. lunulatum	Local; Central & north of Ireland	Small bog pools
Aeshna juncea	Common & widespread	Heath & moorland pools
A. caerulea	Local; widespread in Scottish highlands	Small, shallow bog pools
Somatochlora arctica	Rare; Scottish highlands	Small, shallow bog pools
Orthetrum coerulescens	Local; South & west Britain	Pools, streams & seepages
Libellula quadrimaculata	Common & widespread	Heath & moorland pools
Sympetrum danae	Common & widespread	Heath & moorland pools
Leucorrhinia dubia	Local; England & Scotland	Bog pools

Finally, dragonflies can exploit their position as top predator in bog pools. Unproductive, oligotrophic bog pools, in which food resources appear to be limiting, can nevertheless support many odonate species at high density (Pajunen, 1962). This may be attributable to the absence of fish which are important predators of dragonfly larvae (Corbet, 1962). Alternatively, or in addition, it may be due to the contribution of mycobacteria which are common in *Sphagnum* pools (Kazda, 1977). These bacteria are first concentrated by Cladocera which are in turn consumed by dragonfly larvae. The amino acids and vitamins of the bacteria may be taken up by the dragonflies or the bacteria may maintain some sort of symbiosis with them (Soeffing, 1988).

The continued survival of a large proportion of the British dragonfly fauna is dependent on the conservation and careful management of peatlands. A list of the 11 species largely restricted to acid waters, with notes on their distribution and broad ecological requirements, is given in Table 12.1. Moore (1986) noted that the major reasons for the decline of acid water dragonflies in East Anglia were the cessation of small-scale peat cutting, pond drainage, pollution, scrub encroachment and agricultural reclamation. In the following account, I describe the ecology of three dragonfly species that can be regarded as bog specialists. *Leucorrhinia dubia* is distributed throughout England and Scotland; *Ceriagrion tenellum* occurs only in southern England and Wales, and *Aeshna caerulea* only in Scotland. This account will illustrate the ways in which different species use the particular habitats present in peatlands. It will also provide an insight into the possible impact of conservation management or commercial exploitation of peatlands on dragonfly faunas. Distributional information is taken from Hammond (1983) and Askew (1988).

Bog specialists

Leucorrhinia dubia (Vander Linden) (white-faced darter)

This is essentially a northern European species, at the western edge of its range in Britain where it is rare in England but more widespread in Scotland.

Larvae of *L. dubia* occur in acidic pools on bogs and heathland. At Chartley Moss, Staffordshire, Bailey (1992) noted that females seldom oviposited in water deeper than 7.5 cm. However, in northern Scotland the species inhabits deeper pools that may be 50-100 cm or more in depth (R.W.J. and E.M. Smith, pers. comm.). Adults can apparently distinguish acid bog water from neutral tap water using antennal receptors (Steiner, 1948). Larvae live amongst submerged *Sphagnum* in shallow water (Bailey, 1992) or cling to the underside of floating *Sphagnum* rafts (R. Kemp, pers. comm.) and tolerate acidic waters with pH 3.4 (Beynon, 1995). Larvae also occur in depressions in the *Sphagnum* lawn amongst water-logged *Sphagnum*, which though devoid of standing water, are submerged during the oviposition season. Larval densities in these depressions may be relatively high: Bailey (1992) reported 16 larvae in one 2 m^2 depression. Larvae take 3 years to complete development (Gardner, 1953), so continuity of habitat is essential for the survival of a population.

Larvae emerge on sunny margins of pools during May and early June and use plants such as *Eriophorum* (common cotton grass) as emergence supports but they will also emerge horizontally directly onto the *Sphagnum* lawn (Beynon, 1995). Shaded banks are avoided during emergence (Bailey, 1992). Newly emerged adults seek shelter-belts of trees or woodland away from the pool during a maturation period which may last 2-3 weeks (Pajunen, 1962; Beynon, 1995). Males encounter females at the water when they come to oviposit. Copulation occurs on the ground and usually amongst fringing emergent vegetation. Adults often settle on pale surfaces which radiate heat such as dead, dry *Sphagnum* or rocks (Schiemenz, 1954).

Bailey (1992) notes that the small pools favoured by *L. dubia* become overgrown by *Sphagnum* mats within a few years. In order to maintain the species on a site, new pools can be created after trees are removed from the bog by winching. Pools can also be dredged, but since any *Sphagnum* which is removed is likely to contain larvae, the spoil should be placed in another newly excavated pool. At Thursley Common, pools have been created with the co-operation of the Ministry of Defence by using explosives to remove shallow areas of *Sphagnum*. Nevertheless, *L. dubia* is apparently in decline at this site, its southernmost outpost in Britain, and few individuals have been seen in the last few years (D. Tagg, pers. comm.). The cause of this is unclear but, since the species is predominately northern in distribution, it may be linked to climatic changes.

A large population of *L. dubia* occurs on Fenns and Whixall Moss, Shropshire. For decades this site was subject to small-scale peat extraction and the dragonfly population thrived in the small pools created after peat was removed. However, in the late-1980s mechanised, intensive extraction began and there were fears for the continued survival of this important colony. Many of the pools which supported *L. dubia* began to dry out as the hydrological integrity of the Moss was damaged and the species was eventually restricted to one small corner of the site. However, after a few years of intensive peat extraction, the site was purchased by English Nature and peat-cutting ceased. Since then the site has been carefully monitored and managed. New pools have been created and the *Sphagnum* is growing back. The *L. dubia* population is now large and continuing to increase.

A population of *L. dubia* once occurred on Thorne Moors but is now extinct following peat extraction and drainage. Unfortunately, the species is unlikely to recolonise the site naturally because there are no other populations nearby.

Ceriagrion tenellum (Villers) (small red damselfly)

C. tenellum is locally common on the wet heaths and moors of southern England and west Wales, and in Britain is at the extreme north-western limits of its European distribution. Parr and Parr (1979) suggested that lack of habitat - rather than unsuitability of climate - restricted the range of this species in Britain. However, its absence from suitable bogs in northern Britain and Ireland make climatic factors seem a likely factor governing its current distribution.

Female *C. tenellum* oviposit endophytically into the stems of submerged plants and in the New Forest are associated with *Hypericum elodes* (Parr and Parr, 1979). They favour shallow, unshaded bog pools and seepages which are often highly acidic. The species will also breed in the margins of large well-vegetated ponds, with almost neutral water, where these are close to heathlands (Brooks, 1994). However, such sites seldom support large populations, possibly due to competition with other odonate species. The species does particularly well where it can exploit tiny bog pools devoid of other species of dragonfly, and with very shallow water overlying *Sphagnum* mats. For this reason, the species is especially threatened by the drying out of bogs following drainage, water abstraction or afforestation. For example, the species was lost from Esher Common, a wet acid heath in Surrey, when its breeding sites dried out following the encroachment of planted Scots pine. The situation was exacerbated when low-lying parts of the heath were drained in preparation for the building of a road across the Common. Larger ponds on Esher Common still remain and support a rich dragonfly fauna but these ponds are apparently unsuitable for *C. tenellum*.

The larvae, which live amongst *Sphagnum* mats, take two years to develop (Corbet, 1957). Adults seldom move far from the breeding sites and so are slow to colonise new sites or recolonise sites from which they have become extinct. Moore (1986) describes five unsuccessful attempts to re-establish the species in three sites in East Anglia and it has not recolonised sites in the Somerset Levels following peat extraction (M. Parr, pers comm.).

Aeshna caerulea (Ström) (azure hawker)

This species has a boreo-alpine distribution in Europe and is common in Scandinavia and mountainous regions of central Europe. Sternberg (1990) showed that adult *A. caerulea* are cold adapted whereas the larvae require the warm micro-climate of bog pools to complete development. This requirement for cool conditions would help to explain its distribution in Britain where the species occurs only in Scotland, there being scattered populations throughout the Highlands and one isolated colony in the South-west. However, the species may be under-recorded in the Sutherland/Caithness Flow Country.

Suitable breeding habitats can be found on open moorland, mires and blanket bog. Females usually select small pools at the periphery of bog pool complexes in which to oviposit and will ignore seepages. Typically the pools have a surface area of 5-25 m^2 and a water depth of 15-30 cm. They are not grown over with *Sphagnum* but may include floating *Sphagnum* mats, and have a soft mud or loose *Sphagnum* substratum; they are usually fringed with emergent vegetation, such as *Eriophorum*, but are unshaded by trees. Females settle in bare wet peaty areas at the pool side and oviposit beneath the upper layers of the *Sphagnum* carpet or into floating *Sphagnum* mats (Clarke *et al.*, 1990). Floating *Sphagnum* is also used as a refuge by small, early instar larvae. The species usually shares breeding sites with *A. juncea* but larvae of *A. caerulea* are usually more numerous than those of *A. juncea* in the shallow bog pools favoured by *A. caerulea*. Larval development may take 4-5 years (Clarke, 1994).

Sunlit rocks, bare peaty areas and the grey hummocks of *Racomitrium* are frequently used by adults as basking sites. It is only after basking that the bright blue coloration of the males becomes fully developed. In cool, overcast conditions males remain slaty grey. Copulating pairs settle on *Sphagnum* hummocks. Woodland provides an important component of the habitat for *A. caerulea* and feeding groups will congregate in rides and woodland edges. Woodland provides shelter and roosting sites for adults in poor weather and at night. Although *A. caerulea* is widespread in Scotland, afforestation and drainage pose a threat and have already destroyed some sites (Clarke *et al.*, 1990).

Conclusions

From this brief review of peatland dragonflies the importance of this habitat for many of the British species is apparent. Peatlands provide a complex mixture of different aquatic and terrestrial biotopes that can support a diverse dragonfly fauna. Many species that are peatland specialists have a regional distribution in Britain which may be due largely to climatic restrictions. For this reason the conservation of a matrix of bogs and wet heaths throughout the country is essential for the conservation of a large part of the British dragonfly fauna, including some of our rarest species.

Acknowledgements

I am grateful to Professor Philip Corbet, University of Edinburgh, and Betty and Bob Smith for their helpful comments on this chapter.

Part Four

Functioning Peatlands

An Introduction

Dr Rob Stoneman
Scottish Wildlife Trust, UK.

The Convention (and this volume) tours across the wonder of peatlands dipping into various facets along the way. In this session, a taster of functioning peatlands was given. This can be viewed in two ways.

At its most elemental, we look at the ways peatlands operate. How hydrology, peat formation, plants etc. interact with each other and the surrounding biosphere - the atmosphere, the catchment and so on. Our tasters here were provided by Drs Gilman and Brown who illuminate bog hydrology and the microbiology of methane production respectively. Our interest is captured by the implications of these discussions. Ninety per cent of Scotland's drinking water, for example, is supplied from peat-dominated catchments. The importance of hydrology cannot be overstated. The role of microbes in the production of methane is also critical. Not just critical for the peatland - it may be the ability of peat to store methane or recycle it that allows peat formation to proceed - but also for our own continued sustainable survival as the release of methane in peatlands (through human caused degradation) could dramatically enhance global warming.

The concept of sustainable living links to the understanding of peatland functions. The wider view of peatland functioning, explored by Professor Maltby and Mr Immirzi, allows identification of the full significance of peatlands. This Convention illustrates the many functions peatlands provide: biodiversity, archaeology, palaeoecology, landscape, hydrology, carbon-cycling, flood control systems, pollution filtering, fisheries (in tropical peatlands) and consumptive uses (peat extraction, afforestation). It is a long list and a long way from the nineteenth century view of peatlands as wastelands (Smout, Chapter 17, this volume). Maltby and Immirzi argue that by analysing these functions, the case for conservation can be strengthened because some of these functions can be valued against standard economic criteria. Functions such as diversity, rarity, size etc. - all commonly used for assessing nature conservation value (e.g. Ratcliffe, 1977) - mean less to developers and economists than the economic losses accrued from peatland destruction (e.g. the cost of replacement flood prevention measures).

Perhaps, as conservationists, it is our job to educate economists to value these functions not just in cash terms but in the terms of a sustainable future. Is functional analysis the key?

Chapter thirteen:

Peatlands: The Science Case for Conservation and Sound Management

Professor Edward Maltby
University of London, UK.

Introduction - defining our goal

The term 'peatland' is used to describe the physiographic, geomorphological and biogeographical setting of peat. It includes areas which may no longer carry peat-forming communities such as agricultural land. Systems which are still peat-forming are called mires.

Let me start with an attempt to identify a final goal in relation to peatland conservation and management. It is clear that we must recognise a wide range of possible views and perspectives. These have been categorised by Holdgate (1994) into four main groupings of people: Gaians, Cornucopians, Armageddonists and Arenostruthios. Whilst the first three groups have strong pessimistic, optimistic or resigned views, the last of these have no scientifically informed opinion at all on environmental issues. Since they probably represent the majority it is imperative that a sufficiently forceful and popularly appealing case is presented if the conservation of peatlands is to become a compelling part of society's agenda.

Change is a natural feature of the natural environment and so we are not here concerned simply with preservation of the *status quo*. Indeed many of the peatlands of Western Europe, particularly those associated with the uplands of the oceanic margins, are the result of pre-historic human intervention (Maltby and Crabtree, 1976; Maltby and Caseldine, 1982). Also, most have already suffered more or less significant degradation due to direct or indirect human action.

Conservation is the management of change, which generally means avoiding rapid changes and effects which are undesirable. Whilst this raises an immense number of issues of social value, economic need and sustainability of natural resource use, the fundamental goal in natural resource management is to maintain the integrity and balance of protective, as well as productive, ecosystems for the welfare of both people and wildlife and for the maintenance of environmental quality. Such an ethos and final goal applies well to peatland conservation.

Changing emphasis

It has long been my contention that alone the traditional approach to conservation, endorsing the virtues of uniqueness, rarity and particularly good examples of ecosystems, has not served the peatland stock well. Future policy must be underpinned by a new approach which extends our traditional conservation ethic into one for which public and political support can be more effectively mobilised. Historically, peatlands have been lost to so called 'more productive uses' such as agriculture, forestry or extraction for energy. This has occurred throughout Europe with examples such as drainage of the English fens, the afforestation of Caithness and Sutherland and the mining of Ireland's bogs being particularly significant in Britain and Ireland.

The analysis and decision-making process which has enabled such change to take place has been based on simplistic, short-term and highly sectoral economic assessments. Decisions have ignored the fact that peatland ecosystems may provide a whole range of goods and services as a result of natural ecosystem processes (Fig. 13.1). The trouble is that these goods and services such as fisheries support, wildlife habitat and water quality do not have a market place - they are not traded on the Stock Exchange floor; they are not part of a commodity exchange process - yet they are valuable to people who depend on them either directly or indirectly. The value of these support functions, whether for ecology, for environment, or even for people's economic and social livelihood is only really appreciated when they are destroyed or adversely impacted - such is the need for integrity of the ecosystem - the flora, fauna and the structure, the micro-relief, the hydrological regime and the nutrient balance of the peat bog, fen or other mire type.

Consideration of the natural values of peatlands has been hindered by:

1. lack of scientific information at the ecosystem, as opposed to habitat or species, level, and;
2. the often overriding effects of economics distorted by artificial price support mechanisms, subsidies, beneficial tax structures or other external factors.

Yet we ignore the human dimension of the world's peatlands at our peril. Should we view peatlands as protective or productive? The answer frequently is both, and herein lies much of the problem and dilemma facing society and decision makers.

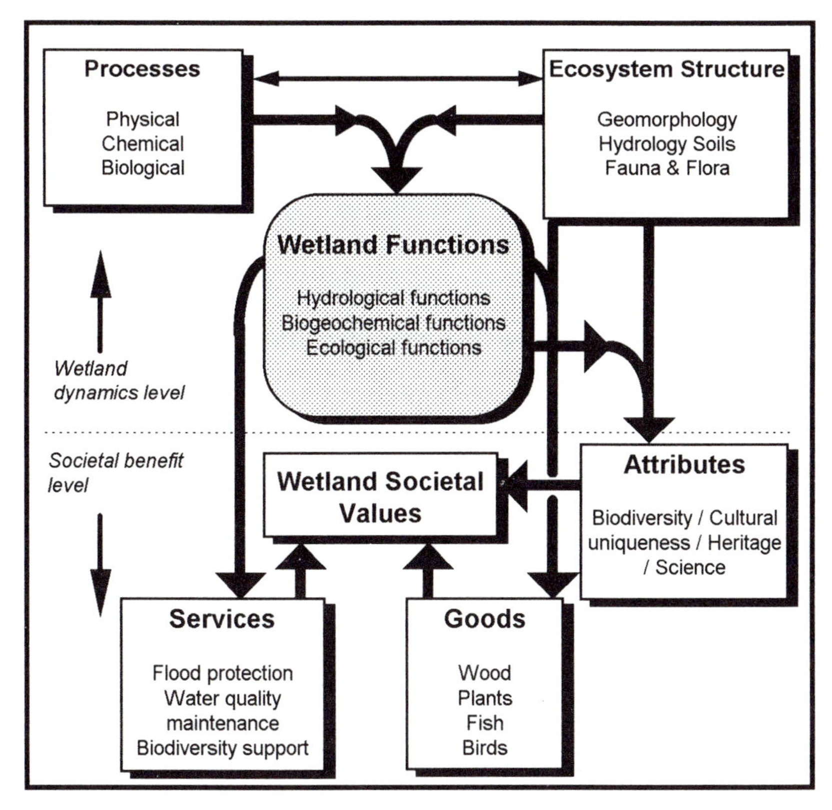

Fig. 13.1. Relationship between wetland (including peatland) processes, functions, attributes and values (after Maltby *et al*, 1996).

Intact, peat systems may provide a wide range of biological diversity functions and ecological services which are valuable but which are naturally taken for granted and not separately accounted for by society (see Fig. 13.2). In general, these benefits decline as peatlands are converted to uses such as agriculture, forestry and mining which are viewed as more directly beneficial. Resource optimisation involves striking the right balance between these effects. Whilst the ideal solution will maintain relatively high levels of biological diversity and thus management options, the general result of sectoral development of peatlands is degradation and loss of biodiversity (Table 13.1). One of the scientists' greater challenges is to modify sectoral use patterns to retain diversity and services.

Nature of the resource

In addressing the science case for conservation and management, it is useful to examine some of the background relationships among peat-based ecosystems, environment and human activity. Peat develops because there is an imbalance between the processes of production and those of decomposition. Various conditions promote this imbalance and throughout both the Boreal zone as well as the humid tropics, local factors of climate, hydrology, vegetation, relief and parent material have been conducive to peat accumulation at least since Pleistocene glacial retreat (and possibly much earlier in lower latitudes). The progressive build-up of the peat mass is eventually limited by the same balance but under some special conditions such as subsiding tropical coastal areas, it may continue unrestricted for millennia, resulting in the accumulation of very deep deposits which may reach several tens of metres. Such is the case in parts of South East Asia and some of these current coastal peat deposits may be modern day analogues of our Carboniferous coal seams.

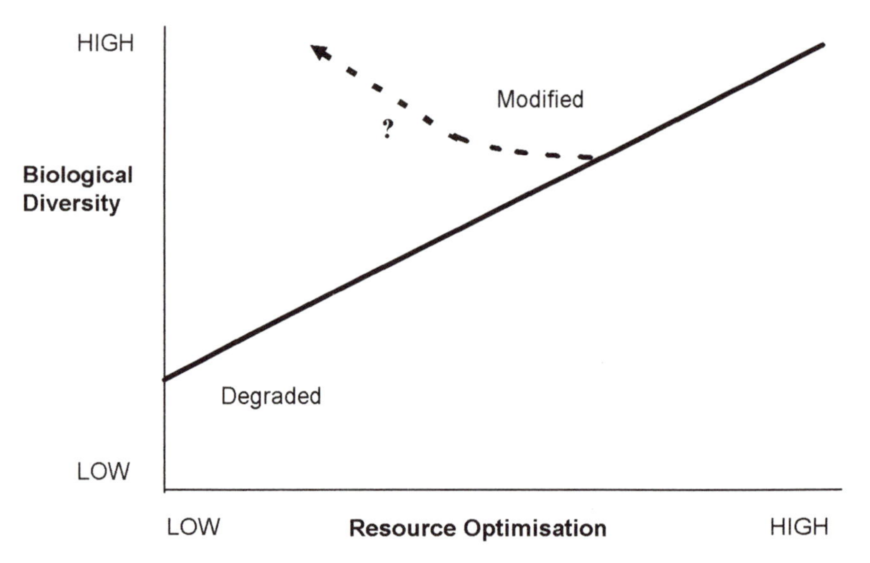

Fig. 13.2. Conversion of peatlands to 'productive' elements gives modification resulting in loss of biodiversity and ecological services. [? -not yet proven].

Peatlands have a long, but episodic and spatially highly variable place in the global environment. The environmental relationships are neither simple, nor necessarily consistent. Overall, they comprise some 390 Mha or about 3% of the globe's land surface, mostly in the northern hemisphere. We have moved a long way since it was generally contended that peatlands were almost exclusively a feature of high latitudes. More than 15% of the world's peat resource may occur in the tropics (Immirzi and Maltby, 1992; Maltby and Immirzi, 1993). In oceanic Europe, we associate the deepest peat accumulations with relatively high rainfall or waterlogged topographic hollows, yet in the Falkland Islands we find peat banks exceeding 5 m on nearly level or only gently sloping terrain where precipitation is less than 500 mm. The peat building species are different, and so may be the causal agents of peat development (Clymo and Lewis Smith, 1984). In this case, it has been suggested that high levels of sodium derived from sea spray may inhibit the decomposition process whereas elsewhere acidity, water-logging, litter type and temperature may be more or less important factors.

Table 13.1. General relationship between resource optimisation and biological diversity.

Biological diversity	Resource Optimisation
Intact • Evolution • Crop Breeders • Medicines	Intact ⟶ Modified Ecological services decline • Biogeochemical Cycles • Hydrological Cycles • Erosion protection • Microclimate 'Productive' Elements increase • Agriculture • Forestry • Mining

There is an enormous variation in the genesis, pattern of development, functioning and present-day character of the world's peatlands. This makes the formulation of standard policy and management guidelines very difficult. The raised bogs of Scotland represent just one part of this range but the present volume testifies to the growing realisation of the need for specific and well-targeted actions to ensure their survival.

Human interactions

Considerable research effort has been devoted to elucidating the relative importance of human activity - as opposed to climate and other natural environmental factors - in leading to the initiation of peat accumulation particularly in upland Britain and in the determination of the course of development of the peat mass. Whilst there is still considerable debate on the subject it is seemingly clear that - in some locations at least - peat may start to form as a result of man-induced as well as natural factors. It is certainly the case on many of the uplands of the United Kingdom that prehistoric forest clearance (especially during the Bronze Age) may have been an extremely important factor initiating peat accumulation. This effect was neither planned nor deliberate and I have described it previously as being one of the early environmental planning disasters of human society. Whilst this may be stretching our interpretation of the transformation of upland wooded and mineral soil landscapes to the present blanket peat dominated moorland, it is nevertheless still ironic that we now value in nature conservation terms an ecosystem which would have been viewed as degradation by our prehistoric ancestors.

If we are going to be able to maintain peat systems or use them as barometers or predictors of environmental relationships and change, it is essential that we continue to improve our knowledge of those factors which lead to the development of these peculiar ecological systems. I say peculiar, because peat systems represent a most remarkable accumulation of organic material. The amount of carbon stored per unit area may considerably exceed that of any other ecosystem type. This storage of organic matter allows the storage also of water, nutrients, contaminants and a whole range of markers of environmental change which we are only just beginning to be able to interpret and understand. Heavy metal distribution (e.g. of caesium 137) and variations in magnetic profiles are just some of the examples which have been utilised in recent years (e.g. Livett *et al.*, 1979; Damman, 1986). Who knows what other wealth of information is contained within the chronological sequence of accumulating peat?

The peat body is also a remarkable store of water and energy. As a large store, the system is highly sensitive to change and degradation. Resilience is an extremely important property of ecosystems which are being utilised in one form or another by human communities. Peatland ecosystems are not very resilient to change. They are not very resilient to stress. They are particularly vulnerable to grazing, fire and especially to drainage. The timescale over which they have developed means that recovery following major disturbance may require centuries or millennia - it can certainly never be instantaneous. What may take thousands of years to accumulate can be lost in a very short period of time. The question is does degradation of peat systems actually matter when they may have developed as a bi-product of human exploitation of the landscape or may represent a temporary phase of environmental conditions such as water-logging or acidification within a particular landscape unit?

We can see from our experimental as well as our management work on the blanket peats of upland Britain, that a whole range of ecosystems may occur in the same unit of space at different times (e.g. Maltby and Gabriel, 1986; Maltby *et al.*, 1990). We may have forest, grassland, heather moorland or bog. We may also have severely eroded landscapes resulting from misuse of the ecosystem. Which of these systems will occur is largely the result of human interference e.g. land 'reclamation' and planting, agricultural intensification, liming, fertiliser application, grazing pressure and use and intensity of fire. The crucial test for planners is in the ability of the modified ecological state to support desirable or essential aspects of ecosystem functioning or to contribute to the maintenance of adjacent ecosystems. Such a test requires society to make decisions. The scientist can guide such decisions.

The science base and rationale

The importance of science is in trying to direct conservation and management so that we retain the optimum balance of ecosystems in order to sustain essential environmental support functions. Peatlands are an essential part of this assemblage. We can divide the rationale for such maintenance along the lines of nature protection and in provision for sustainable development. This is particularly important since there is always going to be much more of the resource located outside any system of strict protection. The first of these categories is probably the easiest to highlight and is associated with the maintenance of biodiversity, the provision of palaeo-environmental information and the maintenance of an important carbon store. The information about the wider importance of peatlands for sustainable utilisation has been less well researched; however, it clearly revolves around such features as the maintenance of hydrological integrity, the maintenance of non-eroding or otherwise non-degrading landscapes and the provision of 'wild' open spaces for public enjoyment.

Biodiversity

Biodiversity is dependent on ecosystem structure, pattern and scale. It is promoted by the occurrence of peat-based ecosystems in a wide range of global environments and biogeographical regions. There is a wide range of species which are responsible for the build up of peat - not just *Sphagnum*, *Eriophorum*, sedges, and grasses, but for example, *Astelia*, a member of Lily family, which is a significant peat-former in the Southern Hemisphere such as in Tasmania and the Falkland Islands. The special ecosystem structure provides special habitat conditions - the stress of waterlogging, of extreme acidity and nutrient deficiency are conducive to the establishment of carnivorous species (such as sundews, pitcher plants and bladderworts) but the variation in nutrients and pH from bog to fen results in enormous variation in species lists and community assemblages. The distribution of invertebrates and higher animal species is frequently linked to the spatial variation of plants such as in the hummock-hollow complexes of bog surfaces. Such is the subtlety of the relationship between micro-habitat conditions especially water table and species distribution, that the biodiversity assemblage is highly vulnerable to perturbation.

Palaeo-environmental information

Preservational quality, incremental pattern of accumulation and relative stability of the deposit make the peat bog an exceptional environmental record of change at local and regional scales. The association of buried cultural objects and *in situ* macro-remains or structures have been important elements in palaeo-environmental detective work. No other system offers such direct access to such important information but the full significance of this record has yet to be exploited - new techniques and analytical developments have long been the curse of the archaeologist and the same is probably true of the bog palaeo-environmentalist - who knows what important information may yet be discovered within the cellular contents of buried leaf material comprising the partly decomposed peat profile?

Carbon store

Despite its global significance, this store of carbon has generally been ignored in models of climate change. Disturbance of the existing peatland carbon store may be sufficiently rapid, and of sufficient magnitude, to materially alter the atmospheric carbon balance, removing or diminishing a vital climate buffer and leading to significant environmental change. The pre-disturbance mass of fixed carbon represented by the world's peatlands is estimated at 329-520 x 103 Mt contained within an area of some 390 Mha (Immirzi and Maltby, 1992).

Peatlands, the global carbon budget and global environmental change

The estimated peatland carbon may constitute 20%, or more, of all the carbon in world soils; it is probably more than three times the carbon stored in the world's tropical rainforests, and in the region of a half of the estimated total carbon in all terrestrial biota (737 x 103 Mt; Matthews, 1984).

The world's peatlands are highly concentrated carbon stores. Due to the long time period over which peat accumulation may occur, non-peat-forming ecosystems are generally unable to match the storage capacity of peatlands. This is particularly true of the tropics which, although only comprising c.10% of the areal extent, may store 70 x 103 Mt, or 15% of the peat carbon pool (Maltby and Immirzi, 1993).

Numerous estimates of carbon fluxes to and from peatlands have been proposed, and these are reviewed in detail by Immirzi and Maltby (1992). It is clear that under natural conditions mires accumulate carbon derived from atmospheric CO_2 and, early in their life, are appreciable carbon sinks. Had all mires escaped human alteration, they would have been a sink for perhaps 100 Mt C $year^{-1}$. As mires age, they become less effective as carbon sinks and increasingly become sources of methane. Thus, even without human intervention, the world's peatlands could have changed from being greenhouse-negative to becoming greenhouse-positive as part of a natural evolutionary process.

However, the last 200 years has witnessed major human modification, especially by drainage for agriculture and forestry, and by peat mining for energy or horticultural products. Mining can lead to the emission of 180-225 t CO_2 ha^{-1}

year^{-1}. Arable cropping may yield 41 t CO_2 ha^{-1} year^{-1} in temperate regions, but 154 t CO_2 ha^{-1} year^{-1} in the tropics (Immirzi and Maltby, 1992). Forestry and pasture result in much smaller emissions (1-10 t CO_2 ha^{-1} year^{-1} in temperate regions). Exploitation of peatlands for agriculture and forestry has affected c. 30 Mha (7.5% of the global total), and is expanding at some 550,000 ha year^{-1}.

The consequences are (i) loss of C-sink capacity, and (ii) release to the atmosphere of accumulated carbon as CO_2. The second is progressive and rapid, often resulting in the release within decades of a store of carbon which has accumulated over millennia. As a net result, there is a major shift in carbon balance; in the case of agriculture alone, this amounts to 426-730 Mt CO_2 year^{-1}, equivalent to about 3.5% of fossil fuel emissions and more than 12% of that resulting from tropical deforestation. Cumulatively since the end of the 18th Century some 30,000 Mt CO_2 may have been released by drainage and peat mining activities (Immirzi and Maltby, 1992). Peat mining alone may be exporting 242-268 Mt CO_2 year^{-1}, equivalent to the annual emissions of 40 100 MW coal-fired power stations.

Hydrology

Peat hydrology is complex and a key element in the development of different types of peat formations (Ingram, 1978; Ivanov, 1981). Its disruption such as by drainage and vegetation change can have far-reaching ecological and environmental effects. Reduction of the water table generally results in physical compaction and increased decomposition producing the well-documented effects of subsidence. In parts of the English Fens the peat has disappeared completely, exposing underlying relatively infertile sands and shallow mineral soils. In historical terms, the rich fertility of the fen peat for agriculture has been short-lived. Economic yields can now only be obtained by high rates of fertiliser application. Leaching of fertiliser nitrate gives rise to contamination of groundwater previously protected by the buffering effect of the peat. In the case of the Florida Everglades the rapid mineralisation of drained peat in the Everglades Agriculture Area has resulted not only in structural instability of many buildings but in the release of sufficient levels of phosphate to cause significant ecological change downstream. In particular this takes the form of expansion of *Typha* spp. as a dense community replacing the natural diversity of *Cladium jamaicense*, *Eleocharis,* Periphyton and open water communities with various aquatic and floating species such as *Nuphar*. The dense *Typha* stands have a much reduced wildlife and fisheries value compared with the unimpacted system and, through the increase of anaerobic conditions, may be responsible for enhanced availability of mercury by methylation eventually resulting in toxic concentrations higher up the food chain.

The consequences of degradation

Protection of the existing peatland resource is important to avoid the adverse environmental and human consequences which result from their disturbance. The effects may be off-site as well as on-site and result in consequences at local, regional and global scales. Examples of degradation underline the importance of conservation in maintaining ecological systems well beyond the peatland boundary. The increase in mercury in a forest lake in Finland was attributed by Simola and Lodenius (1982) to peatland drainage. Increased lead levels were found by Renberg and Sergerstrom (1981) in Swedish lakes adjacent to drained peat. The release of contaminant materials from drained or otherwise disrupted peatlands may extend not only to rivers and lakes but also to coastal estuaries. Accumulation through the food chain gives rise to toxic concentrations in algae, shellfish and other invertebrates and in fish (Winkler and de Witt, 1985). The far-reaching adverse effects of altered peat hydrology on the coastal fisheries of North Carolina have been identified by Street and McClees (1981).

Disruption of the continuity of extensive peatlands or the linkages between isolated systems may have severe effects on regional biodiversity. The possible consequences on atmospheric carbon dioxide levels by peatland alteration, however, emphasises the role of the resource as a global common. In this case, decisions in one part of the world may have far-reaching results quite remote from the place of impact. For example, sea level rise due to the warming effects of raised CO_2 levels produced by peatland loss in high latitudes may produce disastrous economic consequences for tropical island communities.

Resilience - recognising sensitivity

Mires, with their limited capacity for self-repair, are arguably amongst the most sensitive ecosystems on the globe.

- Mires get old - they may collapse, degenerate or dry out as part of a natural cycle - yet there is little evidence for natural recovery after such collapse.
- Natural environmental change may produce surface drying or excessive wetting - causing the development of non-peat-forming vegetation on the one hand or a recurrence surface on the other.
- Human interference may redirect the course of ecosystem development between extremes with end-points which are fundamentally different.

If effective management strategies are to be developed in the future, it is very important to know the critical tolerance levels (or threshold points) for physical pressures, hydrological change, temperature, pH, acidity and pollution, and what effect these have on ecosystem functioning rather than just on appearance. Research on these tolerance levels and their implications for system functioning both on- and off-site is a high priority for the science agenda.

Conclusion

Peatland resources worldwide continue to be degraded and lost, generally on the basis of inadequately informed decisions together with a failure of existing institutional mechanisms to properly evaluate them.

Wise use of these resources depends on new attitudes and new socio-economic, political and planning structures. The SWT plan (Stoneman, Chapter 45, this volume) must address these issues and develop appropriate follow-up actions to ensure implementations of changes which will prevent their progressive loss.

Peatlands cannot be viewed in isolation - they are part of river basins, coastal zones and biogeochemical cycles; they are part of landscapes used by people. In the Third World this use of the landscape might mean survival and the dependence of many communities on 'natural' peatland goods and services is only now being appreciated more fully.

The professional challenge to the peatland or mire scientist is to interpret the ecosystem as a functioning unit within the complex human, and normally dynamic, natural environment, to evaluate their tolerance of various uses, if any, and advise on optimum strategies to maintain or restore functional integrity. His or her more difficult task is to collaborate with other professionals and lobbyists to bring about major changes in the institutional structures required at local, regional, national and international levels to ensure the sound stewardship of some of the globe's most precious yet fragile resources. The time is ripe - the SWT plan, the possibility of strengthening Ramsar with regard to peatlands, the Biodiversity Convention and the new IUCN Commission on Ecosystem Management all offer mechanisms to support this impetus.

Chapter fourteen:

Bog Hydrology: The Water Budget of the Mire Surface

Mr Kevin Gilman
Institute of Hydrology (Plynlimon), UK.

Introduction

Inevitably, any discussion of mires or peatlands must start with classification. Botanists in particular could be excused for being puzzled at this, but the reason is that habitats, unlike species, are not always clearly and unambiguously defined, and not only are there intergrades and regional differences in terminology, but also some quite well-defined mire types can include elements of others.

On an active mire, the process of peat development is intimately connected with hydrology and with ecology, and the classification of mires has a long and complex history, drawing on stratigraphy, hydrology (as indicated by hydrochemistry) and floristic analysis. The simplest division, into bogs and fens, can be made on the basis of either landform or chemistry. Classifications based on shape, dating largely from the work of Osvald (1925), have shown remarkable persistence, because landform is well correlated with the hydrological, hydrochemical and ecological processes of mire development. Tansley (1953) used hydrochemistry to characterise mires as bogs (acid and nutrient-poor) and fens (less acid and richer in nutrients). Moore and Bellamy (1974) supported this simple classification, relating the chemistry of the mire to its form and to the water source. In most hydrological work on mires, it is sufficient to use the term bog to indicate ombrotrophic mires, such as raised mires and blanket mire, and fen to indicate mires fed by allogenic waters such as groundwater or soil seepage.

The central importance of hydrology in peatlands

The pivotal position of landform and water chemistry in defining the character of a mire is a direct consequence of hydrology. The development of peat depends on the maintenance of saturated conditions close to the surface, and a water source with only a limited load of mineral sediment. It is the divergent, radially outward flow pattern of a raised mire that ensures that most nutrients and bases come from the atmosphere, and restricts base-rich soil water and groundwater to a peripheral lagg fen. This domed shape is also present, though usually less dramatic, in blanket mire and some basin mires.

The first step towards characterising the water regime of a wetland site is to appreciate the relative importance of the components of the water budget (see Fig. 14.1). Interactions between the six main storage units differ according to local geology and the wetland type: peatland sites tend to be relatively little affected by surface water inputs, and bogs have little or no interaction with groundwater bodies in the underlying bedrock. Because of the difficulties in assigning boundaries to the various components - mires never conform to the outlines of conventional catchments - the scientific investigation of the water budget is only made possible by some sweeping simplifications. The most important of these is to concentrate on the storage of water, as indicated by water level measurements.

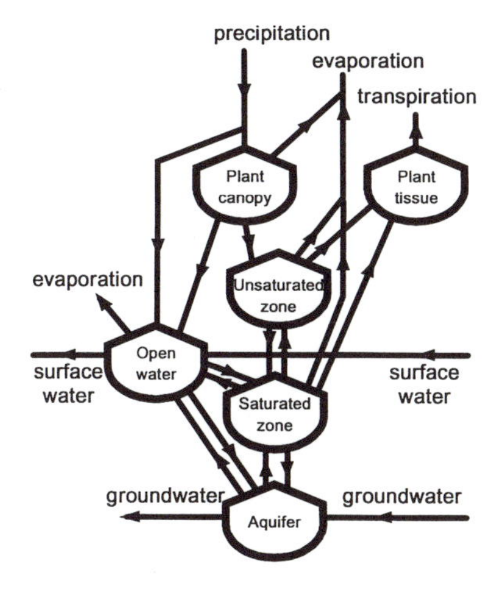

Fig. 14.1. Components of the water budget for a wetland site.

The performance of water budget components on various time-scales is hinted at by the patterns of water level variation (see Fig. 14.2). On a seasonal basis, groundwater levels, for instance measured on a fortnightly time-scale, reflect the distribution of evaporation, but there is usually considerable year-to-year variation in the minimum water level. Particularly low levels, resulting usually from a summer rainfall deficit, can cause concern, especially where there is some perceived external threat such as drainage. An understanding of natural fluctuations can be helpful in assessing whether the threat is real, and in evaluating the effectiveness of mitigating measures. On a finer time scale, sensitive water level recorders can detect diurnal changes relating to the daily pattern of evaporation. Transpiration draws down the water level between late morning and mid-afternoon, and there may or may not be a recovery overnight, as a result of lateral flow from open water bodies (Godwin, 1931). Diurnal fluctuations can be very sensitive and detailed: in addition to transpiration, Laine (1984) identified the effects of dew which raised water levels slightly between dawn and noon.

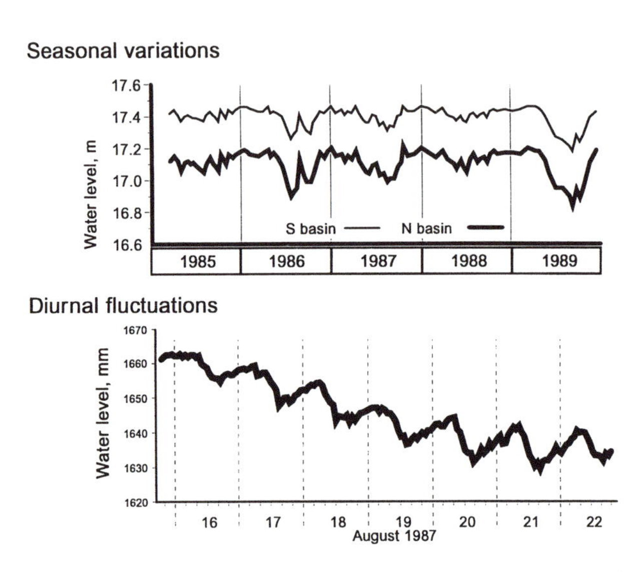

Fig. 14.2. Patterns of seasonal and diurnal changes in water levels. At Cors-y-Llyn, there is an annual cycle, but the amplitude changes from year to year as a result of changes in climate variables. At Skew Bridge Bog, also in mid-Wales, diurnal fluctuations show the influence of transpiration and lateral flow.

Hydrology at work in a small basin mire

The Countryside Council for Wales (and its predecessor the Nature Conservancy Council) has instrumented several important bog sites, among them Cors-y-Llyn (Llyn Mire) near Newbridge-on-Wye in mid-Wales. Llyn is a small basin mire which developed from a lake in a hollow, with final closure of the vegetation mat possibly within historical times (Moore and Beckett, 1971). Recently, some unwise attempts at drainage of the margins have been reversed by the installation of a clay bund, pine trees that were invading have been ring-barked, and the mire expanse is now relatively healthy (though there are still some reservations about the pool and hummock structure and the liquid core). There is a network of 16 dipwells, each accompanied by a ground anchor penetrating to the underlying silt, and two of these wells had continuous water level recorders operating from 1983 to 1994. A boardwalk gives access to the main dipwells. The site divides into two unequal parts, the northern basin containing ten

Using diurnal fluctuations measured by the continuous water level recorder in the north basin at Llyn, the net lateral inflow was calculated for a large number of days between November 1983 and October 1989. The results, shown in Fig. 14.5, indicate that:

1. From this location on the mire expanse, lateral flow is predominantly radially outwards, i.e. the net inflow is negative. In the context of Llyn, this confirms that the mire is behaving as it should.
2. There is more lateral water movement when the water table is near the surface. This fits in with what is known of the variation of permeability with depth in bogs in general.

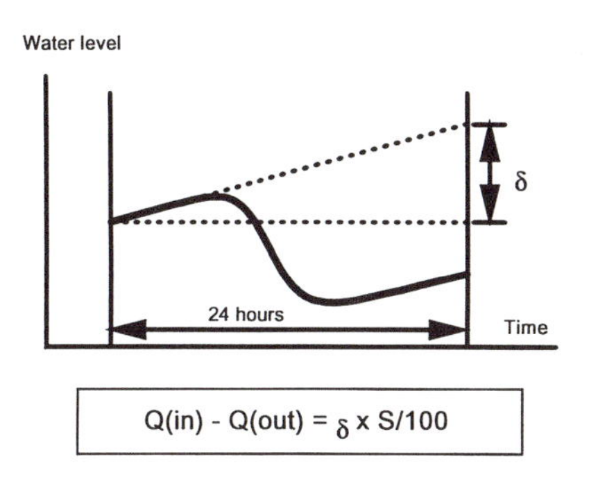

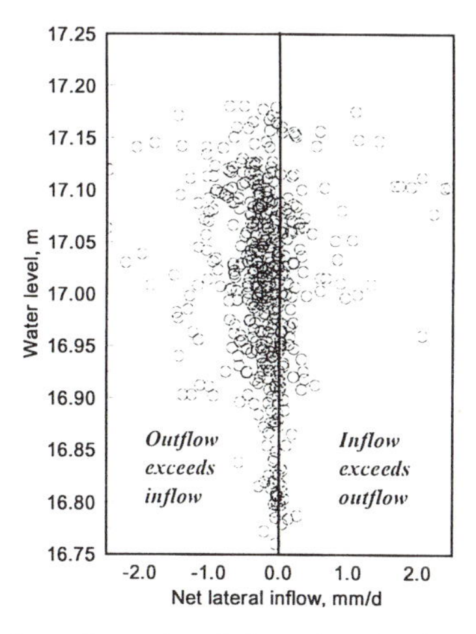

Fig. 14.5. Net lateral flow can be estimated from the rate of rise of water levels during the night. At Cors-y-Llyn, a net lateral outflow is more common, as is consistent with the domed shape of the mire expanse.

Conclusions

The application of science to wetlands is often approached by analysing the various 'functions' of wetlands, i.e. how sites operate as components of the drainage network, the landscape and the larger ecosystem. Wetland 'values', on the other hand, can be regarded as measures of the worth of wetlands to human society. It is widely accepted that the wise use of wetlands can yield such benefits as flood storage, water quality improvement and recreation. This approach is equally valid for peatlands, for the public perception of the values of mires is an essential precursor for wise use, including conservation and restoration.

Hydrology is of central significance in the development, continuation and conservation of peatlands, and moreover many of the functions of mires depend on the detail of the water budget, and its temporal and spatial organisation. Though mires do not yield to the conventional catchment approach to hydrology, it is possible to draw important conclusions about bog hydrology from relatively simple measurements. The future management of mires will need to be based on a better understanding of the role and significance of the water regime, and there will be a continuing role for hydrologists in mire research.

Chapter fifteen:

Microbiology of Methane Production in Peatlands

Dr D. Ann Brown
University of Manitoba, Canada.

Introduction

Peatlands play an essential role in the natural environment, since they have acted as a major carbon sink for at least the past 10,000 years. Ombrotrophic bogs are complex wetland ecosystems where the integrity of the surface, the acrotelm, is essential for the maintenance of the body of the bog, the catotelm. The high water table of this system allows bog pools to be supported on considerable hill slopes, but if the acrotelm is breached either from such insults as overgrazing, or acid rain, or it is removed by mining, or damaged by drainage and afforestation, the water table will fall and the catotelm will become susceptible to aerobic microbial degradation allowing the carbon that is stored in the bog to be dissipated into the atmosphere.

There is a saying that you are what you eat; a similar description may also be suggested for wetlands. Peatland flora, such as sundew or pitcher plants, require a specialised environment in which to grow, and this is determined mainly by the activity of the microorganisms that are indigenous to peat bogs. When surface plants die their decay is mediated by microorganisms. Peat is formed where primary production of plant vegetation exceeds its consumption; this generally occurs in stagnant waterlogged environments where the microbial degradation of plant biomass is incomplete.

The main component of vascular plants is cellulose, but it is enveloped in a network of hemicellulose and lignin (Brown, 1985). Non-vascular plants such as *Sphagnum* contain polyphenols instead of lignin. In both cases the aromatic component of the biomass is recalcitrant to microbial degradation, particularly in the absence of air. However, since cellulose is the main source of microbial nutrition, the lignin coating has to be breached before the cellulose can become available. The cellulose of dead plants is fairly rapidly degraded aerobically to carbon dioxide and water, but in waterlogged wetlands once the oxygen is consumed the environment becomes anaerobic. A consortium of anaerobic bacteria are now necessary to slowly degrade the biomass to methane (Fig. 15.1) (Brown and Overend, 1993). This methane may be re-oxidised to carbon dioxide (Conrad, 1989).

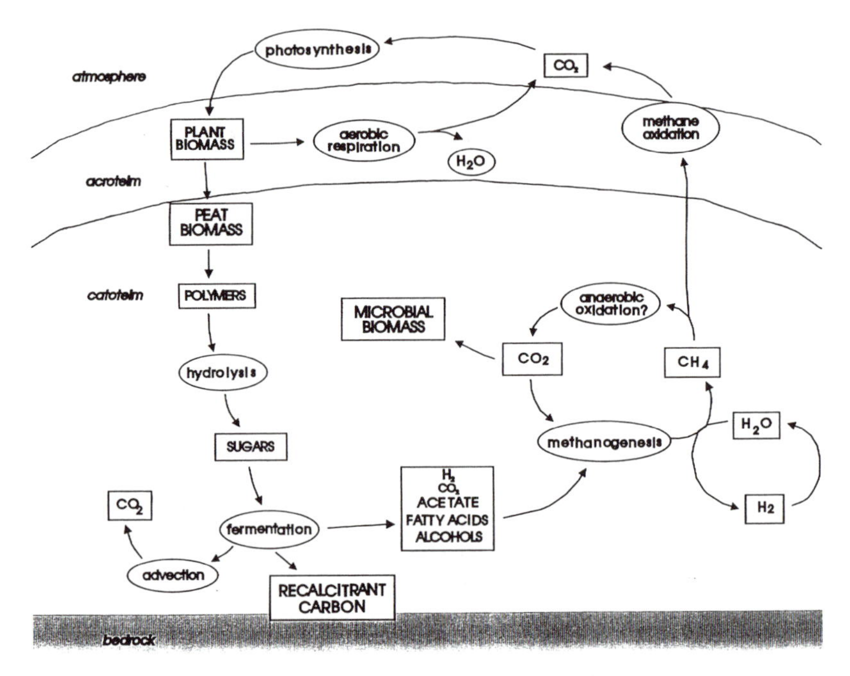

Fig. 15.1. Diagram of the metabolic pathways of carbon in peat.

Peat formation

The type of wetland which develops is determined to a large extent by the hydrologic regime, and this can be differentiated by the fate of the organic production. In highly flushed tidal salt marshes, nearly half of the net primary production may be exported as plants or as animals near the top of the food chain. In fens where there is sufficient water flow for nutrients to be imported and waste products removed, much of the production is in the form of gaseous carbon dioxide and methane. However, if there is little water flow, as in rain-fed ombrotrophic bogs, then the organic production accrues as peat (Moore, 1987).

Wetland ecosystems can progress through many different stages as any process which causes water to accumulate can produce a wetland, many of which may form peat. However, it is only in bogs that there is a significant peat accumulation. Historically, the initiation of ombrotrophic bogs has been from the accumulation of biomass in opportunistic wetlands. Although in the early stages of development a mire is often fed by minerotrophic groundwater where the dissolved ions are derived from sediments and from rock weathering, as the peat accumulates the surface vegetation is gradually lifted above the normal water table until eventually precipitation is the only source of water and nutrients for the living biomass (Maltby, 1986).

Three types of wetland are shown in Fig. 15.2: a fen, where there is sufficient minerotrophic water flow to provide adequate nutrition with no biomass build up; a swamp, where the water flow is in the process of being blocked so that peat begins to aggregate; and a bog, where there is little water flow and the nutrition is minimal (or oligotrophic), and where the wetland can form either blanket (covering the surface) or domed (ombrotrophic) bogs that accumulate peat and act as carbon sinks.

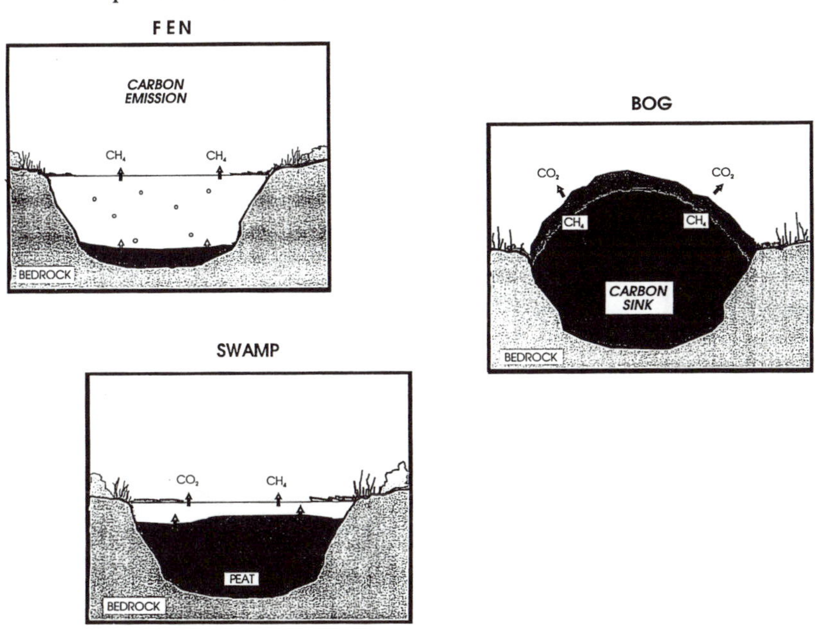

Fig. 15.2. Diagrams of the biogeochemistry of fen, swamp and bog wetland types.

Ombrotrophic bogs have two distinct environments. The acrotelm typically grows *Sphagnum* which forms a fairly open matrix of living and partially decayed vegetation. Here the meteoric water largely flows transversely without penetrating the body of the bog, it produces a variable water table with a fairly aerobic environment but with distinct anaerobic pockets able to produce methane. The catotelm is impermeable and completely anaerobic with little water or gas exchange between it and the acrotelm or the outside environment.

The degradation of plant biomass to peat in the acrotelm is mainly aerobic (Fig. 15.1), and any methane produced here (or which has escaped from the catotelm) is mostly oxidised to carbon dioxide; hence any loss of carbon from the bog is mainly in the gaseous form (although a certain amount will also be lost as bicarbonate in the water). In the catotelm, the degradation to methane is by a consortium of anaerobic microorganisms However, there is little loss of carbon, for although cellulose is degraded, as is shown by the decrease in the cellulose concentration with depth (Brown *et al.*, 1989), the low hydraulic conductivity effectively prevents much diffusion into the acrotelm or dissipation by advection. Since peat bogs have been accumulating for millennia, it would seem that the carbon must be recycled within the catotelm enabling them to become significant carbon sinks (Brown and Overend, 1993).

Methane production

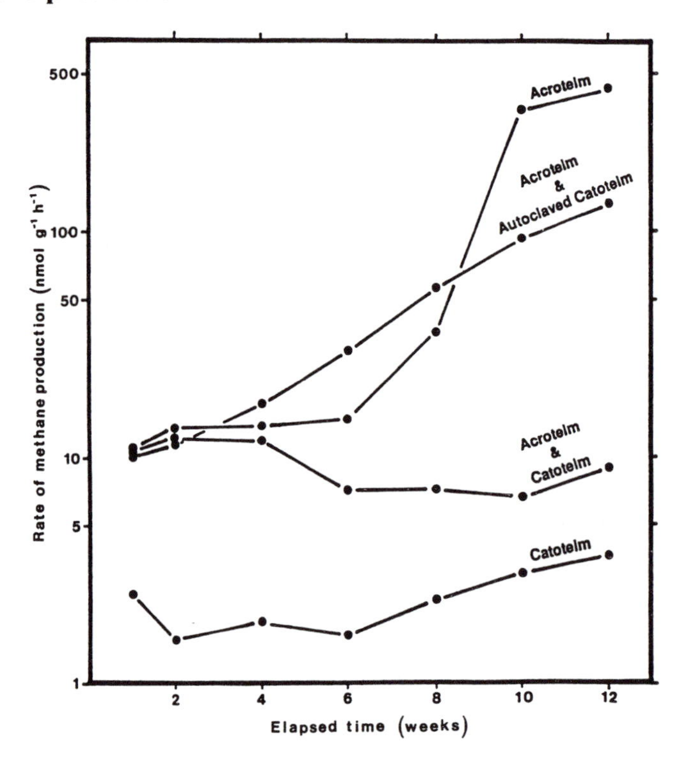

Fig. 15.3. Incubations of 10 g of acrotelm and 10 g catotelm peat, mixed incubations with 5 g of each peat, one with the catotelm peat previously autoclaved (© American Geophysical Union).

Laboratory incubations of peat, taken mainly from the ombrotrophic bog Mer Bleue near Ottawa (Brown *et al.*, 1989), clearly show two populations. The rate of methane production from acrotelm peat is two orders of magnitude greater than that from the catotelm (Fig. 15.3), showing that surface peat is much more active than that in the body of the bog. However if both peats are incubated together, the methane produced is only twice that which would be expected from the catotelm alone. If, prior to joint incubation, the catotelm peat is autoclaved (inactivated), then the methane produced is approximately half that which would have been produced by the acrotelm alone (which is to be expected, since only half of the peat in this incubation will be active). From these results is appears that, not only is the catotelm peat inhibited, but that of the acrotelm is also inactivated.

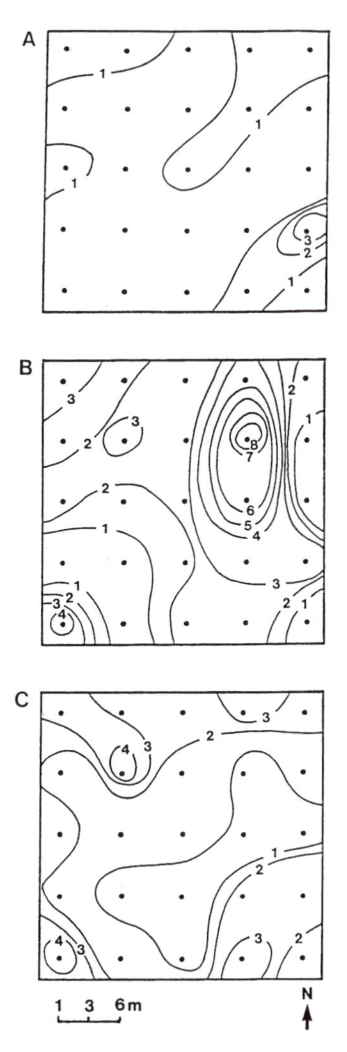

Fig. 15.4. Concentration of methane, in mmol l^{-1} , extracted from 25 stations in Mer Bleue, Ottawa, Canada. A. 60 cm depth; B. 90 cm depth; and C. 120 cm depth (© American Geophysical Union).

Statistical examination of all the incubation results show two distinct populations: nearly half of the total have a fast rate of methane production while the rest are much slower. In general, the active samples are from the acrotelm while the rest are from the catotelm, but due to the heterogeneity of the bog this is not invariably true. The question remains however, what is it that inhibits the rate of methane production in the catotelm peat? A whole range of ideas have been suggested (Brown, 1995), from an inhibitory metabolite, anaerobic methane oxidation, to lack of hydrogen. Whatever the cause, either methane is not initially produced, or it is removed once it has been formed.

We have found a considerable reservoir of gaseous methane within Mer Bleue (Dinel *et al.*, 1988; Brown *et al.*, 1989). The distribution again is quite heterogeneous, both in the horizontal and vertical sense (Fig. 15.4). When the methane concentrations are taken only from the catotelm peat (below 50 cm in this bog) they were found to form one statistical population; but if the readings (much smaller) from the acrotelm are added they show a second population. Confirming once again that there are two distinct environments to be found in ombrotrophic bogs. In the acrotelm there is active production of methane (shown by incubation measurements) and little entrapped methane (shown by the extractions), whereas in the catotelm, although the rate of methane production is much slower there is a considerable reservoir of entrapped gaseous methane.

Methane is produced by the indigenous bacteria of peat, and since this gas is only sparingly soluble in water much of it is thought to be trapped as minute bubbles (~70 μm in diameter) within the peat matrix, thus reducing both the water flow and the total volume of water. Continuous flow laboratory columns using Mer Bleue peat were used to establish the effect of gaseous methane on the hydraulic conductivity of peat. As the methane concentration increased over time so the water flow and the moisture content of the peat decreased, however, when the gas was removed from the column the readings returned to near their original values (Reynolds *et al.*, 1992; Brown and Overend, 1993). It does not appear to matter what type of gas is present, for when the control column was sterilised by irradiation a considerable quantity of carbon dioxide was formed. This gas was quickly washed out of the column by the water flow (as it is considerably more soluble than methane), while at the same time the hydraulic conductivity and moisture content of the peat increased (see Fig. 15.5).

The impeded drainage suggested by Ingram (1982) could thus be explained by the gaseous methane trapped within the peat matrix of the catotelm. In the acrotelm where there is little enmeshed gas, the hydraulic conductivity is correspondingly greater. Although peat bogs are commonly considered to be water saturated, this is not necessarily the case, for they may contain an appreciable reservoir of gaseous methane. This would account for the impermeability of the catotelm (Päivänen, 1973), and may also explain both the heightened water table found in raised bogs and the difficulty experienced in draining them (Brown and Overend, 1993). The two environments of ombrotrophic bogs are thus determined by their hydrologic regime, which in turn is controlled by the microbial production of gas. It is the indigenous microbial consortium of the peat that largely establishes the environment that supports the local cover of vegetation.

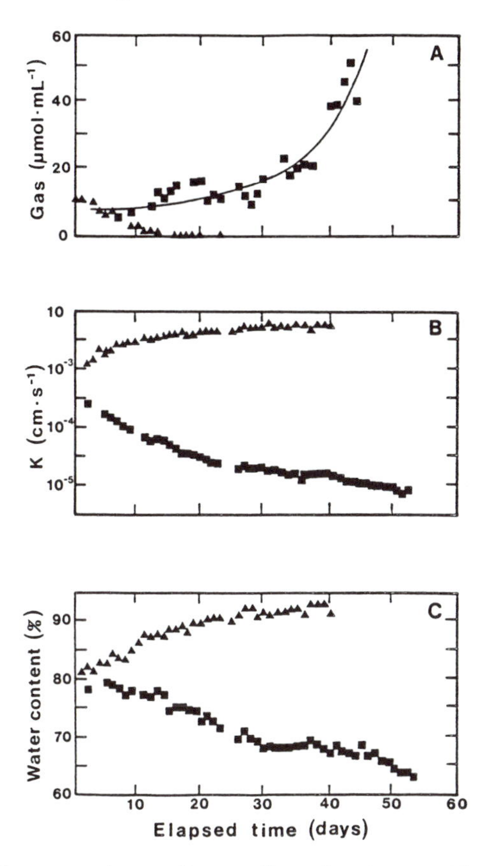

Fig. 15.5. Peat columns under continuous flow showing weekly means of rate of water flow, moisture content and methane concentration plotted against time. ■ active column, ▲ sterilised column (Brown, 1995).

Methane emissions

Methane emissions obtained during the 1990 Northern Wetlands Study of the Hudson Bay lowland (Roulet *et al.*, 1992; 1994) show relatively low methane fluxes which are related to wetland development of the area. Net methane emissions depend on a balance between its production and oxidation. In standing water with a rich organic source (such as beaver ponds) methane production is sufficiently rapid to form gas bubbles which may ebullate before they can be oxidised. Wet environments without standing water, such as peat bogs and forest floors, appear to have the best potential for the oxidation of the methane, although in years of drought with lowered water table the oxidation capacity may be greatly reduced. Undisturbed bog ecosystems emit only small amounts of methane to the atmosphere, as any loss of carbon from them is mainly in the form of carbon dioxide.

There has recently been considerable controversy in Canada over the environmental effects of flooding large areas of wetlands for electricity generation, since this flooding induces the emission of large fluxes of both methane and carbon dioxide (Rudd *et al.*, 1993). Flooding will increase the microbial activity and hence allow a greater production of gas. There is presumably also a considerable amount of gas already trapped within the semi-waterlogged peat matrix, so that when it is flooded the microscopic gas bubbles are able to coalesce and escape to the surface.

Conclusion

Wetlands, notably bogs, are fragile ecosystems that depend on steady hydrological regimes for their conservation. If the acrotelm is not maintained in ombrotrophic bogs the whole ecosystem is liable to collapse and, not only will the entrapped methane be released, but the influx of air (oxygen) will also allow rapid degradation of the peat to carbon dioxide, resulting in the loss of carbon to the atmosphere with a concomitant increase in radiative gases and enhancement of global warming. Destruction of bogs threatens to mobilise this carbon sink, which may be as significant in extent as that of the tropical rain forests.

Although the basis of peat microbiology is becoming clearer, there are still many areas where we have little data. The gross rates of biomass accumulation and loss of carbon from wetlands are unknown, which in part, is due to the lack of understanding of the various metabolic reactions that are involved in the two peat ecosystems. We do not know the amount of carbon that is incorporated from the acrotelm into catotelm, nor whether the catotelm peat is capable of fixing gaseous carbon. To realistically model wetlands it would seem that we need more information on the microbial metabolism of peat bogs.

Chapter sixteen:

Wasted Assets or Wasting Assets: Peatland Functions and Values

Mr Philip Immirzi
Scottish Natural Heritage, UK.

Conflicting notions of worth

Land in nature conservation management is often seen as possessing rather abstract worth. The values arising from agriculture and forestry on peatland generate tradable commodities and income; wilderness, wildlife and science values cannot usually be expressed simply in monetary terms. Conserving peatlands for their wildlife or their palaeoenvironmental archive generates at best a modest, and at worst no, financial return. Whilst the diverse values of peatlands and wetlands are slowly beginning to be appreciated, they are still imperfectly built into the economic calculus which determines decisions affecting them (Barbier *et al.*, 1994). Consequently, dilemmas over their use are still commonplace, with economic sector differences, opposing perceptions of worth and partial, or at times ambiguous, protection afforded by nature conservation law, entrenching sector positions further. Peatlands have continued to attract attention in debates over use and are still capable of precipitating major disagreements (Steel, Chapter 40, this volume). While some conservation 'successes' have been registered, these are attributed to stronger legislation and enforcement (Bain; Raeymaekers; Chapters 36 and 38, this volume). Yet the criteria and legislation that brought this about, are still by and large driven by nature conservation (Ward *et al.*; Bain; Raeymaekers, Chapters 35, 36 and 38, this volume). Notions of decline in environmental quality as a consequence of wetland development, on the other hand, have only played a narrative role in wetlands protection. The legislation that exists can protect many wetlands from damaging interventions but it does not explicitly protect their positive functions, services and attributes. Consequently, the crusade for wildlife has been unsteady and it seems many more failures will be registered so long as the notions of worth remain polarised.

The representation in Table 16.1 illustrates the difficulty of capturing the value of peatlands in economic terms. Whilst no one would disagree that some aspects are highly valued by society, because they do not enter the market process in a straightforward way they remain largely unpriced; the scientific and heritage value for example is not easily translated into monetary value. The economic worth attached to peatlands is strictly utilitarian and boils down to how the human species uses them. In analyses of the costs and benefits of development in peatland, most direct uses (e.g. forest products, sporting uses) are accounted for, whilst indirect uses (e.g. flood regulation) and non-uses rarely figure at all.

The traditional sectoral carve-up of land fails to conserve these broader values even when nature conservation criteria are built-in, well-articulated and supported by legislation. Conventional planning and management of land resources assigns priorities to important economic sectors and promotes or supports sectoral interests (usually to the mutual exclusion of one another). Such specialisation has driven economic development and cramped the ecological diversity and plethora of associated environmental values of wetlands.

Table 16.1. Schematic representation of the on-site and off-site significance of marketed and non-marketed peatland goods, services and attributes.

	Location of goods and services	
	On-site	Off-site
Marketed	*Usually included:* e.g. agriculture, sporting, forestry, industrial extraction	*May be included:* e.g. angling, tourism
Non-marketed	*Seldom included:* e.g. science, viewing & studying wildlife, wilderness, subsistence uses (domestic peat extraction)	*Usually ignored:* e.g. flood desynchronisation, water supply, water quality

In economic terms, the challenge to the narrow utilitarian functions of peatland might come from looking at these ecosystems as providers of a wider spectrum of goods and services most of which lie outside the market place. Alternatively framed, the question might be whether nature conservationists need to seek new alliances in the defence of nature? If conservation of peatlands and wetlands is ever going to be part of a new land ethic, will it have to embrace a notion of their greater worth? While environmental economists have led the way in developing elaborate techniques to capture the value of conservation-orientated management, the traditional nature conservation approach is still shackled with concepts of rarity, uniqueness, and diversity as its normative and operational criteria. Whilst there are clearly fundamental problems with reducing the value of land and its properties by way of an economic equation - where economic instruments or incentives may be necessary - the economist's discipline can expose who should pay and who is paying, who stands to benefit and who will lose.

Peatlands function - services, products, uses and attributes

Table 16.2. Functions, services, products uses and values of peatlands.

Conservative	Consumptive
flood regulation *(floodpeak reduction)* *	commercial forestry **
water supply *(quantity, quality, baseflows)* **	grazing & extensive agriculture **
ion and biomass retention/export *	intensive agriculture **
biogeochemical cycling *	domestic peat extraction **
food chain support *	industrial-scale peat extraction **
sporting estate management **	
fisheries support *	
global biogeochemical cycles *e.g.* CO_2 *sink &* CH_4 *source* *	
wildlife habitat *	
wildlife **	
palaeoenvironmental archive ***	
archaeological resource ***	
forage **	
genetic resources **	
biological diversity ***	
cultural - *wildscape/wilderness* ***	
science ***	
erosion prevention *	

Key: *services; ** products and uses; *** attributes

Functions are the physical, chemical and biological processes that characterise wetland ecosystems (Dugan, 1990). The processes are vital to the functioning of the wetland and operate whether or not human use is actually derived. Thus, a function is an aspect or property of a wetland that may support or protect a human activity, but not all wetland functions are necessarily useful or important to society. The dynamic interactions which drive the hydrological, biogeochemical and ecological machinery yield services, products/uses and attributes on which society places at least some value.

Functional analysis is the study of the interactions of these processes. By unravelling the interconnections and quantifying the flows, there is a basis for determining which of the natural functions provide useful services, products and attributes. When demonstrated, the sum of these can be far greater, and - if conserved - more sustainable, than many proposed developments. The most frequently cited case of an intervention which stood to increase public expenditure, is that of the reclamation of flood-plain wetlands in Charles River basin in Massachusetts. Preservation of c.4000 ha of wetland may have saved US$ 3-17 million each year in flood prevention measures and damage (USACE, 1972).

Frequently cited functions and values of peatland are summarised below (Table 16.2) contrasting those which are conservation-based and those which are consumptive. The functional philosophy can clearly be applied widely to non-wetland terrestrial and aquatic resources (Winpenney, 1991; Swanson and Barbier, 1992) and the simple framework will appeal to economists who see matters in a similar vein (Table 16.3).

Table 16.3. Ecological and economic system characteristics.

Characteristics		
General system	Ecological system	Economic system
Stocks	Structural components	Assets
Flows	Environmental functions	Services
Organisation	Biological/cultural diversity	Attributes

If we explore the range of natural values of extensive peatland systems, we begin to see that we can list in narrative terms such things as habitat for wildlife, and express this as some measure of biodiversity. Landscape values such as wilderness are examples of other important attributes of Scottish peatlands. As far as the public may be concerned the value of other attributes of peatlands may be more obscure. Examples include the so-called peat archive, which includes records as disparate as archaeology (Coles; Chapter 4, this volume), palaeoenvironmental information such as climate change (e.g. Barber, 1981), atmospheric pollution, natural chronometer, past ecology and past land use (Lindsay, 1995).

Peatlands can, of course, be of considerable value in supporting economic activities such as forestry, intensive agricultural activities in the lowlands, and extractive industries (such as those for fuel and horticultural peat). These are examples of consumptive uses and products. In the case of peatlands, certain activities such as drainage and burning may alter the peat substrate in such a way that future options are narrowed. The rehabilitation prospects of afforested peatland are not well known and the restoration prospects of eroding hill peatland from over-grazing are daunting, demanding re-examination of the policies which have promoted such changes.

By virtue of their areal scale, vegetation, soil properties, and role in hydrological cycles peatlands also provide a form of natural infrastructure. Their role in regulating hydrological and biogeochemical cycles are among the least well understood. Where peatlands occupy large parts of catchments, they will indisputably have a role in flood flow regulation, water supply and ion and biomass retention or export Whether or not the functions can be construed as a service remains unclear. On this basis, however, it is easy to appreciate that large transfers from one land-use to another can have catchment-scale effects beyond the mere alteration of habitat.

What does a functional approach offer over traditional conservation methods?

A functional approach does not deny the importance of traditional nature conservation criteria. Traditional conservation criteria can and must be built in and generally should strengthen the conservation case. It is not that the fundamental conservation arguments are necessarily inadequate, but - being constrained - the conservation lobby is the weak relation in the economic development process. The altruism of conservation is seen as a burden both on the State and on private landowners. Nature conservation often deliberately excludes other uses, although perhaps less so than other sectors. In the UK, it is immensely difficult to establish unassailable criteria for the defence of wildlands outside Sites of Special Scientific Interest and Areas of Special Scientific Interest. Even land that is designated is not sacrosanct and over-riding public interest is frequently invoked to damage or destroy the interest. When weighing up the trade-offs between development and conservation, the vexed question remains - what is tradable and on what terms? The introduction of the notion 'critical natural capital', the economic analysts' term for irreplaceable or non-tradable habitats, is fraught with difficulties in its practical application given the wide perception of precisely what should remain inviolable (Buckley, 1996). In the USA, functional analysis has played a prominent role in mitigation by extending the range of properties to be compensated (Ainslie, 1994).

The appreciation of potential damage to hydrological processes which may extend well beyond the boundary of designated sites has forced conservation agencies to extend the compass of their interest (Newson, 1992). A functional approach can make a valuable contribution provided it is rigorous and scientifically based. Moreover the approach recognises the importance of wetlands to other sector interests and offers a potentially strong scientific basis for making land-use allocations. Many choices could enhance natural heritage features and vice versa. The analysis and assessment of values need not be sectorally organised. Theoretically this offers a means of spreading the burden of costs and begins to redress imbalances in the distribution of the benefits of conservation options. Another allure is that, having identified functions and assessed their responses to different interventions, it becomes possible to assign monetary values to different land-use scenarios and enhance the cost-benefit analysis (CBA). Whilst there is considerable resistance to the wide application of economic valuation methods to certain intangible values of the environment, and widespread distrust in CBA, there is potentially considerable merit in demonstrating costs and benefits of development versus conservation. This analysis would be particularly illuminating for those values which are hydrologically mediated or support economic activities down-stream or otherwise off-site.

How might the functional approach be applied?

Table 16.4. Possible applications of wetland functional analysis.

	Application
normative level	research tool - experimental design, hypotheses generation & testing
policy development	strategic decision making physical planning
	SEA (Strategic Environmental Assessment)
	Integrated Catchment Management (ICM) or Catchment Management Planning (CMP) etc.
operational delivery	regulatory assessment tool
	Environmental Impact Assessment (EIA), impact prediction
	speed processing of consents for potentially developing operations (PDO)

There are potentially a number of applications for functional analysis. Some of these are summarised in Table 16.4. It is worth taking a look across the Atlantic, to the USA where the longest experience of functional analysis (FA) exists and an array of functional assessment tools (FAT) have been developed. Such tools employ rapid information appraisal methods and infer function using complex variables termed predictors. Relative performance is established using reference wetlands. FAT have evolved in the context of Section 404 regulatory programme under the Clean Water Act (Ainslie, 1994). In the USA, wetlands are defined as waters and permits are required for both fill and dredge materials (which are regarded as pollutants). In fact regulation of wetlands is so important that developers are as concerned about their presence as they would be of hazardous wastes (Stokes, 1990). FAT play a multiple role, they are used in the determination of significant degradation of waters, have a part in analysis of alternatives, compensatory mitigation and veto actions. Although by 1980 some 40 FAT had been developed (Lonard and Clairain, 1983), it may come as a surprise that few of the methodologies are extensively used (Adamus, 1992). Furthermore, in-depth assessments are rarely performed and only 2% of permits reviewed by the EPA have FA applied (Ainslie, 1994). There appear to be several reasons for this including concerns over technical validity, usefulness of the results in the permitting scenario, training requirements, time and resource constraints, and the assessment of relative value (Smith, 1992 cited in Ainslie, 1994).

Much of the USA experience has been preoccupied with assessment of permits in the context of large numbers of routine applications to dredge and fill wetlands. There is little in the way of their application in a strategic sense, for example in the identification of policy areas for development which would have obvious applications in the reduction of site-specific conflicts. Neither has there been as much research directed towards FA as might be expected given the importance of the regulatory programmes. In a European context there is limited institutional appreciation of the need for anything quite as holistic as FA, however some research has been carried out (Maltby *et al.*, 1994). Large-scale peatlands have not been studied so far.

At a strategic level, partial application of FA could help identify the suitability of areas for a single or restricted range of activity/ies. At a larger scale, despite many merits, ranking or determination of protection areas could suffer from a variety of difficulties associated with protective designations and the need for compensation. So, the application of FAT in strategic impact assessment may bring with it major problems. Clearly there is an opportunity for reducing the risk of attrition and making the deployment of tactical FAT more effective. This could make the overall costs of implementation more manageable. All types of impact, however, would have to be assessed, posing considerable analytical complexity. As less information would be available, such analysis would be less exact; however, at a strategic level, impact predictions need not be precise. On the other hand, an initial lack of practical experience and a poor literature base would hamper consideration of its early use in anything but an experimental way.

Requirements for taking functional assessment forward

While there are a number of examples where loss of functional capacity is plausible, there are few, if any, fully worked examples of the way that peatlands provide benefits. As long as this remains the case, it seems unlikely that exhortations for peatland protection will lead to action. Wider acceptance of the FA approach requires the development of a sound proven science base. Coupling wetland function to societal benefits will be a challenge for researchers and economists, whilst development of transparent decision-making aids may be onerous. FAT may also face more problems than just an inadequate scientific basis (see Table 16.5), though none of these are necessarily insurmountable.

To date, applications have not been widely tested or proven (e.g. Butcher, 1995) and they are likely to be received with a degree of customer resistance. One reason for this is that any such procedures are likely to be eclectic, making a full analysis redundant for many types of agency casework. Essentially practical agency applications are likely to be required with special attention being paid to constraints such as time, resources and expertise. Given the complexity and interconnectedness of the questions being posed, are such tools necessarily going to enhance best professional judgement? Are bad procedures worse than no procedures at all, as is the case in so much CBA (Bateman and Bryan, 1994)?

Table 16.5. Anticipated problems with taking functional analysis forward.

- paucity of examples where functional capacity demonstrated
- inadequate research and development base
- lack of recognition of functions, products, attributes
- little or no quantification of the societal benefits
- poor understanding of the tolerance to impacts
- performance of natural systems may differ significantly
- essentially practical planning and decision-making tools would be required
- increasing reliance on layers of national and international legislation & hence bureaucracy
- would require a major policy review
- must understand the social dimension and implications for stakeholders
- application demands inter sector compliance
- could require changes in legislation

The absence of any regulatory framework in Europe could be a hindrance or a help. The US experience of regulation shows a statutory process believed to be slowed by many layers of bureaucracy. While the Environmental Protection Agencies' (EPA) hand has been measurably strengthened, it is not altogether clear just how much FAT for wetlands have helped. Many activities that alter functional capacity have escaped regulation in the US (e.g. diversion of water from a wetland, diversion of sediment, flooding, change in nutrient concentrations, grazing, alteration of adjacent land (NRC, 1995)). But is it realistic or sensible to regulate everything? The lack of any compensatory mechanisms in the USA leads to protracted and acrimonious litigation. Is it possible that legislation here will merely replicate the problems registered in the USA? The adoption of principles of wise use of wetlands in a non-regulatory framework on the other hand, risks them being relegated to anodyne political rhetoric.

We need to ask whether FA has a natural locus? Whilst there is usually more than one agency for dealing with, say, the freshwater environments, for peatlands and other wetlands the responsibilities and stakeholders are fragmented, if any exist at all. In Europe, institutional frameworks for managing wetlands are non-existent. Whilst in some European countries integrated catchment management (ICM) is facilitating the development of integrated strategies (with a focus on river quality and other related variables), there is still no common mechanism. FA on its own does not necessarily offer such a mechanism, though it differs from the concept of ICM. ICM is also holistic in philosophy, strategy and operational levels, but usually has a much wider remit. There may well be difficulties in translating FA into the current planning context, some of these difficulties are constrained by organisational culture and a reluctance to give up influence or take on extra duties. The context for limited applications exists and may be provided by *ad hoc* stimuli for action: be they flooding, erosion, eutrophication, fishery decline, decline in nature conservation value etc.

There are a number of inherent antagonistic tendencies including: competing ideologies (e.g. ICM and its many clones), economic pressures (force majeur), resourcing aspects (not clear who stakeholders are), financial and administrative arrangements which need to be recognised. If some kind of functional analysis is ever to become a regulatory requirement (namely through an EC directive, see CEC, 1995), the goals and objectives of the participating interests would need to be identified, their respective roles clarified, rules of engagement agreed, and intervention and arbitration mechanisms confirmed at both national and operational levels. Whilst some of these are perhaps already enshrined in statute or government commitments, further statutes and additional tiers of bureaucracy are likely to be required.

Prior to FA and FAT being taken forward, a number of steps can be identified some of which will help frame government and agency responses to any impending notions of legislation. These are:

1. authoritative scientific review of ecosystem and hydrological functions, the societal benefits and the effects of interventions on wetland functioning;
2. review of successes and failures of current regulations and functional assessment applications - can such tools be relied upon to deliver simple answers to complex problems ?
3. precise identification of targets/goals *vis-à-vis* interactions with surrounding land and the overlap with ICM;
4. analysis of constraints - legal, administrative, resource;
- identify respective roles of current agencies and the need for new ones,
- examine participatory aspects - the imposition of further constraints on development will be carefully scrutinised,
- the resourcing initiatives required to fund research and implementation.

Conclusions

By looking at functions we can begin to see how multiple sectors of society might benefit from wetland conservation without compromising or foreclosing the needs of others. In the development of less favoured regions (climatically, edaphically or economically), it becomes even more important to afford greater protection to natural assets and natural infrastructure. Peatlands of north western Europe would appear to represent a prima-facie case. Debates in the last two decades have identified that the overriding economic imperatives for the development of wetlands can no longer go unquestioned and that economic interest balances uncomfortably with public interest. There is also a paradigm shift in the way natural resources are being considered. The policies and contributions that various sectors of the economy can make towards achieving a sustainable future are being re-defined. Governments still need to carry out a fresh analysis of the costs and benefits of development in wetlands, facing up to the challenges of the failures of the neo-classical economic model. Only then will the terms and shape of new policies emerge. It is easy to promote the desirability of sustainable, or wise, use of wetlands. Delivery of this aim is likely to be hampered by significant gaps in the understanding of the many ways wetland and peatland systems work.

The development of new policy lines is being encouraged by the European Commission (EC) where the idea of 'no further wetland loss' has permeated (CEC, 1995). However, the EC believes that no frameworks for integrating environmental concerns about wetlands (at either national or supranational level) are sufficiently well developed. Paradoxically, it expects the Habitats and Species Directive (92/43/EEC) - a conservation instrument - to provide the momentum. Since 1984, the EC has provided 60-100 million ECUs towards wetlands related conservation and restoration (CEC, 1995). On the one hand, this has been a measurable incentive but on the other, it has not matched the expenditure on detrimental impacts (e.g. dams, polders, aquaculture). New regional development plans will have to comply with European Union environmental policies (article 7 of the new Framework Regulation for the Structural Funds).

A number of alternatives exist to help shape wiser use of wetlands, e.g. strong enforceable legislation, new tools for assessment of wetlands values, and/or more enlightened and innovative policies. It seems that conservation generally can be most effective when it is the shared responsibility of the various social and economic players. A fairer distribution of the costs and benefits might be contrived by traditional measures without resort to legislation. Specific subsidies, grants or fiscal incentives applied through taxation could be developed. Each of the options has its rewards, but has the added obligation of policing, monitoring and justification of the benefits. Whatever the solution, while one state sector alone is responsible, it becomes inevitable that problems arise. One dilemma faced today is that the costs of restoration and rehabilitation of peatlands cannot be borne by a single sector.

There was a time when wetland conservation was the exclusive domain of nature conservationists. An alternative view is emerging. Legislation on protection of the functional importance of wetlands may - in the long term - be the only viable policy response. It does seem however, that much more public debate, research, and education are necessary. New policy initiatives and financial fixes should be exhausted before new legislation is considered because a hasty or precipitous regulation could soon turn into an administrative nightmare.

Acknowledgements

I should like to thank Dr Helen Armstrong and Dr Des Thompson for allowing me the intellectual space and time to prepare this chapter. The ideas developed in the chapter benefited from discussion with Bill Ainslie (EPA), Rob McInnes, David Hogan and Ed Maltby (RHER).

Part Five

Mires as Cultural Landscapes

An Introduction

Dr Hugh A.P. Ingram
University of Dundee, UK

The consideration of mires in a cultural context has severely practical implications for the way we manage mire sites, especially peat bogs and particularly those which it is thought suitable or necessary to conserve. By conservation I refer to the objective of enabling our descendants to study and enjoy these places. This in turn implies that their unique ecological processes shall continue to function so that they will support the range of living organisms and organic communities which are characteristic of mires.

This may seem an unremarkable objective. Why should we worry about it? Two observations suggest that all is not well. One is that, both in the United Kingdom and elsewhere in N.W. Europe (e.g. Åberg, 1992), various prime bog sites are becoming colonised by trees. These profoundly affect the mire surface and its wildlife in ways that diminish the scientific interest of both; that destroy the characteristic organic communities; and that, in an oceanic region of predominantly treeless bogs, have operated seldom if ever during our current interglacial. Two bog sites owned by the Scottish Wildlife Trust, one at Bankhead Moss in Fife and the other at Flanders Moss in the Carse of Stirling are both affected and control of scrub invasion at these sites is costing heavily in money and human effort.

The other problem may or may not be linked with the first one. Some bog surfaces appear to be drying out and are doing so in the absence of any obvious cause, such as drainage or the extraction of peat. Even without tree invasion this causes changes in the surface which are profound enough to be a source of worry. At Dun Moss in East Perthshire, dwarf shrubs seem to be increasing at the expense of the *Sphagnum* cover, so an array of plants that are interesting and unusual is giving way to *Calluna*: a plant that already covers most of the surrounding hill-slopes.

Why are these changes occurring? I doubt if we can be certain; but some interesting evidence is relevant.

Many of our mire sites have had a long association with human cultural activity. There are Neolithic hut-circles within 500 m of Dun Moss (RCAHMS, 1990) and the adjacent mineral soils show evidence of disturbed profiles (E.L. Birse, pers. comm.). At Offerance Moss in the Carse of Stirling an extensive type of Mediaeval earthwork called a homestead moat lies closely adjacent to the edge of the mire (NCC, 1983). About 200 years ago in the same area the peat bogs became so important to the local economy that those wishing to 'live above the shop' contrived dwellings by carving them out of the very peat itself (Fenton and Walker, 1981).

To suppose that the people associated with these artefacts ignored their local bogs would be absurd, and not only in the last of these instances which was associated with extremely intensive destruction of the mire. Britain is an overcrowded island, and there is abundant evidence (RCAHMS, 1990) that even the area round Dun Moss, which lies at about 350 m a.s.l. in a part of the S.E. Grampians that is now sparsely populated, was formerly quite densely settled; and that this was so at various times over a period of millennia. The area is indeed a 'multi-period archaeological site' (Halliday, 1993) as well as being a Site of Special Scientific Interest.

The only reasonable inference is that these sites were managed by our ancestors through a combination of burning and grazing. Interpretation of midden evidence for domestic livestock is the province of the archaeologist; but burning to rejuvenate heather and enhance its palatability is a method of increasing livestock productivity that is well understood in the north and west of these islands (Symon, 1959). Burning and grazing both suppress tree colonisation. When the fire travels with (rather than against) the wind, fresh *Sphagnum* colonies become re-established (e.g. at Dun Moss) well within two decades, following an initial period of dominance by *Eriophorum vaginatum*. Furthermore, Walton Moss in Cumbria is part of a common-grazing for sheep. Their presence maintains a thriving and diverse *Sphagnum* carpet together with an insignificant cover of dwarf shrubs.

It therefore seems probable that many bogs in the British Isles have developed under a regime of human intervention that must be judged as moderately intense. They are, in fact, 'cultural landscapes', to use the fashionable terminology. Even if this were not so, we are aware from evidence of charcoal layers in peat, gathered on both sides of the Atlantic (Tallis and Livett, 1994; Kuhry, 1994), that fire, at least, has formed part of the complex of environmental influences under which these systems have undergone most of their development.

Thus, many of the most notable mire sites are in part the product of deliberate human activity. But the management prescriptions that have been written for many of our more interesting peat bogs seem to ignore this. Their authors appear to regard them as 'natural' systems in which *Homo sapiens* and its livestock is an alien component. To them, the challenge is to show that their desire to regulate grazing and to reduce the use of fire as a management tool is not contributing to changes in these sites that conflict with the remit of wildlife conservation. For the scientific community, the challenge is to help land managers of all kinds by showing how cultural influences actually affect the ecology of bogs and other mires.

Chapter seventeen:

Bogs and People Since 1600

Professor T.C. Smout
University of St. Andrews, UK.

When the first detailed geographical accounts of Scotland began to be written, generally in the years between 1630 and 1730, the reputation of bogs as a useful resource stood high. Thus at Fetteresso in Kincardineshire the parish was said to be supplied with 'inexhaustible mosses, wherein are digged the best of peats, very little if anything inferior to coals', from which the inhabitants supplied not only themselves and the neighbouring communities of Dunnottar, Inverbervie and Stonehaven, but also Aberdeen some 15 miles away (Mitchell, 1906, p. 248). At Cortachy in Angus 'the hills and glens of this county abound with excellent moss and muir for feuell, with wild fowl of different kinds, and sometimes with deer and roe' (Mitchell, 1906, p.284). A description of Aberdeen and Banff by Robert Gordon of Straloch, probably of the 1630s, says that 'there is no occasion here for stoves; the hearths are well supplied with peat, which is dug out of the ground, and is black and bituminous, not light and spongy, but heavy and firm' (Mitchell, 1906, p.268). A parish with 'moss ground' was blessed, like Keith: it had 'great plenty of fir under ground, which the people thereabouts dig up some two fathoms deep, and by this they are served with winter light and timber for their houses. In this hill is a large peat bank about six or seven foot deep and near two miles long' (Mitchell, 1906, p.89). A parish without such resources was cursed: of Cushnie it was said 'it is a poor countrey both for corn and pasture, and exceeding scarce of Fewel' (Mitchell, 1906, p.31). Of the lower ground of Morayshire near the coast it was observed that 'they suffer from scarcity of peats for fuel, which is the only inconvenience felt by this highly favoured region, but even that in few places, and they remedy it by hard drinking in company, for this also must be admitted' (Mitchell, 1906, p.457). The implication seems to have been that if you could not warm yourself with peat you needed to warm yourself with whisky.

This generally cheerful and positive attitude towards bogs may be contrasted with the attitudes that came to rule in the following century, the age of the agricultural improvers, when the old assumption that natural resources were a given changed to one where they were regarded as capable of betterment. To the late eighteenth and early nineteenth century improvers, peat bogs cried out for money and effort to transform them from waste into arable land, even though locally in parts of the countryside the exhaustion of peat supplies was already said in the 1790s to be leading to depopulation.

Reclamation of mossland was not itself new: as early as 1724 it was observed that parts of Flanders Moss 'by casting, pareing and burning' had been in some places 'cut quite thorow and made arable ground' (Mitchell, 1906, p.341). The great scheme to drain Blairdrummond Moss nearby, by the notable improver Lord Kames, was hailed as a national benefit, greater even than that conveyed by David Dale in founding his famous cotton manufactory at New Lanark. Both had employed displaced Highlanders, wrote William Aiton in 1811, but whereas at Blair Drummond 'the moss colony remain healthy and happy, delighting in their situations, warmly attached to their patron, and to the Government, daily increasing in wealth, and rearing a numerous offspring, ready to extend their brawny arms, in the cultivation of the dreary wastes, or to repel their country's foes', those in the cotton mill 'became discontented with their situation, and soon abandoned it'. That 'several hundreds of ignorant and indolent Highlanders' went on Aiton, were 'converted into active, industrious, and virtuous cultivators, and many hundreds of acres of the least possible value rendered equal to the best land in Scotland are matters of the highest national interest, to which I can discover no parallel in the cotton mill colony' (Aiton, 1811, pp. 341-2).

These tones were quite characteristic of early nineteenth century commentators. Thus the Rev. Robert Rennie of Kilsyth, a pioneer in the systematic study of bog formation, wrote in 1807 that 'innumerable millions of acres lie as a useless waste, nay, a nuisance to these nations. The benefits that might accrue to Europe by a slight attention to this subject, are above all calculation. It is impossible for numbers to express, or the imagination to conceive, correctly, the extent of these' (Rennie, 1807, p. 6; see also Rennie, 1810). Aiton himself believed that 'the intrusion of Moss earth has been attended with two evils of great magnitude; first, the loss, or at least the reduction, of the value of an immense extent of soil, and secondly, its pernicious effects on the atmosphere' (Aiton, 1811). He estimated the amount of ground under bog in Scotland at over 14,000,000 acres, and believed that the accumulation of moss over so much of the original soil since (he thought) the time of the Roman invasion, had led to a decline in the temperature. Andrew Steele, in 1826, in the opening chapter to his treatise on peat moss, spoke of the bogs as 'immense deserts . . . a blot upon the beauty, and a derision to the agriculture of the British Isles'. He also commented that 'the only animals found on these grounds are a few grouse, lizards, and serpents' (Steele 1826, pp. 38-40).

The agricultural experts received general support in their view of bogs as dreary encumbrances from the ever-growing band of Romantic tourists, who came to Scotland to view the glens and to obtain a frisson of excitement from the 'picturesque', 'the sublime' or the 'terrific', but who found nothing attractive in bogs as they floundered through them. Thus John MacCulloch, Walter Scott's friend, had a memorable diatribe against the Moor of Rannoch: 'hideous, interminable . . . a huge and dreary Serbonian bog, a desert of blackness and vacuity and solitude and death . . . an ocean of blackness and bogs, a world before chaos; not so good as chaos . . . even the crow shunned it . . . if there was a blade of grass anywhere it was concealed by the dark stems of the black, black muddy sedges, and by the yellow melancholy rush of the bogs' (MacCulloch, 1824, pp. 317-20).

When, however, the train replaced the pony as a way of crossing the bog, the traveller could view them more dispassionately and in greater comfort. The first thoroughly appreciative description of the aesthetics of the Moor of Rannoch came from the anonymous author of a public relations book for the new line. The moor becomes in winter 'a study in sepia', in summer 'one colossal Turkey carpet, so rich and oriental'. Even: 'to see a sunset on Rannoch Moor is as essential as to see Loch Lomond by moonlight' (Anonymous, 1894).

Nevertheless, until far into the second half of the twentieth century, the idea that the peat bog was a desert that needed reclamation, or at least was a wasted resource that could legitimately be used up for some economically productive purpose, ruled most thinking on the matter. Economic and political changes helped to reinforce the view. Following the great expansion of coal mining in the eighteenth and nineteenth centuries, and the transport revolution enabling coal to be more readily moved, the role of the local peat bog as a fuel resource declined to extinction except in parts of the northern and western Highlands. The coming of the national grid, and Thomas Johnston's insistence, as head of the first Hydro Board after 1945, that it should reach the remotest areas, further emancipated even the crofting population from absolute slavery to the peat spade. Meanwhile in the twentieth century the rise of the agricultural subsidy encouraged the transformation of many peatlands into farmed pasture, after 1945 the coming of new ploughs and tractors made it possible to get conifers into the hillsides and flows, materially assisted by tax concessions, and the rise of the garden centre gave further encouragement to cut out the bogs. If we were spared the depredations of the Irish peat-fired power stations, we were not spared much else. In the immediate post-war period there was even a prototype peat-fired power station set up in Caithness, but the experiment was not successful.

The ecological movement itself, with its talk of wet deserts, was not friendly to the notion of peatlands. In 1973, H.A. Maxwell, writing in highly favourable terms about modern forestry plantations, quoted a personal communication from Frank Fraser Darling: 'The Sitka spruce has been a godsend in re-afforestation of Scottish hill ground of peaty character . . . . To recreate a forest biome after a long period of soil degradation is inevitably a slow process (I think myself in terms of one or two centuries) . . . as ground and shelter conditions ameliorate we can confidently expect improved appearance of the plantations gradually becoming forest' (Tivy, 1973, p. 182).

In the post-war decades the call to reclaim marginal land of all descriptions reached a peak of frenetic zeal quite analogous to that of the early nineteenth-century improvers. In a pamphlet entitled *Reclamation!* the Scottish Peat and Land Development Association (c. 1962), called for a Land Development Board to encourage the transformation of the waste: 'we can no longer afford to have so much marginal land put to so little use, or deteriorating through misuse'. Among its proposals were to establish experimental farms on bogs, and to initiate reclamation and improvement by concentrating drainage machines in groups 'in an area that forms a natural entity - e.g. the whole of a glen or a major bog'. J.M. Bannerman, the well-known Liberal, called for 'widespread arterial drainage schemes', especially in Strathspey 'where the land to be reclaimed runs into scores of thousands of acres'. He also proposed lowering Loch Lomond by four feet by removing the silted sandbanks at the outflow. Others called for an onslaught on the 4 million acres of partially productive land by 'chemical ploughing' (i.e. by massive application of herbicide). These people were visionaries but they were not cranks: apart from Bannerman, they contained respected MPs like Tom Fraser and the future Conservative Secretary of State for Scotland, Michael Noble (Scottish Peat and Landowners Association, c.1962). The most useful outcome of these years was a national survey of peat-bogs begun by the Scottish Office after 1947 and taken over by the Macaulay Institute. Although intended as the basis for commercial exploitation of the resource, its greatest utility in the long run has been to aid nature conservation in accurate assessment of the bogs.

Changes within the last ten years have, ultimately, been towards much greater appreciation of the nature conservation value of all kinds of bogland, and towards a reluctance by government to allow public money to be spent on subsidising drainage ventures that made little sense in the realities of the world economy of the late twentieth century. The crunch came in the late 1980s, when the controversy over the Flow Country between forestry interests and conservationists, in which the conservationists were enthusiastically supported by the *Daily Telegraph*, led to a reduction in tax concessions to the foresters' rich backers in Nigel Lawson's budget. If the threats to the bog are still numerous, its friends have never been so numerous either.

So far in this chapter, we have considered how outsiders and largely self-defined 'experts' regarded bogs. This concluding section touches briefly upon the rather more obscure topic of how those who had bogs upon their land actually managed them on a day-to-day basis. Peatlands and bogs were, of course, of many kinds: for practical use the most important variation was the spectrum from wet to dry. Most occupiers, however, saw them as a source for two things - fodder, and extracted material for building, fertiliser and (above all) fuel. The chapter discusses grazing and bogs, but I would here emphasise two aspects: bogs as sources of hay, and of spring bite. These have to be seen in the context of earlier grazing regimes, where the main constraint on stocking levels was the ability to keep animals alive between November and May without artificial foodstuffs, turnips, or silage. 'Until well into the last century', writes John Mitchell, 'cattle being over-wintered were fed almost exclusively on 'bog-hay', a mixture of wet meadowland plants scythed from undrained land': it could yield between 100 and 150 stone of hay per acre, and some of the bog hay meadows were of great extent. What was said to be the largest in Scotland was the Carron Bog, four miles long and a minimum of one mile wide: it almost all now lies beneath the Carron Reservoir, but in its heyday in the late eighteenth century was described by the local minister as adding 'great liveliness and beauty to the general face of the country. The scene it exhibits during the months of July and August, of twenty or thirty different groups of people employed in haymaking, is certainly very cheerful'. You may still see bog hay cut today in Co. Donegal and Sligo, and no doubt elsewhere. The Aber Bogs on Loch Lomondside (not themselves in the least peaty) were last so cut in 1952, but the practice must now be unfamiliar at least in mainland Scotland (Mitchell, 1984). On the other hand, most upland farmers still know of the value of bogs for spring bite, when the emerging sprouts of bog cotton and other plants provide a richness of early protein that is still welcome for animals coming out of winter quarters.

The use of the peat bog for opencast extraction by the occupier was obviously important, but it is significant that in some societies - Ireland, Denmark - the generic term 'turf' (qualified in different ways) is used for anything cut from the ground. The Danes distinguish between 'grass-turf', 'heather-turf', 'bog-turf' (or peat) and 'sand-turf' (peat under sand) (Hove, 1983). The point is that all were obtained by skinning the ground, and the difference to the farmer was in degree rather than kind. They were all suitable for cutting sods for burning, and subsequent use as fertiliser, a practice that in Scotland was viewed sometimes with favour, and sometimes with hostility by experts, but - at least until the middle of the nineteenth century - was widely practised by farmers. Turf, or 'grass-turf' to the Danes was also used as a building material, for 'feall dykes' (the Scottish term) and for 'divots', or scale-shaped roofing turves (Fenton, 1970). Peat, no doubt depending on its fibrous qualities, could also be used for building, but it would have a much shorter life, particularly in frost: it would therefore be avoided except for buildings that were essentially temporary, like shieling huts. The most significant use of peat, however, was for fuel, though in districts where peat was scarce - as in parts of Perthshire - turf was used in its place: the disadvantage of 'grassturf' was that it burnt too fast. Even good quality peat was needed in very large quantities. Fraser Darling reckoned that a family of four in the Western Highlands needed to cut 15,000 peats a year for cooking and warmth: a good man could cut 1000 a day, but a month's work for the family was involved in winning peats, drying and carrying them (Darling, 1945, p. 6).

It was therefore very important to the farmer and crofter that cutting peat (and indeed 'grass-turf') should be as simple and quick as possible, and the best way to ensure that was to burn off the overburden of vegetation. Since this was often also likely to improve grazing, burning on moors and peat bogs was a regular event: presumably this, along with direct grazing itself, was what kept the drier ones from instantly regenerating with birch and other trees as so many are doing today. Because it was convenient, the burning itself often took place at unlawful seasons when the bog was driest, in late spring or summer, with attendant risks: a fact that perhaps accounts for much of the Scottish legislation limiting muirburn from the fifteenth century onwards.

For the farmer peat-cutter, convenience was more important than the law. This was how a cutter in Co. Antrim described his experiences around 1932: 'I had to get a winter's firing for myself, and having been used to watching my father when I was a young fellow, and seeing what he had done, I followed his footsteps. He always picked out a nice spot in the bog where he would start and cut. So I burned a nice wee bit first. That was to make the turf easier cut with the spade. Burning was against the law. I was even caught by the police myself one time and it cost me £3-10-0.' He went on to explain how a friend who worked alongside was determined not to be caught by the police in the same way, and always lit the moss before he went to dinner, so that he could pretend it was an accident if the police came and he was not there: 'one day when he got back and the moss was burning he saw Mr Wilton [the policeman] watching and he ran as fast as he could and he got his shovel and he started beating out the fire and throwing water on it from the drain . . . . Wilton got over and demanded what he was doing and he said, 'Well you might ask. This was all right when I went for my dinner and look at it now - and even some of my turf has been burned.' He had the presence of mind to bluff Wilton that somebody else had lit the moss on him and Wilton just threw off his tunic and fell to and helped him to put it out and said nothing' (Smyth, 1991, pp.10-11). No doubt the village constabulary in Scotland had many equally trying experiences when people's treatment of the bog did not quite keep within the law.

Acknowledgements

I am most grateful to Dr Hugh Ingram and Professor Alexander Fenton for drawing my attention to several important texts.

Chapter eighteen:

Bogs as Treeless Wastes: The Myth and the Implications for Conservation

Professor Frank M. Chambers
Centre for Environmental Change and Quaternary Research, C&GCHE, UK.

Introduction

The Quaternary period, comprising the last c. 2 million years, has been witness to remarkable environmental upheavals, with climate in mid-high latitudes apparently alternating between warm temperate stages and cold stages. It has become apparent that the rate of change between these states has been exceedingly rapid (Broecker, 1992; Taylor *et al.,* 1993); the magnitude of change has been such that, during the late-Quaternary, major ice sheets have repeatedly built and decayed in northern latitudes.

During the Late-glacial at the close of the Pleistocene, there were frequent, rapid temperature changes of high magnitude (Taylor *et al.*, 1993; Lowe *et al.*, 1994), which culminated in a rapid phase of 'global warming' at the start of the Holocene. By comparison, the climate of the Holocene has been relatively stable (Broecker, 1992; Lehman and Keigwin, 1992), although recent research shows that it has nevertheless varied appreciably (Briffa, 1994). Most northern bogs have developed during the Holocene, and so are less than 10,000 (radiocarbon) years old; some have developed extensively in areas formerly occupied by late-Pleistocene ice sheets. If the pattern of environmental changes of the late-Quaternary is repeated, most of these mires will be obliterated several thousand years from now, during the next cold stage. On a geological timescale then, not only have most of the bogs of north-temperate lands had a relatively short lifespan, but it must be recognised also that they are ephemeral features of the landscape, and that their present or recent vegetation may not be fully representative of their potential in the varying climate of the late-Holocene.

Ombrotrophic mires of the British Isles

In significant parts of the northern hemisphere the Holocene climate has been conducive to paludification of soils and to the extensive growth of mires, which have developed and then persisted for thousands of years. By AD 1800, peatlands covered some 6% of the landscape of Britain and some 16% of Ireland (Taylor, 1983). There were extensive upland peats in Scotland, northern England, Wales and south-west England, with even more extensive peatlands in Ireland, not only at altitude but also in lowland areas, including large parts of the west and of the Central Irish Plain. Much of the peatland in Ireland and in western and northern Britain supported bog (as opposed to fen) vegetation. Bogs that apparently depend upon rainfall for their sustenance, both in terms of moisture and nutrients, are termed ombrotrophic; in the British Isles, there are two principal types: raised mires and blanket mires.

Bogs in the British Isles have been variously classified: by form, genesis, nutrient status, nutrient source, vegetation type or vegetational formation. Irrespective of the method of classification is the common perception amongst conservationists that ombrotrophic mires in the British Isles are, by nature, treeless. This derives, at least in part, from Tansley's (1939) descriptions of the 'vegetational formations' of raised and blanket bogs. Characteristically, these accounts are treeless, but that does not necessarily mean either (i) that these mires are *always* treeless; or (ii) that these mires are *naturally* treeless. Indeed, Tansley and his contemporaries recognised that the vegetation of the British Isles was significantly influenced by human activity (Ingram, p. 1, this volume).

This perception - of treeless mires - has, implicitly, guided conservation aims and objectives (though arguably not the management practice) on remaining relatively untouched bogs and it is strongly influencing management objectives on those cutover bogs where conservation is aimed at restoring the conditions of a living, treeless mire. Conservation practice may involve 'trashing' the existing trees (note that trees *are* there) and raising the water level, which is aimed both to re-establish a living bog moss vegetation and to prevent (or to inhibit) further tree colonisation. This chapter questions the common perception of bogs as naturally treeless, and examines the implications of this perception for mire conservation.

Bog history

The history of individual bogs can be investigated by coring through their accumulated peat, and peat stratigraphy can be investigated further in the laboratory through study of the contained microfossils and plant macrofossils. Research confirms that, for the majority of their history, a treeless vegetation has prevailed on the ombrotrophic mires of the British Isles; their surfaces appear to have been dominated for millennia by Sphagna, or by sedges, or by ericaceous shrubs. In this, they differ from some of their cousins on the continent, where, for example in Fennoscandia, a rather different type of mire is prevalent, which is characteristically part-dominated by trees. The differences are attributed to different climatic regimes: the more continental climate of Fennoscandia favours the growth of trees on mires. This begs the question as to what degree of climate change, either in the past, or in the future, might favour tree growth on bogs in Britain and Ireland.

Climate change and peatlands

Whilst in the 1960s and early 1970s there was considerable speculation amongst Quaternary scientists as to the future direction of climate, with opinion seemingly anticipating imminent global cooling (West, 1984), current concern amongst climatologists is less about a future ice age, and more about so-called 'global warming'. Temperatures recorded over the past century apparently show mean global temperatures increasing by 0.5°C over the period, allegedly as a result of human activities. Prognostications from contemporary climatologists show temperatures continuing to rise by a further 1-2°C by AD 2050 (see Fig. 18.1), with a greater temperature rise in mid-high latitudes (cf. Houghton *et al.*, 1990).

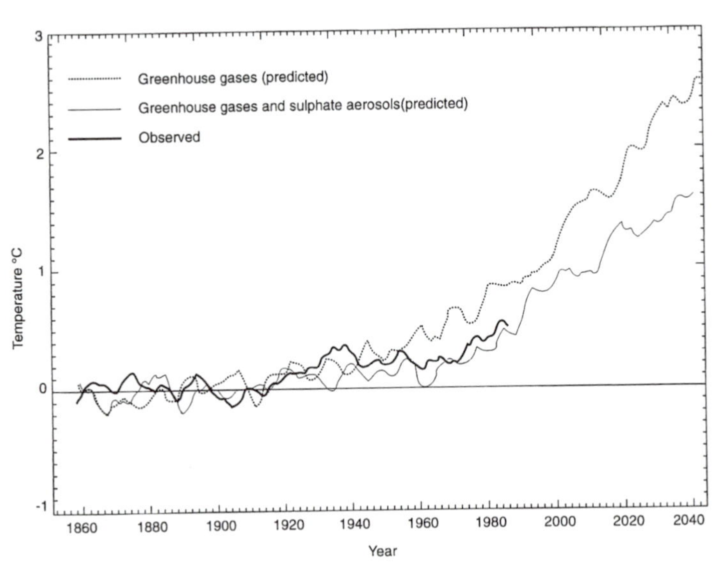

Fig. 18.1. 'Global' surface air temperatures from 1860-present (based on average values recorded at stations located in both hemispheres) and predicted temperature change to AD 2050, compared with past and projected emissions of greenhouse gases. The simulations show a predicted temperature rise in excess of observed values, but a closer match is obtained when both greenhouse gases and sulphate aerosols are included (re-drawn, after Bennetts, 1995). Note the projected warmth for next century, which has implications for bog conservation.

This 'consensus' amongst an inter-governmental panel of scientists has led some conservationists to be concerned as to how existing biomes will fare in the face of such changes. In northern latitudes, these concerns range from the fate of boreal forests to the melting of permafrost. There is now concern over the future of northern bogs in the face of the rate and magnitude of predicted climate change.

Climate history

To ascertain what has happened to British bogs during previous episodes of warming, it is necessary to reconstruct past climates, and this can be attempted from detailed research (Barber, 1981; 1982; Blackford, 1993). Bogs are valuable archives of past climatic conditions. Contained within their peat stratigraphy, and often clearly visible, are the major climatic and vegetational changes that occurred during their lifetime. Many bog plants have restricted water-level tolerances; examination of their relative abundance in a core of sediment will indicate changing bog-surface wetness through time. This type of work has been pursued by Barber (1981, 1982) and associates (Haslam, 1987; Barber, Chambers and Maddy, 1994a; Barber *et al.*, 1994b) and is the basis of most detailed palaeoclimatic data from raised mires. However, detailed records of stratigraphic change in peat bogs were not exploited for climate reconstruction until the 1970s and 1980s, largely owing to a prevailing misconception as to how peat bogs grow (Backeus, 1990). Pool-and-hummock topography of raised mires was thought to be in cyclic evolution, in which pool species grew luxuriantly, peat accreted faster, and pools overtopped hummocks, and so themselves became hummocks, with the hummocks becoming pools. This notion of cyclic regeneration held sway for c.50 years, from the 1920s, before being refuted finally by Barber in 1978 (see Barber, 1981). So, it is only really within the last 20 years that the climatic signal from peat bogs has been seriously investigated; new findings are still emerging (see, for example, Blackford and Chambers, 1995). The history of bogs, and their future conservation, needs to be reinterpreted in the light of these palaeoclimate studies.

Example 1: plant macrofossil data

A continuous proxy record of late-Holocene climate change comes from a core taken from a raised mire in northern England, at Bolton Fell Moss, in Cumbria (Barber *et al.*, 1994a; 1994b). There, analysis of plant macrofossils has given a sensitive, high-resolution, proxy-climate record. The macrofossil data down the core can be summarised using ordination techniques to provide a time series of past bog-surface wetness, and the summary data (Fig. 18.2) show an oscillating record of change. There is a particularly marked shift c. 4000 BP (uncalibrated; about 4500 years ago in calendar years), which might be interpreted as the main climatic event of the latter half of the Holocene in northern Britain. A summary of the principal findings of the Cumbrian study is given in Table 18.1.

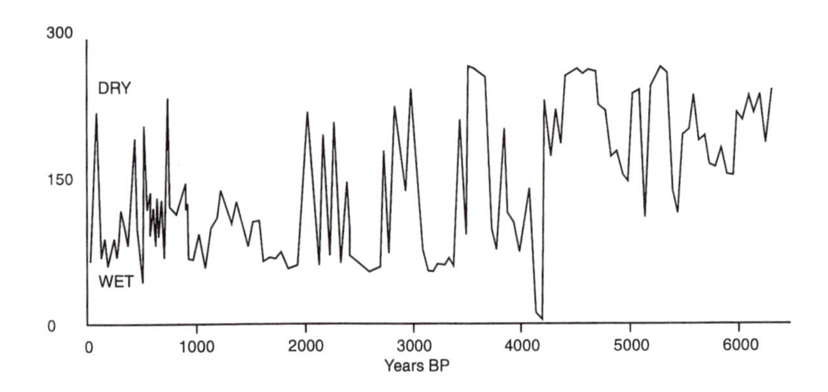

Fig. 18.2. Changes in bog-surface wetness for the past 6300 radiocarbon years recorded in a core from Bolton Fell Moss, Cumbria: axis 1 DCA plotted against years BP (redrawn from Barber *et al.* 1994a, with corrections and additions).

Table 18.1. Summary of principal findings of the palaeoclimate data from Bolton Fell Moss, Cumbria.

The plant macrofossil data show:
(a) that there have been marked climatic fluctuations in northern Britain during the Holocene;
(b) that an oscillating climate is the norm;
(c) that a shift at c. 4000 BP may be the major climate change in the latter half of the Holocene;
(d) that the records for the present century may not adequately reflect prevailing climate, owing to human interference with the mire.

Source: (Barber *et al.*, 1994b)

Example 2: Analysis of sub-fossil pines in mires

In Scotland, fine-resolution pollen analyses of peats, combined with tree-ring data from subfossil pines, point to a climatic excursion c. 4000 BP, when conditions apparently ameliorated to allow *Pinus sylvestris* L. (Scots pine) to grow in the far north of Scotland for 300-400 years (Gear and Huntley, 1991). The record of pine macrofossils and pine pollen suggests an expansion and retreat over some 80 km of latitude (that is, north to south, and back) in a period of four centuries, indicating significant changes in climatic conditions, at least regionally, c.4000 BP.

In Ireland, subfossil pines have been found stratified in raised bogs, indicating temporarily dry surfaces of such mires in prehistory. In some cases the pines seem to have been particularly abundant, and persistent. At Sluggan Bog, Co. Antrim, where three generations of pine stumps have been found dating from c. 6000 Cal. BC (c. 7200 BP), Pilcher *et al.* (1995 p. 666) observed: 'The stratigraphy gives the impression that there was a relatively dense pine forest covering most of the surface of the bog over an extended time period'. The overall chronology of these subfossil pines lasts some 809 years, and probably spans the period from c. 6300-5500 Cal. BC. A more recent phase of pine growth on many raised bogs in the north of Ireland centres on 3000 BC. Cross-dating with the Belfast bog-oak chronology shows that it spans the period 3451-2569 BC (Pilcher *et al.*, 1995), although on individual bogs, the period of pine growth may have been for only part of this time range. This period of pine growth apparently ceases c. 4000 BP (i.e. c. 2500 Cal. BC), which is similar to records for pines on raised mires in central Ireland (McNally and Doyle, 1984) and beneath blanket mires in Scotland (Bridge *et al.*, 1990; see also Gear and Huntley, 1991); its culmination with increased surface wetness finds echoes in the palaeoclimatic data from Cumbria.

The palaeoenvironmental record

In the British Isles then, studies of peat stratigraphy through raised mires and blanket mires show that many bogs seem to have been treeless for several thousand years; indeed, for some, they may have been treeless either throughout their existence, or for the ombrotrophic part of their lifespan. However, that is not true for all bogs. Studies in mire palaeoecology have long shown that, in the early-mid or mid-late Holocene, some bogs in Britain and Ireland apparently dried sufficiently for their surface to colonised by cohorts of trees, notably of *Betula* (birch) or of *Pinus sylvestris* (Scots pine). At Fenns and Whixall Moss in the Welsh borders, a major episode of pine colonisation of part of the mire surface has been dated to c. 3200-2900 BP (Grant, 1995; Chambers *et al.*, 1996), which is one of the more recent pine episodes recorded in England and Wales. Recent research into Holocene palaeoclimates suggests that pine colonisation may have occurred during episodes of regional drying or warming. Bogs then apparently changed, or reverted, to a treeless state after increases in bog-surface wetness. The palaeoenvironmental data imply that ombrotrophic mires in Britain and Ireland may not naturally be treeless during episodes of significant regional or global warming. This surely has implications for conservation of remaining raised and blanket mires in the context of current and projected climate change.

Human influence upon mires

There is, however, another aspect to this treelessness: implicit in much of the above is a presumption that the vegetation of British mires is 'natural' and uninfluenced by human activity. That view can be challenged (H. Ingram, unpublished). For example: 'In the past Cors Fochno [Borth Bog] has been the scene of annual fires, especially as winter turned to spring and east winds desiccated the vegetation. Such burning of bogs, now a part of agricultural practice may date back to... [the] Dark Ages...' (Condry, 1981 p. 202). This extract implies that, far from being unused, bogs were grazed at particular times of the year, as part of agricultural (pastoral) practice (see Smout, Chapter 17, this volume), and might be regarded as 'waste' *sensu* Rackham (1976) - land used for communal pasture.

Clearly, burning, and grazing, might have an effect upon the bog flora; both are particularly damaging to trees. Any trees that might naturally have colonised bogs would be seriously disadvantaged - destroyed even - by repeated fires or by grazing. This is not readily recognised by some contemporary bog conservationists who have inherited bogs that have been lightly grazed or fired in the past, but who now exclude livestock and burning from their management regimes. The practice of burning has long been recognised by other naturalists: 'A legend of Cors Fochno was that ague [malaria] was spread by an old witch (Yr Hen Wrach) who came out of the marshes by night to curse people with the disease while they slept. So every February or March the witch's bogland haunts were fired in revenge, no doubt with considerable ecological results. So when in our time peat bogs become nature reserves and fire is excluded, some interesting changes in the vegetation can be looked for' (Condry, 1981 p. 202). Examples of these changes can be seen in the Central Lowlands of Scotland at Flanders Moss and at Cander Moss: the former has recently become heavily infested with young birch; the latter supports both 20-year-old pine and a plethora of young birch where other pines have recently been cleared. Of course, for many bogs the 'problem' of tree growth is exacerbated by marginal peat cutting, shrinkage and drainage, which makes the bog surface drier and more amenable to tree colonisation. However, attempts to restore a damaged bog to some supposed 'natural' vegetational state, begs the question as to what that state should be.

Conclusions

Bogs are ephemeral features of the landscape. Northern bogs have only been around for a few thousand years, and (global warming notwithstanding) they would be likely to disappear at the end of the present interglacial. In Britain and Ireland, raised bogs have been viewed as naturally treeless, and conservation efforts presently involve removing trees and preventing their re-growth. Yet, for significant parts of their existence, sometimes for hundreds of years, they have naturally carried populations of pine. These episodes, equating to almost 20% of the duration of the Holocene, may correlate with episodes of regional, or global warming.

Since late-prehistoric times, bogs have been part of the cultural landscape - perhaps as 'waste' (common pasture); some may have been lightly grazed by domestic stock, and sometimes deliberately fired. This may well have inhibited or prevented re-colonisation of bogs by trees during historical episodes of warming (as might otherwise have occurred during the early medieval period). So, if we look at present-day bogs and think of their future, then whether one accepts greenhouse warming, or whether one prefers solar variability (Fig. 18.3) as being responsible for the past-century rise in global temperatures (Friis-Christensen and Lassen, 1991; Kerr, 1991), one must accept that some ombrotrophic bogs in the British Isles are not necessarily naturally treeless, and particularly not during periods of regional or global warming.

The view of ombrotrophic mires as naturally treeless - or as ungrazed land - is a myth. This has profound implications for their conservation in the face of actual, or anticipated, regional or global warming.

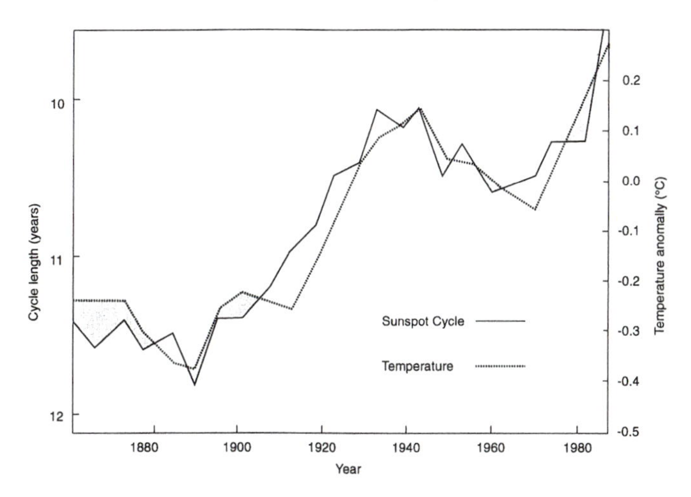

Fig. 18.3. An alternative explanation for the past-century recorded rise in global temperatures (re-drawn from Friis-Christensen and Lassen, 1991): the temperature rise apparently matches changes in solar-cycle length more closely than do greenhouse emissions (compare with Fig. 18.1). Either way, continuing warmth would favour tree growth on mires in the British Isles.

Acknowledgements

Thanks to Kathryn Sharp for re-drawing Figs 18.1-18.3.

Chapter nineteen:

The Impact of Large Herbivores on North British Peatlands

Mr David Welch
Institute of Terrestrial Ecology, UK.

Introduction

There is good evidence that large herbivores of various sorts have been present in the British uplands since the last Ice Age (Ritchie, 1920; Simmons, 1974). Wild cattle were hunted there in Mesolithic times as proved by finds of bones and microliths in North Pennine peat deposits (Johnson and Dunham, 1963), and until a few hundred years ago red deer were widespread in the upland districts of England, Wales, Scotland and Ireland (Whitehead, 1964; Doyle, 1982). Between the twelfth and eighteenth centuries hunting gradually restricted red deer to remnant woodlands and remote mountain plateaux, but many uplands became grazed by domestic livestock, particularly sheep and cattle and also goats and horses (Symon, 1959; Donkin, 1962; Campbell, 1965; Hughes *et al.*, 1973). For example, a remote tract of the North Pennines with extensive blanket bog was being grazed by sheep and cattle as early as the fourteenth century (Welch, 1975).

However the numbers and densities of large herbivores in the uplands are known only for the last one or two centuries. Before this period the existence of herbivore impact has to be inferred from a combined knowledge of the past condition of the vegetation and the present-day consequences of particular stocking rates. As it is now established that a comparatively light stocking can prevent the regeneration of trees, the past absence of dense forest in the uplands below the altitudinal limit of trees may well have been caused by the herbivores present. To judge this, and to assess present-day effects on peatlands, I will first describe how red deer and sheep currently behave and affect their range in upland Britain.

Behaviour and impacts of red deer and sheep

When red deer and sheep graze freely, uncontrolled by managers, there is great variation in their occupance (defined as accumulated presence) within their overall range (Hunter, 1962; Rawes and Welch, 1969; Colquhoun, 1971; Gordon, 1989). Grassy communities are preferred to dwarf-shrub heath, and sheltered places to cool windswept ground, so that often there is ten-fold or even hundred-fold variation between occupance on different vegetation units within 100-ha blocks of ground. In experiments comparing the behaviour of Scottish Blackface sheep and red deer exposed to a mosaic of *Agrosto-Festucetum* and *Callunetum*, both herbivores spent more time on the *Agrosto-Festucetum* than expected from the extent of its patches, but the sheep selected more strongly (Clarke *et al.*, 1995).

Once present in a stand of vegetation, herbivores can affect it by grazing, dunging, urinating, trampling, lying and various minor activities such as fraying with the antlers and wallowing in pools. In most of these actions red deer and sheep are selective, i.e. they prefer to eat certain plants or plant parts, and they choose to rest in drier or more sheltered positions or even specific 'camp sites', so impacts are not merely proportional to presence in an area. Moreover, the resulting impacts are often constrained by attributes of the swards, e.g. on preferred grasslands the low sward height may cause intake rates to be lower than on bogs and heath which have greater standing crops of foods such as *Eriophorum vaginatum* L. (cotton-grass) and *Calluna vulgaris* (L.) Hull (heather). Because defecation and urination occur fairly regularly over time, there will tend to be a net transfer of nutrients from swards that allow higher intake rates to swards that only allow low intake rates. A greater effect on the nutrient balance of bogs would result if the animals avoided resting there.

The size differences of red deer and sheep affect their impact. Sheep with smaller mouths can graze more selectively than red deer, for example they are better able to avoid the unpalatable *Erica tetralix* L. (cross-leaved heath) when it is growing mixed with *Calluna*, and so may have a greater influence on the composition of bogs for a given occupance rate. Red deer, with longer legs, move faster and range further than sheep, so they can spread out greater distances from the most preferred swards in search of food items such as saplings of pine and birch scattered over bogs and open moorland.

Effects on botanical composition

Bog and wet-heath communities

Under moderately heavy grazing pressures *Calluna* and other ericoid species are reduced in cover and height, and are replaced by graminoid species such as *Eriophorum vaginatum, Juncus squarrosus* L., *Molinia caerulea* (L.) Moench and *Scirpus cespitosus* L.. Graminoids withstand defoliation better than ericoids because they grow from the shoot bases rather than the shoot tips. *Calluna* is the potential dominant of most North British peatlands in the absence of trees, and its performance and palatability are of crucial importance in controlling botanical composition. For *Calluna* growing well, the threshold stocking density for decline in cover has been estimated to be 2.7 sheep ha^{-1} (Welch, 1984), but at higher altitudes and on waterlogged badly-aerated soils the growth of *Calluna* is sharply reduced, hence the threshold densities are much lower. For blanket peat at 550 m in the North Pennines a year-round stocking of 0.6 sheep ha^{-1} was found to be checking *Calluna* (Welch and Rawes, 1966), whereas grazing at 0.35 sheep ha^{-1} gave no measurable differences compared with ungrazed controls (Rawes and Hobbs, 1979); Grant *et al.* (1976) from observations of utilisation on blanket bog at 250 m in Argyll judged that a year-round stocking of 2.2 sheep ha^{-1} was causing decline but 1.4 sheep ha^{-1} was tolerated. Similarly, at a peaty gley site in Northumberland with vegetation related to NVC M15 (*Scirpus-Erica tetralix* heath) (Rodwell, 1991), stocking at 1.2 sheep ha^{-1} allowed a slow increase in *Calluna* cover (Nolan *et al.*, 1995).

The graminoid communities that develop on heavily grazed bogs and wet heaths vary considerably in botanical composition, the dominance of particular species being related to peat depth, degree of wetness, aeration and climatic factors (Fig. 19.1). On wet deeper peats, *Eriophorum vaginatum* mire (M20 (Rodwell, 1991)) is produced, and on peaty gleys *Juncus squarrosus-Festuca ovina* grassland (U6 (Rodwell, 1992)). Occasionally this latter community develops on peat, usually where there is some lateral water movement, although, on slopes at lower altitudes, where peat soils are to some extent aerated, *Molinia caerulea-Potentilla erecta* mire (M25 (Rodwell, 1991)) generally occurs.

Schemes similar to Fig. 19.1 relating community occurrence to site factors have been presented by Ratcliffe (1959), Rodwell (1991), Rodwell (1992), Thompson *et al.* (1995) and Thompson and Miles (1995). All agree on the broad relationships outlined above, but there are differences in detail between schemes - especially regarding whether grazing together with burning produces wet heath M15 and M16 communities from *Scirpus-Eriophorum* (M17) and *Calluna-Eriophorum* (M19) bogs. These discrepancies are not surprising since for peatlands, known herbivore densities have only rarely been monitored long enough for successions between communities to occur, and the schemes have been built up largely from intuitive observations at sites differing in grazing intensity and management. However, there is good evidence from the North Pennines for a shift from *Calluna-Eriophorum* (M19) bog to *Eriophorum* (M20) bog caused by a stocking of 3.4 sheep ha^{-1} (Rawes and Hobbs, 1979), and that *Calluna* and *Eriophorum* quickly replaced *Juncus squarrosus* in a wet grassland protected from grazing (Marrs *et al.*, 1988).

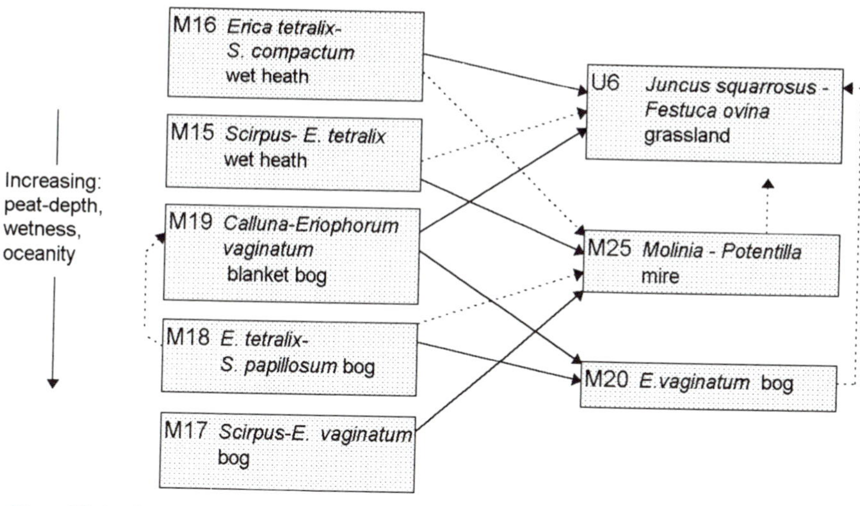

Fig. 19.1. Successional inter-relationships of mire and wet-heath NVC units induced by moderate to heavy grazing. Main lines show usual successions, dotted lines show occasional successions occurring in a narrow range of circumstances. Reverse successions under light or no grazing may be delayed by absence of potential dominants.

The effects of total prevention of grazing by large herbivores have been inferred from the composition and structure of vegetation inside long-term exclosures, observations from several sites across Britain being brought together by Marrs and Welch (1991). Generally *Calluna* cover and height were greater inside the exclosures than outside, and the protected vegetation was richer in lichens, especially *Cladonia impexa* Harm.. But *Calluna* spread little at some sites where it initially had negligible or no cover e.g. it reached only 10% cover after 24 years of protection of *Eriophorum* bog at the Silverband exclosure at 685 m in the North Pennines. Sometimes only very minor differences were apparent between ungrazed and grazed vegetation, e.g. for *Calluna-Molinia* wet heath on Rhum after 12 years (Ball, 1974), which probably indicates very light usage from herbivores. Often dead bushes or branches of *Calluna* were evident in the protected vegetation, giving up to 10% of the cover (Marrs and Welch, 1991). At Coom Rigg Moss a build-up of standing dead material of litter from *Calluna, Eriophorum, Molinia* and *Deschampsia flexuosa* (L.) Trin. was thought the likely cause of a decline in *Sphagnum* spp. and *Andromeda polifolia* L. from 1958 to 1986 (Chapman and Rose, 1991). Formerly this moss had been lightly grazed by sheep, and because the cover trends and loss of pools resulted in the vegetation changing from M18 to M19, I indicate a partial relationship between these two types in Fig. 19.1.

My conclusions from these various observations are that the impact of herbivores on bogs, even with low occupance, can build up over a period to become important. When grazing is light, and lichens and tall bushes of *Calluna* make a greater contribution to the sward, trampling by herbivores is likely to have a greater impact than defoliation. Red deer, being heavier than sheep and ranging further from preferred vegetation types, may well have a greater effect on outlying bogs than sheep pastured in the same areas. Thus Ratcliffe and Walker (1958) considered that sheep rarely visited the Silver Flowe pool-complex bogs in S.W. Scotland, but deer tracks were visible. Even sporadic visits can affect floral composition because many bog plants are slow growing e.g. lichens, which can take years to recover from damage, whilst, in the case of trees such as *Betula pubescens* Ehrh. and *Pinus sylvestris* L., time is required for individuals to grow sufficiently tall as to be safe from browsing. Colonisation by trees will be considered in the next section.

Forest communities

Large herbivores damage trees by browsing the leading shoots of seedlings and saplings, and by bark-stripping the trunks of taller trees. Because red deer and sheep select between tree species, they modify the composition of both regenerating tree communities and existing woodlands; the effects vary somewhat between regenerating and adult stands because the preference rankings for leader browsing and bark-stripping are rather different. There are also marked seasonal changes in browsing preference: in summer, red deer preferentially select deciduous trees such as *Betula* L. spp. (birch), *Salix* L. spp. (willow) and *Sorbus aucuparia* L. (rowan) over *Pinus sylvestris* (Scots pine), in winter, when leafless, deciduous trees are less selected than *Pinus* (Cummins and Miller, 1982). Hence the timing of grazing can influence composition.

Colonisation by trees is currently being prevented, either by herbivores or by a lack of seed source, in most bogs and wet heaths in northern Britain which lie below the natural tree line (McVean and Ratcliffe, 1962; Rodwell, 1991; Hester and Miller, 1995; Thompson *et al.*, 1995), and 1 red deer per 10-15 ha has been found to prevent the regeneration of Scots pine (Holloway, 1967; Beaumont *et al.*, 1995). In the absence of large herbivores either in exclosures or on tracts of ungrazed moor, tree establishment is erratic but sometimes abundant (Marrs and Welch, 1991), depending upon proximity of seed sources and the availability of colonisation niches in heather stands. On grazed moors, saplings that have grown taller than the heather or grass swards are generally found to have suffered leader browsing. This suggests that sufficient height cannot be gained to escape above the reach of red deer or sheep in the interval between visits. For some sites, e.g. wet heath in Glen Affric, which Scots pine is colonising (Fenton, 1985), sapling growth rates have been measured, providing definite evidence that deer damage more than counterbalances potential height increment. When growth is poor, as for pines at Coille Coire Chuilc in Perthshire (Sykes, 1992),

30 years' protection was thought necessary to allow a cohort of suppressed saplings to reach a safe height.

In upland districts where tree regeneration is known to be occurring, herbivore stocking is minimal. Thus in Deeside, N.E. Scotland (Watson and Hinge, 1989), Scots pine regeneration was found in roughly half the 1 km grid squares containing grouse moorland in the middle part of the valley, but, was virtually absent in moors stocked with red deer in the upper valley. Grid squares lacking regeneration in mid Deeside were mainly on the fringes of the moors and so were visited by sheep and cattle coming from peripheral farmland. On Hoy, Orkney, low levels of sheep stocking and the absence of red deer were considered by Prentice and Prentice (1975) to have led to *Salix aurita* L. prominence in mires and heaths. Lichen-rich Trichophoro-Callunetum was also widespread.

For mature forests in the British uplands, there are few documented cases of changes in tree species being brought about by herbivores, although severe bark stripping has been observed to cause some trees of vulnerable species to die. For example in two mixed deciduous woods in N.E. Scotland, cattle took 78% of accessible rowan bark but less than 1% of the bark of *Betula* spp., *Fagus sylvatica* L., *Fraxinus excelsior* L., *Prunus avium* (L.) L., and *Pinus sylvestris* (Kinnaird *et al.*, 1979); *Salix cinerea* L. and *Picea abies* (L.) Karsten lost 10% of their accessible bark. Strong selection for rowan and willow bark has also been recorded in red deer in thicket woodland at Beinn Eighe in N.W. Scotland (Mitchell *et al.*, 1982). Here *Alnus glutinosa* (L.) Gaertn., *Betula* and *P. sylvestris* suffered very little bark stripping. Recruitment of trees in mature woodlands is also influenced by large herbivores, whose presence in low numbers boosts overall recruitment rates by creating more niches (Miles and Kinnaird, 1979a; Mitchell and Kirby, 1990). However, selective browsing may also prevent some species regenerating. Thus, the pure pinewoods of the Central Scottish Highlands are thought by Miles and Kinnaird (1979b) to be attributable to a long history of grazing.

By curtailing the regeneration of woodland, and by favouring trees that produce mor humus, e.g. *Pinus,* rather than mull humus, e.g *Betula* spp., *Salix* spp. and *Sorbus aucuparia* (Miles, 1978), large herbivores may have encouraged podzolisation and the subsequent development of blanket bog during the Holocene; they may still be assisting peat accumulation, for example, by stopping birch and willow colonisation on wet moorland. The onset of blanket peat development has been linked with the occupation and use of uplands by man, e.g. Moore (1975) and Birks (1988); the grazing of domestic livestock is thought to have aided the process. However, because wild herbivores - particularly red deer - would be displaced by this occupation, and stockings of wild and domestic herbivores can only be guessed, the arguments for domestic livestock being an important cause for the general initiation of bog are still flimsy, and burning, woodland clearance and climate change seem more likely causes.

Effects on peat erosion

The surface of bogs and wet heaths is prone to break up when trampled by large herbivores. *Sphagnum* moss is especially susceptible to trampling damage (Bayfield, 1979) and this may lead to erosion of the underlying peat. Erosion is now occurring widely in British blanket bogs, and is particularly prevalent in *Calluno-Eriophoretum* mire (NVC 19) and *Eriophoretum* mire (NVC 20). It is most widespread in the South Pennines (Tallis, 1964a) and is also reported from Ireland (Bradshaw and McGee, 1988), North Wales (Ratcliffe, 1959), the Eastern Cairngorms (Huntley, 1979) and Orkney (Prentice and Prentice, 1975). Explanations for erosion have been widely sought (Bower, 1961; Tallis, 1964a; McGreal and Larmour, 1979). The presence of sheep in moderate numbers on some eroded peatlands prompted some workers, e.g. Radley (1962), to claim that sheep are a main cause. Wallowing by red deer has also been blamed for peat erosion in the eastern Cairngorms (Huntley, 1979). However, detailed studies on the timing, zonation and spread of erosion on particular moors, e.g. Tallis (1985a; 1987), suggest that large herbivores are merely exacerbating other causes including: cut back from peripheral streams due to the natural instability of deep peat; bog bursts and slides resulting from very heavy rainfall; intense peat fires; and, atmospheric pollution killing the *Sphagnum* cover.

Concluding remarks

The main impacts of grazing by large herbivores on British peatlands appear to be the loss of heather and the prevention of tree regeneration or colonisation. Given the present state of knowledge, whereby isolated facts from a few investigations are being intuitively used to describe relationships for very widespread vegetation types, we can only suggest that other impacts of herbivores are relatively unimportant. Further examples in this category, besides that of peat erosion discussed above, are: the compaction of the peat surface produced by treading; the depletion of nutrients from bogs caused by herbivores grazing but not resting there; and the drying out of bogs resulting from greater biomasses of heather transpiring more when grazing is light. To properly assess the contribution of these processes requires further detailed studies.

Acknowledgements

Dr Sue Hartley has kindly commented on the draft.

Part Six

The Need and Means for Survey and Inventory

An Introduction

Mr Richard Lindsay
Scottish Natural Heritage, UK.

Conservation is a subjective activity which is based entirely on human values. It is possible to dress it up in all manner of different ways - for example maintenance of the biosphere, or retention of the gene-pool - but these are functions which we, the human species, have chosen to regard as important. The species whose gene-pool we are protecting almost certainly care little, if anything, about the gene-pool or a balanced biosphere. What place does inventory - the dry, cold logic of lists - have in conservation if conservation is so fundamentally driven by subjective human values?

In fact 'value' is a key concept even for inventory or, more accurately, for what we do with the results of inventory. To elaborate and explain, imagine you are an art dealer who has been commissioned by a millionaire to purchase an example each of Leonardo, a Salvador Dali and a Picasso. You have also been given a purchase ceiling of $100 million. Unfortunately the auctioneer is only willing to reveal to bidders a small fragment of each painting on offer, and will reveal neither the title nor artist for these. Your decision to bid, or not, must be based on these fragmentary glimpses. What do you do?

Fortunately Salvador Dali has such a distinctive style that his pictures are easy to spot, although some are much more valuable than others. The problem here lies in judging whether the obvious Dali is one of his extremely valuable major works, or is instead a minor painting of a much lesser value. Perhaps only a small amount of additional information - for example the date of the painting - would be sufficient to make a judgement about the proportion of the millionaire's purchase fund which should be spent on this piece.

Then there is the Picasso. A few fragments have a clear 'Cubist' style and all the hallmarks of a Picasso, but equally they have all the hallmarks of a George Braque. This is because both artists shared in the development of Cubism and copied each other's ideas on many occasions. Only by seeing a significantly greater proportion of the painting would it be possible to make a decision about whether there is a Picasso to bid for, and, if so, how much it might be worth.

Leonardo is even more difficult, potentially, especially as the auctioneer has revealed only small fragments of what appear to be background, possibly taken from the corners of the various paintings. The style of these fragments reveals them to be from the Renaissance period, but this is not sufficient to pinpoint a Leonardo. Comparison might be made with a range of known Leonardo backgrounds, but most such Renaissance paintings tended to have such similar settings and styles that even this cannot be relied upon entirely. How much easier it would be if the auctioneer had happened to reveal a different fragment of the various paintings, and in one there had been a mouth with a faint ghost of a smile. However, to chance upon a fragment containing a clue of such crucial importance that one can be certain it is worth going for broke in the bidding - even from such a famous work as the Mona Lisa - requires a great deal of luck, given the total extent of the painting.

However, having established that this is indeed the Mona Lisa, one then faces another dilemma. This one painting is likely to be so expensive that there would be little or no money left in the purchase-fund to buy the other two paintings. Perhaps, having positively identified the Mona Lisa, one is therefore then obliged to establish whether there is another, rather less expensive Leonardo as an alternative. These minor works would require much more effort to identify because they are generally less well-known. Alternatively, it is perhaps appropriate to discuss this development with the millionaire. The chance to buy the Mona Lisa does not arise often. Having given him the facts, as far as they are known by then, he is then in a position to decide that the opportunity of buying the world's most famous painting outweighs the desire to own a Dali and a Picasso. He might even decide he wishes to buy all three. It is his subjective choice, but it would be based on the information provided by you as the specialist art-dealer, information which consists of the potential pictures involved, their likely values, and an assessment of whether others may be determined to out-bid you.

Thus, the basic decisions of the day are subjective ones made by the millionaire. He wanted a Leonardo, a Picasso and a Dali. He also had to decide whether he wanted the Mona Lisa more than the other two paintings, or indeed whether to spend far more than originally intended because the opportunity presented itself.

The role of the art dealer was very different, but crucial. It involved obtaining sufficient data about the various paintings on offer to help an informed judgement to be made as to what was worth purchasing, and what was not. Where the information was not adequate to make such a judgement, the dealer had somehow to obtain more information from the auctioneer - in some cases only a small amount, in others it might need to be a considerable amount.

In this short introduction I have not discussed much about nature conservation inventory. However, the similarities between the processes of, and dilemmas facing, the art dealer and the conservationist will not, I hope, be lost on the reader. Without adequate knowledge and information, sensible judgements cannot be made. If such knowledge can be obtained, it may reveal unexpected items or opportunities which in turn may bring about a re-assessment of priorities. It certainly enables activities to be correctly focused. This is especially important when resources are limited - and in nature conservation resources are always limited. Inventory helps us to squeeze the last drop of value from the resources we have.

Chapter twenty:

The Status of European Mires by the IMCG and the Swedish Wetland Inventory (VMI)

Mr Michael Löfroth
Environment Protection Agency, Sweden.

The Swedish Wetland Inventory (VMI)

The total area of wetlands (mires, swamps, wet heath/meadows etc.) in Sweden, (excluding lakes and other open water areas) is c. 9.3 Mha.

Wetland exploitation in recent centuries has destroyed about 2.7 Mha (forestry drainage: 1.5 Mha; agricultural drainage: 1.0 Mha), and damaged about two-thirds of the remaining 9.3 Mha.

The majority of Sweden's wetlands are underlain by peat and are thus regarded as 'mires'. These mires show a great diversity of types, from nemoral types in the south, sub-oceanic types in the south-west, alpine types in the mountain chain, aapa mires in the boreal zone and subartic palsa mires in the far north. Existing mire types are, for example, different types of raised bogs, flat bogs, level and sloping fens, string fens, mixed mires and various peaty swamps.

In Europe, only Norway and Russia can show as great a diversity in mire types as Sweden.

Mires and other wetlands play an important role for biodiversity, and, given ongoing exploitation, an inventory was required that could show which mire sites should be conserved.

The Swedish Wetland Inventory (VMI) began in 1981 lead by the National Environmental Protection Agency who were also responsible for the development of methods, classification, computerised data storage, data analyses and education of personnel in the county administrations.

The first stage of the inventory was mapping and data collection on wetland sites (using stereo-analysis of aerial photographs).

Each area of a wetland, or connected wetlands, was regarded as a site. Within each site, one or more sub-sites was distinguished, corresponding to a mire-type (a mire unit type). A whole set of parameters such as size, types, tree-cover, wetness and damages (e.g. ditches, peat-excavations and roads) were registered and stored in a computerised database.

The data from this first step was used, together with, if available, earlier records of fauna and flora, to target 10-15% of the sites for field survey.

The second step of the inventory was to carry out a field survey taking - on average - one day per site. In the field survey, the data from the first step was checked, if needed, corrected and dominating vegetation types in the sub-sites were described using a plant species list.

To date, inventory has been carried out in all counties in Sweden except for the alpine area and the county of Norrbotten. An inventory has recently begun in the latter, and will continue for another three years.

To date, the inventory includes c. 27,000 sites containing c. 70,000 sub-sites and totalling c. 2.4 Mha. 4300 sites have been field surveyed resulting in the description of about 30,000 vegetation relevés (containing about 350,000 recordings of species). To my knowledge, this is the largest wetland-site inventory in Europe.

As the data is computerised, it eases analysis enabling production of maps of mire type distribution and mire regions, analysis of damage status and vegetation types, as well as enabling the primary aim of evaluation of nature values (ecosystems, refuges, functions, species, landscapes etc.). The sites have thus been classified in four 'nature value classes', the highest class including some thousand sites where exploitation should be avoided.

The inventory was also used as a basis for a mire protection plan for Sweden. This plan has recently been published, and identifies 345 mire sites, (covering a mire area of 210,000 ha) requiring protection (mainly via nature reserves designation).

The Status of European Mires, by IMCG (International Mire Conservation Group)

One IMCG project is the production of a report regarding the status of mires in Europe. The project began in 1990 collecting information from Western Europe, but, following political changes, this information gathering was extended to East European countries.

The majority of the work has been carried out on voluntary basis. National mire experts have been contacted and asked to write a summary of the status of the mires in their respective countries.

The goal is to publish a report that for each country presents data about:

1. the mire types and their distribution,
2. the exploitation history and use,
3. current threats,
4. conservation situation,
5. management needs.

The IMCG hopes that producing such a report will provide:

1. an overview of distribution, status, threats etc. on European mires, presented in a common language, in one set of terminology and understandable to a wide audience.
2. the possibility to formulate conservation strategies for Europe.
3. a sound scientific basis for selecting representative mires on the basis of regional distribution and variation, rather than only selecting sites within national borders.

At the moment, data has been collected from 23 European countries. We are compiling, or hoping to receive, data from another five countries. The greatest lack of information is from the Balkan area and some of its neighbouring countries which might not be covered by the overview.

One difficulty is that many countries lack a national ecological mire inventory. Some countries have inventories which are acceptable for this purpose (e.g. UK, Switzerland, Austria and Sweden), but in some countries (e.g. Portugal, Spain, France and Italy) very little information on mires is available.

As results were collated, it was realised that many definitions for mires are used across different countries. The figures for mire areas actually represented different things! A 'mire area' reported from one country might include agricultural peat areas and other dry peat soils utilised by man; whilst such data from another country might only refer to virgin mire-areas still functioning as mires although possibly with a change of vegetation (i.e. excluding damaged mires).

The need for a unified mire definition for this project was clear. This matter was discussed during the IMCG conference in Switzerland and it was agreed on following definition:

1. *Mire*: A mire is a wetland, usually dominated by peat-forming vegetation.
2. The term '*peatland*' was defined as all areas covered with peat, whether they can be defined as mires or not.

There remained a need to define the most fundamental differentiation in mire types; a matter that was discussed in IMCG conferences in Switzerland and Norway. The following definitions for the main mire types were agreed upon:

1. *Bog*: A bog is an ombrotrophic mire; its upper layers receive their entire water supply directly from the atmosphere.
2. *Fen*: A fen is a minerotrophic mire; its upper layers get at least some of their water from surrounding mineral soils, bedrock or open waters.

For most European mires, such definitions classify bogs as systems that are naturally poor in species - the only species which exist are those that can survive in conditions of low nutrient and mineral availability (i.e. that derived solely from precipitation). This changes in areas under the influence of oceanic climate conditions. Precipitation in these areas is enriched with ions like sodium, chloride, magnesium etc., and the flora consequently becomes somewhat richer - species that usually only occur in fens, e.g. *Sphagnum papillosum,* can grow on such bogs.

In special situations, airborne dust, like volcanic ashes (e.g. Iceland) or calcium rich dust (e.g. a few areas in Switzerland) can enrich bog vegetation, making it very similar to intermediate fens.

Mire areas are sometimes differentiated into units according to main hydrological regime. These units are further, internally, differentiated into smaller units like elements or structures. To be clear what level is actually described, a mire classification hierarchy has to be used. The following hierarchical levels will be used in the report: mire complexes, mire units and mire sub-units.

Mire complexes

A hydrologically connected mire area can contain one or more distinguishable 'mire types' e.g. level fen, dome shaped bog, peaty swamp etc. When more than one mire type dominates the vital parts of the mire area it is called a 'mire complex' (Sjörs, 1948). Mire complexes (or macrotopes) have been used in some countries as the basis for mire classification. In Finland, for example, the northern type of mire complex, dominated by fen-vegetation in their central parts, are called 'aapa mires'.

In Europe's bog regions, mire complexes often consist of one or two bogs and level fens. The further south one moves in Europe, the more mires tend to occur as single mire units rather than complexes. On the other hand, to the north (e.g. in Fennoscandia and Russia) the majority of the mire area occurs in complexes consisting of several mire units.

Mire units

A mire unit (or mesotope) normally represents a single, identifiable hydromorphological unit such as a concentric bog, a flat bog, an unpatterned level fen, a string fen etc.

The mire unit seems to be the best level for mire classification and will be the level on which mires are described, analysed and compared in the report. Mire complexes are sometimes used in this context, but they can vary too much across Europe (depending on what types of mire units they contain) and, in reality, some of these combinations of mire-units are more or less accidental depending on two different hydrological regimes close to each other (e.g. when an eccentric bog happens to occur attached to a soligenous fen).

A mire unit can exist by its own or as a part of a mire complex (see above). In Russia a synonymous word is 'mire massif', in the UK a 'mesotope', in Sweden a 'sub-site' and in Norway 'synsite'.

Within a mire unit, several sub-units can occur.

Mire sub units

Within a mire unit, several features can occur, sometimes forming the component that builds up the mire unit and sometimes as a single, smaller, part with a certain morphology/hydrology. These sub-units can form 1-3 (or perhaps more) levels within a hierarchy. Examples of mire sub-units are a lagg, a flark, a flush, a pool, a string or a hollow etc. The mire sub-units can often form a distinguishable surface pattern across a mire.

In the mire-unit type 'eccentric bog', on its sub-unit 'mire expanse', there can exist a further sub-unit of a 'flush' which itself contains smaller sub-units of 'hummocks' - an example of three levels existing below the mire unit. In another mire unit, a hummock may exist as the first sub-unit. Since the same feature may exist at different levels on different mires, it is not useful to name more levels under the mire unit connected to certain types of features. However it can be useful to describe as many levels as needed for a certain mire site, preferably labelling the levels one, two, three etc.

Results

As data on European mires is still being compiled, and many figures are preliminary estimations, no final results are presented here.

However, it might be interesting to consider some of the preliminary estimations. Data available to date indicates that the total area of mires in Europe was originally c. 100 Mha, whilst today only c. 76 Mha remain. It could then be considered that, of the remaining mire sites, probably over 50% are damaged or have an artificial ecology due to human activity.

The total area of 'protected' mires appears to be slightly over 7 Mha. The question is, are they really protected against damage, given that some countries allow certain human activities upon them? Political changes in eastern Europe might also create new situations for previously protected areas.

The main causes of mire destruction in Europe are drainage, especially for agriculture, but more recently also for forestry. In at least 15 European countries, various forms of peat extraction is responsible for serious damage on the mires. Other forms of destructive utilisation are overgrazing, road construction and inundation.

Several of these activities are a current threat to European mires. Even if the damaging activity has stopped, long term changes are continuing (e.g.

degradation of natural mire vegetation). New threats are also notified, one of the more serious ones being eutrophication caused by pollutants (airborne or otherwise).

Preliminary data allows an estimation of which European countries have the largest mire areas today. The top five are Russia (no data yet), Sweden (c. 6 Mha), Finland (c. 4.9 Mha), Norway (c. 2.5 Mha) and Belarus (c. 2.5 Mha).

The preliminary percentages of still virgin/near natural mire area, compared with original mire area, for a selected number of countries are: Former Soviet Union c. 69%, Sweden c. 66%, Finland c. 47%, Ireland c. 16% and UK c. 11%.

Preliminary estimations of the bog area remaining for some countries are as follows: Russia has the largest area, but no data is available at the moment, Sweden c. 1.3 Mha, Estonia c. 250,000 ha, UK c. 170,000 ha and Ireland c. 140,000 ha.

However bogs show a diversity in morphology, topography, hydrology and vegetation. Some of the more common bog-types in Europe are dome shaped bogs, eccentric bogs, and plateau bogs (the so called raised bogs), flat bogs, boreal bog types and blanket bogs. The species composition on bogs vary from south to north and in particular from west to east. The western, oceanic influenced bogs show the most marked difference in vegetation compared to other bogs and are unique for their kind.

UK and Ireland are the main distribution areas in Europe for the western oceanic bog types; blanket bogs and oceanic raised bogs. It is therefore serious, from an ecological point of view, that there is such a great loss of these ecosystems. It is important from an European point of view that efforts are made for preserving and rehabilitating these ecosystems.

Within the project, a modern mire region map will be produced. A draft version illustrates a west-east gradient according to oceanic influences and a north-south gradient from palsa mires and aapa mires in the north, followed by different bog regions and mainly fen regions in the south of Europe, and finally a differentiation according to altitude. Mountainous and alpine areas form a special mire region, containing a number of mire types (mainly fens) that are characterised by an unusually thin peat layer.

Chapter twenty-one:

The Scottish Raised Bog Land Cover Survey

Ms Lucy Parkyn and Dr Rob Stoneman
Scottish Wildlife Trust, UK.

Introduction

The task of conserving Scotland's raised bogs must begin with an assessment of the problem to allow effective conservation strategies to be developed (e.g. Stoneman, Chapter 45, this volume). Gauging the overall condition of any habitat is a difficult task. Conveniently, raised bog was always a rare habitat in Scotland; the original total extent covered 26,949 ha - just a third of one per cent of the total landmass. However, this area is divided amongst 851 sites. Hence, survey of the whole resource is likely to be a complex and expensive task.

The National Peatland Resource Inventory

The process of constructing a national picture of the total raised bog resource began with the setting up of the National Peatland Resource Inventory (NPRI) by the Nature Conservancy Council (NCC). The NPRI used British Geological Survey drift data on peat deposits as the basis of an inventory of all UK peat deposits. This dataset was supplemented, where necessary, using the Soil Survey of Scotland 1:50,000 soil maps.

This baseline dataset has proved to be a reasonably accurate reflection of the peatland resource except where 1:50,000 maps could not be obtained and coarser resolution maps were used (e.g. in the Grampian region of Scotland). The dataset gives a rough approximation of the original extent of peatlands. However, areas in which peat has been removed to the extent that few traces can now be found, are not recorded as peat on drift or soil maps.

Existing site accounts and aerial photographs were then used to assign the peatland areas of Scotland into four classes: blanket bog, intermediate bog, raised bog and fen. The basis of this classification scheme is set out in Lindsay (1995). The scheme has been criticised by Wheeler and Shaw (1995b) who refer to the surface and basal topography of three raised bogs in order to demonstrate some of the scheme's shortcomings. Given that classification is an artificial construct, it is inevitable that classification schemes will unsatisfactorily compartmentalise natural systems. However, since the NPRI is the only dataset which attempts to separate the whole resource into raised and blanket bogs in a - more or less - consistent manner, it was used here.

The category of intermediate bog is often confused with that of raised bog as the former often form 'domes' on hill-tops or ridges; such domes are totally controlled by underlying morphology and, as such, are not included as raised bog in this study. Two sites appeared to be misclassified within the NPRI: Dogden Moss and Moss of Corthie (both large and important sites) which ought to be reclassified as raised and blanket bog respectively. In most other cases, the present classification scheme works well in practice.

Raised bogs in Scotland were then classified according to their major (MLC) and best (BLC) land covers based upon previously published information, unpublished NCC information and aerial photographs (see Lindsay and Immirzi, 1996). Nine classes were used (see Table 21.1)

Table 21.1. Land cover classes.

Code	Description
P1	Primary (uncut) with 'near-natural' (i.e. *Sphagnum* dominated) vegetation
P2	Primary surface where the vegetation had been modified by factors other than drainage
P3	Primary drained bog in which a regular drainage pattern exists
P4	Primary bog supporting open canopy woodland or scrub
P5	Primary bog supporting closed canopy woodland
S1	Secondary (cut-over) bog which has re-vegetated
S2	Secondary bog where the surface is bare or is still being actively worked
A1	Archaic soils used for agriculture
A2	Archaic bog soils beneath built developments

An analysis of the major land cover classes of Scotland's raised bogs gives a reasonable summary of the state of the overall resource, although there will clearly be inaccuracies (e.g. DoE, 1994b). In an effort to get a more detailed and clearer picture of the conservation status of Scotland's raised bogs, two further datasets have been added into the NPRI: McTeague and Watson (1989) and SWT (1995a) - see below.

The Mid-Strathclyde sites

Sites in mid-Strathclyde were surveyed in detail by McTeague and Watson (1989). Both intermediate and raised bogs were surveyed, although only those classified as raised bog are considered here. Aerial photograph interpretations were ground truthed to provide a comprehensive dataset of mid-Strathclyde's bogs. The resulting surveys were digitised and added as an extra coverage of information to the NPRI. Whilst this survey was remarkably comprehensive, some sites in the area were not surveyed - these sites are categorised with the unsurveyed sites above.

McTeague and Watson (1989) classified bogs according to a ten point scheme (see Table 21.2).

Table 21.2. Mid-Strathclyde land cover classifications.

Code	Definition
0	Not peat
1	Intact mire surface
2	Degraded mire surface
3	Old baulk and hollow cuttings
4	Broad-leaved woodland
5	Conifer plantation
6	Agriculture and wet acid grassland
7	Other land-use (urban/industrial etc.)
8	Minerotrophic mire
9	Commercial peat workings

Source: McTeague and Watson (1989)

In retrospect, the data collected in this survey has been under-used. A significant shortcoming is that the attribute data for digitised polygons of peat do not distinguish between primary and secondary bog, nor do they incorporate any polygon-specific identification code. Thus, a revegetated milled peat-field would be given the same class as a primary degraded bog. Additionally, most of the survey information is only stored on paper, thereby limiting its use.

None the less, this remains a valuable dataset providing accurate information about land cover for those bogs surveyed.

The Scottish Raised Bog Land Cover Survey

Using LIFE funding from the European Union, the Scottish Wildlife Trust (SWT) and Scottish Natural Heritage (SNH) set out to extend McTeague and Watson's approach across the rest of Scotland. However, before the survey commenced, the methods used were examined critically and modified. In general, SWT's (1995/6) approach was similar to that of McTeague and Watson (1989). The main differences relate to the land-use categories, data storage and photographs.

Aerial photograph interpretation

Stereo aerial photographs were used to assess the current extent of the peat, to record adjacent land-use, to check whether the peat polygon is in the right place and, most importantly, to subdivide the site into units of similar land cover/use. Only units greater than 0.25 ha are mapped. Each unit is then assigned appropriate codes according to Table 21.3. Land use categories were modified from the previous NPRI categories, P1-5, S1-2 and A. Land classes P1 to P5 represented a continuum of degraded bog rather than separate classes. Accordingly, a simpler system was devised merging classes P2-5 into one class (P2). Added to this simple structure are a series of codes designed to provide detailed information on the conservation status of the peat polygons.

Aerial photographic interpretation (API) is a useful tool for gross characterisation of a site and to accurately map boundaries of different land-covers. However, API does not provide enough information to assign certain mapping codes. These include agricultural use (arable or pastoral), condition (wet or dry), modifiers for secondary or archaic areas, and certain vegetation structure codes. Additionally, some codes, assigned using API, may be subsequently altered during ground-truthing (see below). For example, an aerial photograph may suggest an area is covered in closed canopy woodland (given that the crowns of trees cannot be discerned). Ground-truthing, however, may show that the canopy is actually fairly open and the site still supports vegetation associated more with open bog than woodland. Further, depending upon the time lapse between the photograph being taken and field survey executed, the situation may have changed: a bare field re-vegetated, a plantation felled or scrub expanded etc. A detailed description of the way API was conducted is shown in Parkyn *et al.* (1995, unpublished).

Pre-survey preparation

Sites subdivided and coded by API were then 'ground-truthed' to confirm that the boundaries and land-use codes were correct, to add in land-use modifiers which were not possible to determine using API and to collect further information concerning the site (vegetation data, photographs, damage and structural assessments). However, these surveys are expensive in terms of time, money and personnel. Sites were thus divided into three categories:

1. sites which must be surveyed - have the highest wildlife interest;
2. sites which should preferably be surveyed - usually smaller sites of reasonable wildlife interest.
3. sites where ground-truthing is unnecessary - usually completely afforested or archaic sites.

Table 21.3 Land-use categories used during aerial photograph interpretation and field survey.

LANDUSE TYPE		HUMAN IMPACT		VEGETATION MODIFIER					OTHER	
PRIMARY			***Drainage***	***Tree***	***Scrub***	***Shrub***	***Field***	***Ground***	***Erosion***	***Condition***
Natural	P1			BP, CP	BP, CP,					
Degraded	P2		I	MP	MP				MIC	
			N	BPF, CPF	BPF, CPF			SPH		W
SECONDARY		***Extraction***	M	MPF	MPF	L	H	BRY	Y	D
Vegetated	S1	M, B, S,	W	BPFR	BPFR			BAR		WF
Exposed	S2	DOM	A	CPFR	CPFR			BBA		WS
			U	MPFR	MPFR			BS		
				BS, CS	BS, CS, MS			BAS		
				MS	GS (O or C)					
ARCHAIC		***Development***								
Agriculture	A1	P, A HO								
Built	A3	RR, B, HA								
Utility	A3	AM, R								
OTHER										
Fen	F									
Swamp	SWP									
Open water	OW									
Unknown	U									
Not peat	NP									

Key to Table 21.3:

Drainage Modifiers

I	irregular
N	narrow
M	moderate
W	wide
A	absent
U	unknown

Vegetation Modifiers

BP	broadleaf plantation
CP	coniferous plantation
MP	mixed plantation
?PF	plantation, felled
?PFR	plantation, felled, replanted
BS	broadleaf selfsown
CS	coniferous selfsown
MS	mixed selfsown
GS	gorse, bramble etc.
O or C	open or closed canopy
L	low shrub dominated
H	herb dominated
SPH	*Sphagnum* dominated
BRY	bryophyte dominated
BAR	bare peat dominated
BBA	bryophyte/bare peat co-dominant
BS	bryophyte/*Sphagnum* co-dominant
BAS	bare peat/*Sphagnum* co-dominant

Erosion Modifiers

MIC	micro-broken
Y	gully erosion

Extraction Type

M	milled
B	block-cut (baulk-&-hollow)
S	sausage cut
DOM	domestic, segment

Development Type

P	pasture
A	arable
HO	horticulture
RR	road/railway
B	built
HA	hard-standing
AM	amenity
R	refuse disposal; tipping; quarry etc.

Condition Modifiers

W	wet
D	dry
WS	wet swamp
WF	wet fen

In practice, the majority of class one and two sites were, in fact, surveyed. Once sites were selected for survey, access was arranged by contacting land-owners/occupiers when known or by visiting the nearest farms. Often, no-one knew who owned a site, so the bog was surveyed anyway.

Site survey

For each site, a series of sheets was completed which detailed information on the entire site, primary (un-cut) areas, secondary (cut-over) areas and individual land-use polygons. The results of the survey were divided into nine sections (see Table 21.4).

Table 21.4 Sections of the survey report from the Raised Bog Land Cover Survey.

Section No.	Section Title	Details
One	*Site details*	Site code, site name, surveyor, grid reference, altitude, date of survey, weather conditions, owner information (on paper copy only), access permission, management plan, photograph references, birds or other animals seen during survey
Two	*Damage*	Animal influence, burning, drainage, other
Three	*Mire quality*	Bare peat, *Sphagnum* cover, surface texture
Four	*Detailed microforms*	Microforms present, microform distribution, associated vegetation
Five	*Quadrat data*	Maximum of nine quadrats taken from the 'best' (in nature conservation terms) land-cover/use polygons recorded using the DOMIN scale
Six	*Species list*	Listing of all species recorded on the site
Seven	*Site description*	Describing the site, structure, vegetation, damage and a site assessment
Eight	*Land-cover information*	see Table 21.3
Nine	*Target notes*	Each polygon is briefly described
I	*Site map*	Information is accompanied by a site map showing the different compartments
ii	*Photograph*	Sets of stereo-photographs of landscape views were taken on most of the sites surveyed. Location and direction of shots are recorded on the site map

Source: Parkyn *et al.* (1995)

Whilst this represents a considerable volume of information, skilled surveyors were, on average, able to process one site per day including survey, write-up and data input (see below).

Data storage

Building on the McTeague and Watson (1989) survey and taking cognisance of faults identified subsequent to survey, considerable attention was paid to data storage. The information was stored in four formats: on paper, in a flat-file database, as digitised information within a geographic information system (GIS) and as photographs.

Paper Copy - Sections one to nine and the site map are held as paper copies both by SWT and SNH (central and regional).

Database - Sections one to nine are presently held on a flat-file database (Filemaker Pro). This will later be transferred to a relational database.

GIS Data - The site maps were originally generated by the existing NPRI which plots site boundaries onto AO acetate sheets at 1:10,000 or 1:10,560 to overlay the corresponding OS maps. After ground-truthing, the boundaries are re-drawn (where necessary) and each polygon numbered. These sheets were then digitised into ArcInfo format and attribute data attached (codes shown in Table 21.3).

Photographs - Two sets of photographs were taken and catalogued in another flat-file database. Copies are held at SWT and SNH.

Results and discussion

The resulting database of information provides a powerful tool to answer questions about the state and condition of individual bogs or sites across geographical regions.

The digitised GIS data, in particular, provides detailed information in an easily manipulated format. Reference to Fig. 21.1 demonstrates the potential of this data for a site-by-site analysis.

Flanders Moss East is considered to be one of the largest and most intact raised bogs remaining in the UK. Fig. 21.1a illustrates the basic land cover categories. Clearly, the site is still largely un-cut although much of the area now supports degraded vegetation. Overlaying information on trees and scrub (Fig. 21.1b) clearly shows that tree and scrub encroachment is a severe problem at the Moss. In Fig. 21.1c, the density of drainage has also been overlain. Such information effectively allows the polygons to become conservation management units. Indeed, if the cost of blocking drains and removing scrub and trees can be defined, then accurate cost assessments for conservation management work can be assessed. For example, compare the cost of managing the units shown in Fig. 21.1d - the larger area has an irregular drainage network and little scrub; management costs are approximately £10 ha^{-1} (using voluntary labour) (SWT, 1995b). The smaller area is a closed canopy plantation where rehabilitation costs would be c. £600 ha^{-1}. The data shown gives a conservation management cost of £1590 and £20,400 for the intact and afforested area respectively.

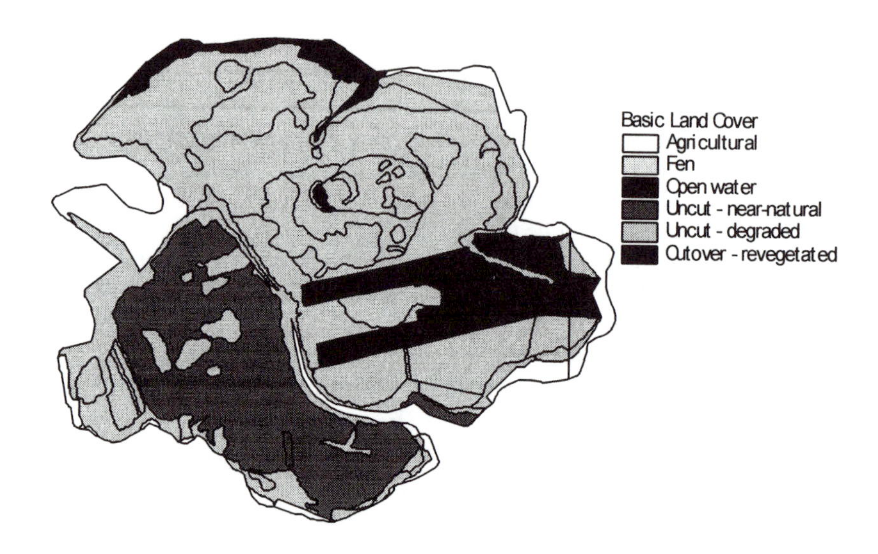

Fig. 21.1a Basic land-cover classes.

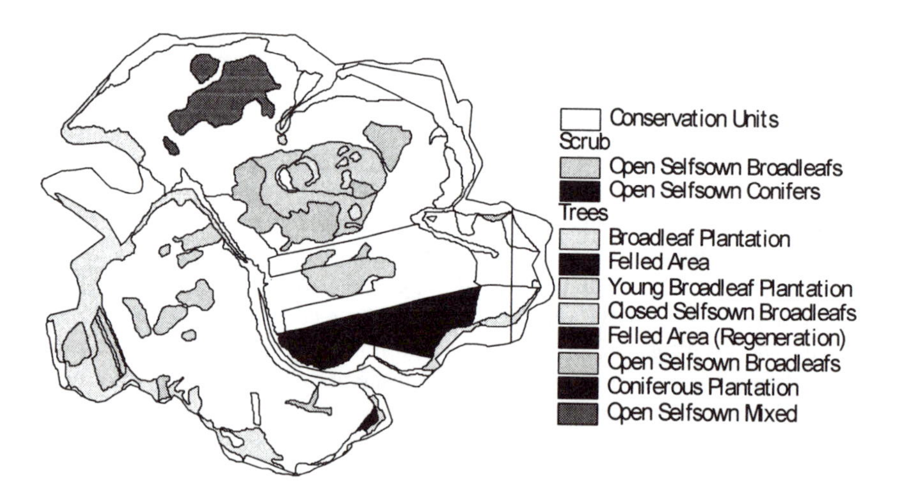

Fig. 21.1b Trees and Scrub.

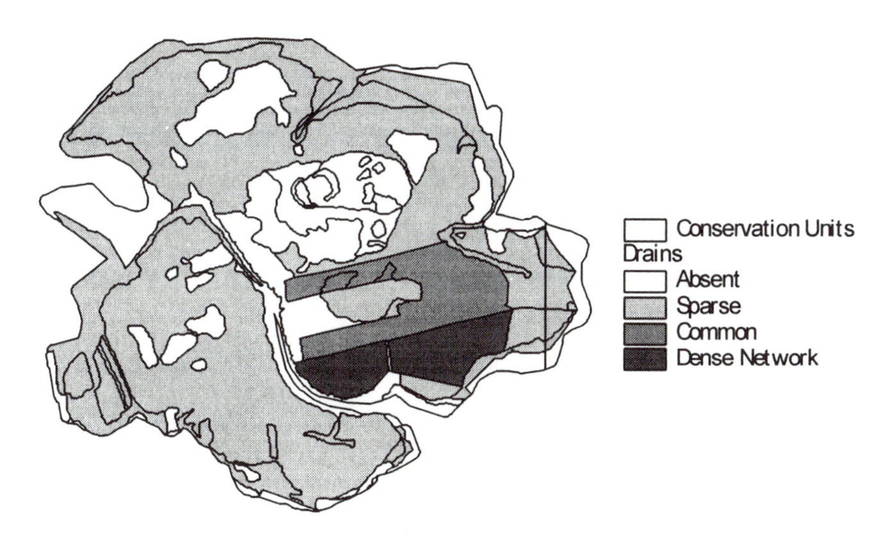

Fig. 21.1c Drainage intensity.

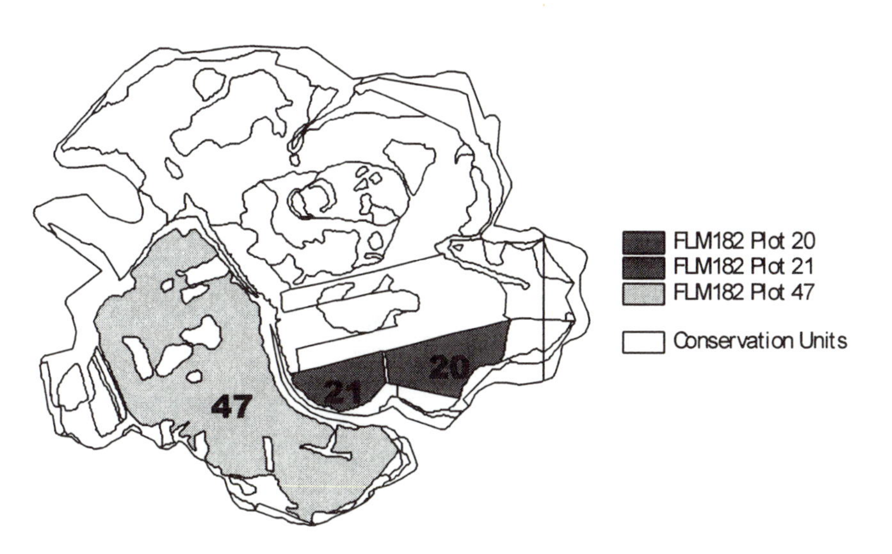

Fig 21.1d Management Units.

Fig. 21.1(a-d). Examples of Land Cover Survey data for Flanders Moss East.

Future surveys of the site using the same methodology would offer a detailed and accurate way of defining both the degree and type of change.

The land cover survey also allows information to be analysed on a regional level by looking at groups of bogs. Fig. 21.2 shows the cluster of bogs surrounding Flanders Moss East. This shows that the Upper Forth Valley bogs are in a rather poor condition requiring considerable management. An analysis across all these sites shows only 10% of the area now supports the 'near-natural' vegetation communities of un-cut surfaces. Fig. 21.3 shows the raised bogs of South Lanarkshire illustrating that many of these sites are in a good condition. Table 21.5 shows the distribution of land-cover here.

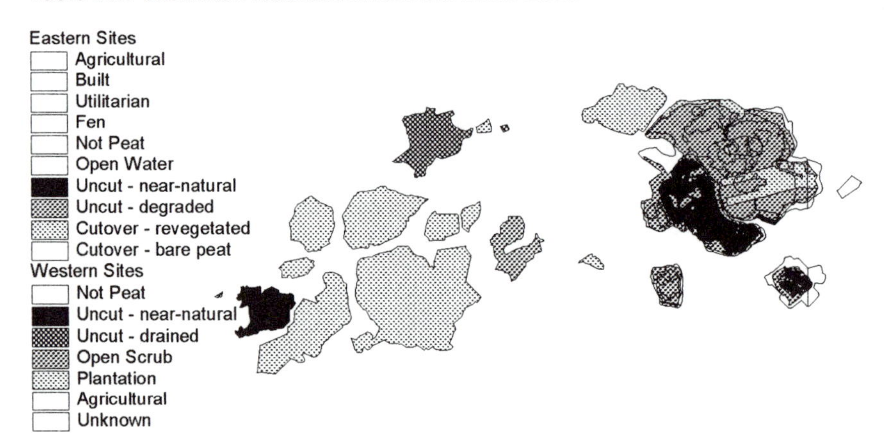

Fig. 21.2. Regional overview of raised bogs surrounding Flanders Moss East.

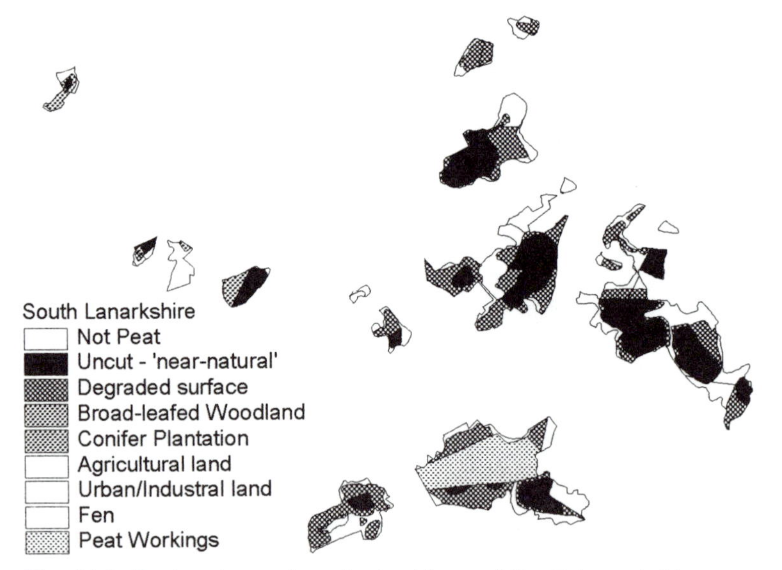

Fig. 21.3. Regional overview of raised bogs of South Lanarkshire.

Table 21.5 Land-use/cover class summaries for raised bogs in the South Lanarkshire new Local Authority.

Cover Class	Totals (ha)	%
Unknown	0.00	0.00
Primary - Natural	627.54	24.14
Open Degraded (Primary/Secondary)	603.69	23.22
Wooded (Primary/Secondary)	413.03	15.89
Active Commercial Workings	300.54	11.56
Developed Areas	21.21	0.82
Agricultural/Fen	633.53	24.37
Total	**2599.54**	

Conclusion

The land-cover survey is still not complete, 11% of the resource remains to be surveyed, and the mid-Strathclyde data needs revising. However, this information still provides an immensely useful dataset to assess the state and condition of Scotland's raised bogs. It allows sensible priorities to be set, areas for action to be defined, costs to be assessed and, most usefully, sets a baseline from which conservation strategies can be developed. In Scotland, the opportunity to reverse the progressive decline of this extremely rare habit has arisen as a direct result of the land cover survey. Conservationists can now act in a coherent and effective manner. The challenge is to turn the opportunity into action (Stoneman, Chapter 45, this volume).

Chapter twenty-two:

From *Sphagnum* to a Satellite: Towards a Comprehensive Inventory of the Blanket Mires of Scotland

Ms Eliane Reid[1], Ms Sarah Y. Ross[2], Dr Des B.A. Thompson[1] and Mr Richard A. Lindsay[1]
[1]*Scottish Natural Heritage, UK.*
[2]*University of Stirling, UK.*

Introduction

There is a growing realisation that biodiversity is intimately connected with the long-term health and vigour of the biosphere (Stoms and Estes, 1993). Indeed, new legislative mandates such as the European Community Habitats Directive (92/43/EEC), and the follow-up to the Rio Conference (key statutory instruments are listed under Table 22.1) reflect this and have led European Union countries to develop specific plans to protect and enhance species and habitat. One important element of this is the adoption of a more strategic approach to inventory of the location, extent and interests of key habitats (Miller, 1994).

Scottish Natural Heritage (SNH), the Statutory Conservation Agency, responsible to the Secretary of State for Scotland for the natural heritage of Scotland, is currently undertaking habitat inventory, monitoring and management activities in order to meet statutory and ministerial priorities.

This chapter describes work in hand to provide a comprehensive inventory of the blanket mire resource of Scotland. This habitat is listed both as 'priority' (active bog) and as non-priority types (degraded) under the EC Habitats Directive. It is the most extensive semi-natural habitat found on land in the UK, presenting formidable challenges for an inventory programme.

Peatland habitat conservation evaluation is inherently multi-disciplinary: the hydrology, morphology, topography, vegetation, species dynamics and management are all key elements in assessing the scientific interest of a site (JNCC, 1994). There are several steps which should be taken in this evaluation however, particularly with reference to the 'active' state of the habitat.

Table 22.1. Some of the key International and National Legislation's, Conventions and Agreements.

Date	Legislation
	National Legislation
1981 *(amended in 1985)*	The Wildlife & Countryside Act
1990	The Environmental Protection Act
1991	The Natural Heritage (Scotland) Act
1994	The Conservation (Natural Habitats etc.) Regulations
	International legislation
1979	Council Directive on the conservation of wild birds (79/409/EEC)
1992	Council Directive on the conservation of natural habitats and of wild fauna and flora (92/43/EEC)
	International Conventions and Agreements
1976	Convention of Wetlands of International Importance as Waterfowl Habitat (The Ramsar Convention - 75/66/EE)
1972 *(UK ratified 1984)*	Convention concerning the Protection of the World Cultural and Natural Heritage (The World Heritage Convention)
1982	Convention on the Conservation of Migratory Species of Wild Animals (The Bonn Convention - 82/461/EEC)
1981/82	The Convention on the Conservation of European Wildlife and Natural Habitats (The Bern Convention - 82/72/EEC)
1992	Convention on Biological Diversity: (The Biodiversity Action Plan - 93/626/EEC)

Inventory of blanket mires - data and techniques

Since 1992, SNH has been evaluating different inventory and survey techniques for assessing the extent, condition and scientific interest of the blanket mire habitat. Blanket mires in Scotland cover approximately one million ha (Lindsay, 1995), and are generally found in harsh, rugged, remote terrain, which is costly to survey by traditional small scale ground survey techniques (Fig. 22.1). Indeed, it is estimated that to fully evaluate the extent and condition of Scotland's blanket mire, a minimum of 25 person years would be required (Lindsay, pers. comm.). In addition, blanket mires form a complex continuum of mire units which often extend over 1000s of hectares and are therefore inherently difficult to define on the ground and to map.

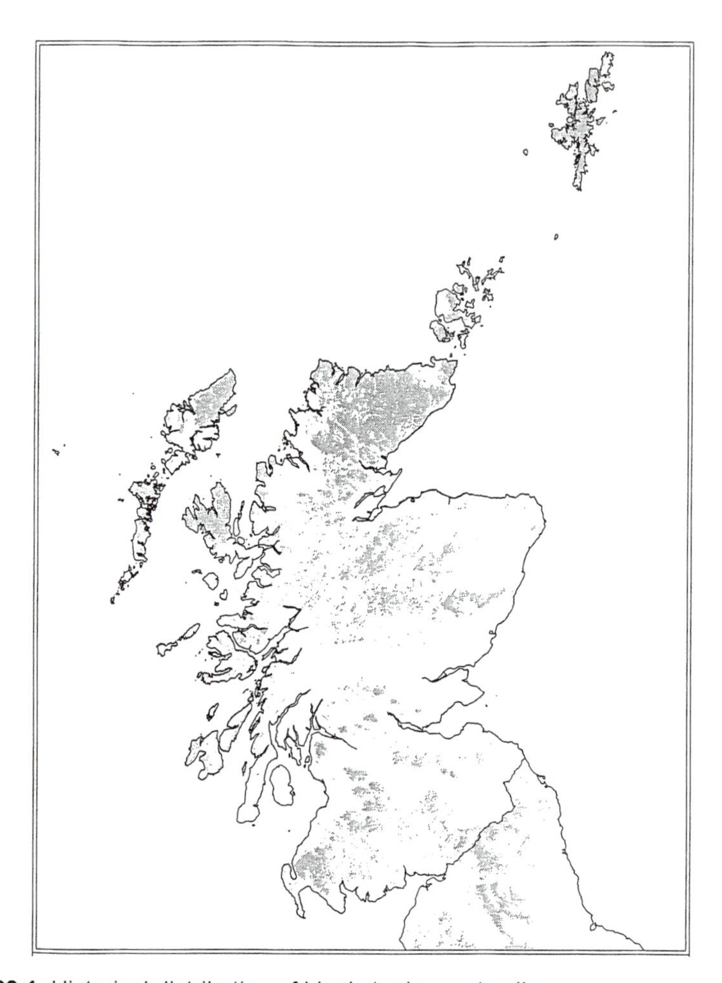

Fig. 22.1. Historical distribution of blanket mire peat soils.

A number of datasets and techniques are available to categorise natural and semi-natural land-cover and habitat condition, extending across a wide range of variability and scale (Fig. 22.2). All provide information at different scales and resolutions and aid local, regional, national and international policy. In outline they are:

1. Small Scale - Traditional Survey
2. Intermediate Scale - Remote Sensing - Air Photograph Interpretation
3. Large Scale - Remote Sensing - Satellite Imagery Interpretation

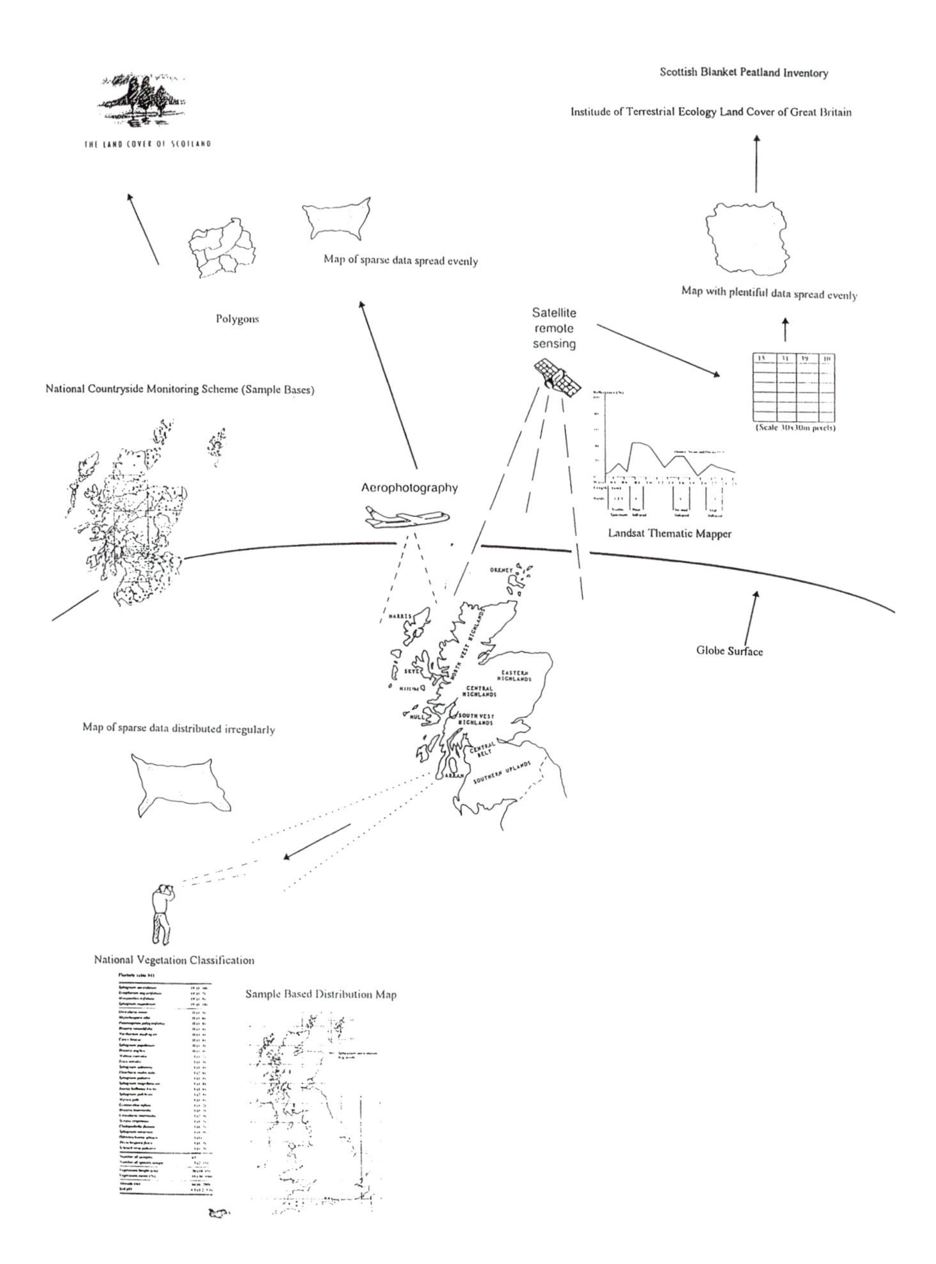

Fig. 22.2. Examples of Mapped data and Mapping scales Available to SNH (after Miller, 1994; Rodwell, 1992; MLURI, 1993; Tudor *et al.*, 1994).

Traditional small scale site survey - the National Vegetation Classification (Rodwell, 1991)

This classification system is based on sample surveys of 2 m^2 (for mires) quadrats throughout Britain. The system is used extensively by SNH for mapping specific Sites of Special Scientific Interest (SSSI) and potential Special Areas of Conservation (SACs). On a site, homogeneous stands of vegetation are identified as typical of given communities (following methods by Rodwell, 1991). These are then mapped on to air photographs which are registered with 1: 10,000 Ordnance Survey maps. Boundaries between communities (or vegetation types) are subsequently drawn onto a map (1:10,000) creating a site habitat map.

Within British peatlands, the National Vegetation Classification system (Rodwell, 1991) is currently used as the standard system to determine EC Habitats Directive 'active' peatland types (Table 22.2).

Table 22.2 'Active Peatland' - National Vegetation Classification Types.

NVC Type	Description	Active raised bog	Active blanket bog
M1	*Sphagnum auriculatum* bog pool community	◆	◆
M3	*Eriophorum angustifolium* bog pool community	◆	X
M15	*Scirpus cespitosus-Erica tetralix* wet heath	X	◆
M17	*Scirpus cespitosus -Eriophorum vaginatum* blanket mire	X	◆
M18	*Erica tetralix - Sphagnum papillosum* raised and blanket mire	X	◆
M19	*Calluna vulgaris - Eriophorum vaginatum* blanket mire	◆	◆
M20	*Eriophorum vaginatum* blanket and raised mire (active ONLY - M20a for raised bogs)	X	◆

◆ = identified as active. (after EC, 1994)

This standard methodology for habitat survey is used by ecologists throughout Britain, and the results are readily comparable. The system, however, requires a substantial capital investment in the form of the personnel and time needed to cover large areas of blanket mire (c. 150 ha day^{-1} $person^{-1}$) and, as it is a scheme based on vegetation sampling, it is only designed to classify vegetation types which appeared in the original samples. It cannot classify, or evaluate, previously unsampled vegetation types, vegetation condition, site hydrology, or microtopography, all of which are important for assessing the scientific and conservation interest of an area. These limitations are particularly important in relation to blanket mire, which appears to be under-represented in the distribution maps given by Rodwell (1991).

Intermediate scale - air photograph interpretation

Discrimination between phytosociological vegetation types continues to be one of the most elusive goals in remote sensing studies of natural and semi-natural vegetation. Studies have more commonly concentrated on distinguishing between, and mapping, vegetation related variables which demonstrate distinct homogeneous features (e.g. arable, heather moorland, unimproved grassland, etc.), or at large regional/national scales (Jewell and Brown, 1988; Grenon, 1989; MLURI, 1993; Fuller *et al.*, 1994). Examples of such datasets used within SNH are as follows:

Large National Scale - Air Photo Interpretation - The Land-cover of Scotland 1988 (MLURI, 1993):

The Land-cover of Scotland survey was proposed by the Under-Secretary of State for Scotland in 1987, with the aim of producing a detailed census of land-cover throughout Scotland. The project was set up in three stages: (i) flying air photography of Scotland (scale = 1:24,000), (ii) interpretation of the photography and computation, and (iii) validation. Land-cover within this system was largely concerned with interpreting 'the vegetation and artificial constructions covering the land surface' (Burley, 1961). The Macaulay Land Use Research Institute (MLURI) system is hierarchical and recognises the principal, major and main land cover features. Within the principal feature 'Semi-natural ground vegetation', heather and dwarf shrub heathland, blanket bog and other peatland vegetation are identified (Table 22.3). These classes are currently used for general casework within SNH. They allow coarse-scale evaluation of the distribution of categories deemed to be of conservation importance within the organisation (e.g. heather moorland - Thompson *et al.*, 1995). It does not, however, give a comprehensive evaluation of the ecological variability within the blanket mire habitat, i.e. the variability between communities dominated by sedges, grasses or mosses (e.g. *Molinia caerulea, Trichophorum cespitosum, Sphagnum papillosum*) and shrub species dominants which are indicative of variations in habitat condition and recent management.

Table 22.3 - LCS88 Classification (MLURI, 1993).

Principal feature	Major feature	Main features	Sub-categories based on
Semi-natural ground vegetation	Heather & dwarf shrub heathland	Dry heather moor Wet heather moor Undifferentiated heather moor	Muir-burn Rock outcrops Scattered trees
	Blanket bog & other peatland vegetation	Blanket bog with dubh lochans Undifferentiated blanket bog	Erosion Scattered trees Undifferentiated blanket bog Mechanised exploitation Domestic exploitation

The National Countryside Monitoring Scheme (NCMS) (SNH):

Air photograph interpretation is also an extremely useful tool in estimating change over time in the appearance of the countryside with changes, for example, in farming structures and other land use. The National Countryside Monitoring Scheme, has evaluated such changes, where sample squares of air photographs have been interpreted from the 1940s, 1970s and 1980s. The dynamics of changing landuse and land-cover are represented as gains and losses amongst feature types (Fig. 22.3) (Mackey and Tudor, 1996). For example, large areas of blanket mire, unimproved grassland, and heather moorland, have been afforested. There has also been a net reduction in the cover of heather moorland and a reduction in blanket mires in favour of heather moorland. Again this scheme is not sensitive to the finer scale changes in habitat condition.

To date, little success has been achieved in distinguishing between vegetation types using remote sensing techniques other than at gross scales, although recent attempts have been made to clarify where this is possible for a few British types (Haines-Young and Bunce, 1995). The dilemma is that many vegetation classification techniques (such as the NVC) are sensitive to differences in vegetation species composition rather than broad land-cover in order to describe, classify, evaluate and subsequently map phytosociological relationships (Davis *et al.*, 1990). Consequently, the techniques utilised by ground-based ecologists in Britain have not been extensively practised within the field of remote-sensing. Likewise the synoptic view presented by remote sensing has not been fully taken advantage of in ecological mapping or as an aid in the long term evaluation and conservation of habitats.

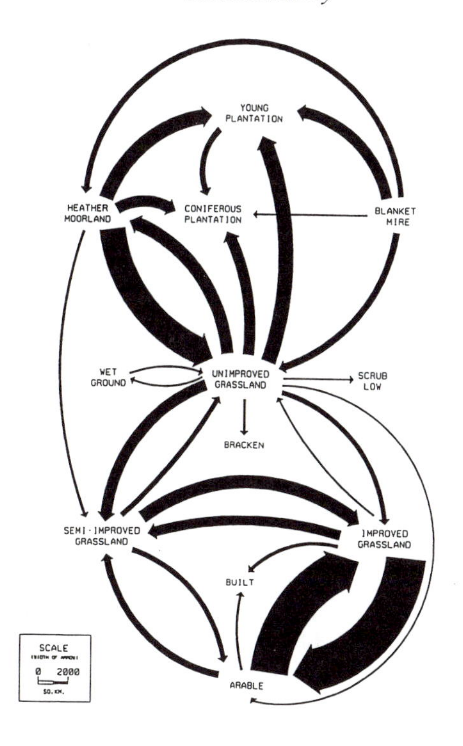

INTERCHANGES GREATER THAN 100 SQ.KM. FOR SCOTLAND

Fig. 22.3. National Countryside Monitoring Scheme - the dynamics of changing land-use and land-cover through time (Mackey and Tudor, 1996).

Some research techniques have however attempted to examine the relationship between botanically classified data and spectral data derived from satellite sensors. This has included specific studies comparing spectral data (from multiple sensors) with data developed by the use of multivariate analysis techniques to predict community type on the basis of spectral variables (Wood and Foody, 1989). Other work has considered the problem of representing the continuum between community types mapped by ecologists and satellite, and have investigated the 'fuzzy sets' method in order to judge the 'end points' of the vegetation continua (Foody and Cox, 1991). None of these approaches have so far developed a thorough, robust and cost effective methodology for operational mapping and monitoring.

Remote sensing - satellite image processing and interpretation. The National Peatland Resource Inventory Phase II - Scottish Blanket Peatland Inventory

Blanket mire has been identified within SNH as being intrinsically suited for mapping at large scale by utilising satellite platforms such as Landsat Thematic Mapper (TM) or SPOT imagery (Pooley and Jones, 1995). Satellite imagery (or datasets derived from satellite imagery) depicting habitats is available to SNH, for example in the form of the Institute of Terrestrial Ecology Landcover of Great Britain satellite data (Fuller *et al.*, 1994). Although this is a valuable and immensely useful dataset for identifying where a particular habitat occurs, it is limited in describing the condition and ecological variability within that habitat as described above. SNH has therefore developed its own methods for evaluating the extent and key elements of habitat variability, using satellite imagery.

The Scottish Blanket Peatland Inventory (SBPI) uses a methodology, developed specifically for mapping the extent, condition and hence conservation interest of the blanket peatland habitat throughout Scotland (Reid *et al.,* 1994). Priorities are currently guided by those areas within previously unmapped or little mapped peatland which have been listed as important within Europe, e.g. Lewis (Outer Hebrides) (Fig. 22.4). This methodology, with small modifications to the 1994 procedure, is described briefly (Fig. 22.5).

The method adopted by the SBPI is specifically designed for mapping blanket peatlands, and makes no attempt to map other habitats. In this respect it differs from most other large-scale mapping datasets available in Britain. Resources and techniques are designed to concentrate on evaluating and interpreting the spectral characteristics of the TM imagery within the spatial locations of blanket peatlands.

1. The first step identifies and prioritises peatland areas of interest. This is achieved by employing local knowledge and by making use of data which have been captured within a Geographic Information System (GIS). This data (polygons) has largely been captured with kind permission from the British Geological Survey (1: 50 000) map series of soil deposits. Peat soil deposits (over one metre in depth) have been classified within the NPRI into four categories: raised bogs, blanket bogs, intermediate bogs and fen. The blanket peat soil category 'polygons' create boundaries which are used to highlight the spatial location of the blanket peat and to then focus image interpretation within these boundaries.

2. TM data are then selected based on the most appropriate time of the year for peatland type characterisation (May, June, July), TM availability and a recent date. Identification of sites of importance and variability are achieved by 'overlaying' the BGS dataset onto the imagery within image processing software, using a ratio of spectral bands. Once sites (approximately 10 km^2 in area) have been identified, and in order to highlight areas of homogeneity within the image, a principal component analysis, and then an unsupervised classification (ISODATA) are run within the system (individual classes - i.e. areas of homogeneity - are automatically based upon the statistical distribution of the pixels within the image). The number of end classes is determined by image analysis of the spectral variation within the BGS boundaries and by local ecological knowledge of the sites. The classification is iterated, within the software, by analysis and visual interpretation, based on ecological knowledge of the area, until a robust classification of the sites is produced. A raster-to-vector conversion algorithm is applied to the image in order to produce vector lines around areas of homogeneity. This map is overlaid onto contrast enhanced bands 5, 4, and 3 of the imagery and produced as hard copy. For survey, areas (blocks) are targeted for ground survey by matching vector classes with the visual identification of blocks that are homogeneous in tone within the bands 5,4, and 3. These blocks are superimposed, at the same scale (by using Ordnance Survey maps), onto air-photographs for ease of navigation and survey in the field.

3. In the field, attribute data are collected for each block where both species composition (NVC) and non-vegetation ground cover are quantified. Other attributes gathered include; dominant shrub and herb species, ratio of shrubs:herb species, height and structure of the vegetation, peat depth, slope, aspect, management/damage, and microtopography.

4. Vegetation classification patterns throughout the survey blocks are analysed with particular reference to linking spectral features with vegetation features (e.g. Table 22.4). Other attribute data are assessed using multivariate statistical techniques based on the pattern identified for each spectral class. The classification of imagery is repeated and manipulated, and final vegetation and environmental (physical) class descriptions and confidence values are calculated for each spectral class.

A confusion matrix is used to test the end-product image within peatland areas. Other datasets (e.g. other vegetation surveys, OS maps, SNH datasets etc.) are used for qualitative image interpretation (where little ground truth is available in the original field dataset).

Fig. 22.4. Satellite image archive and areas of blanket bog.

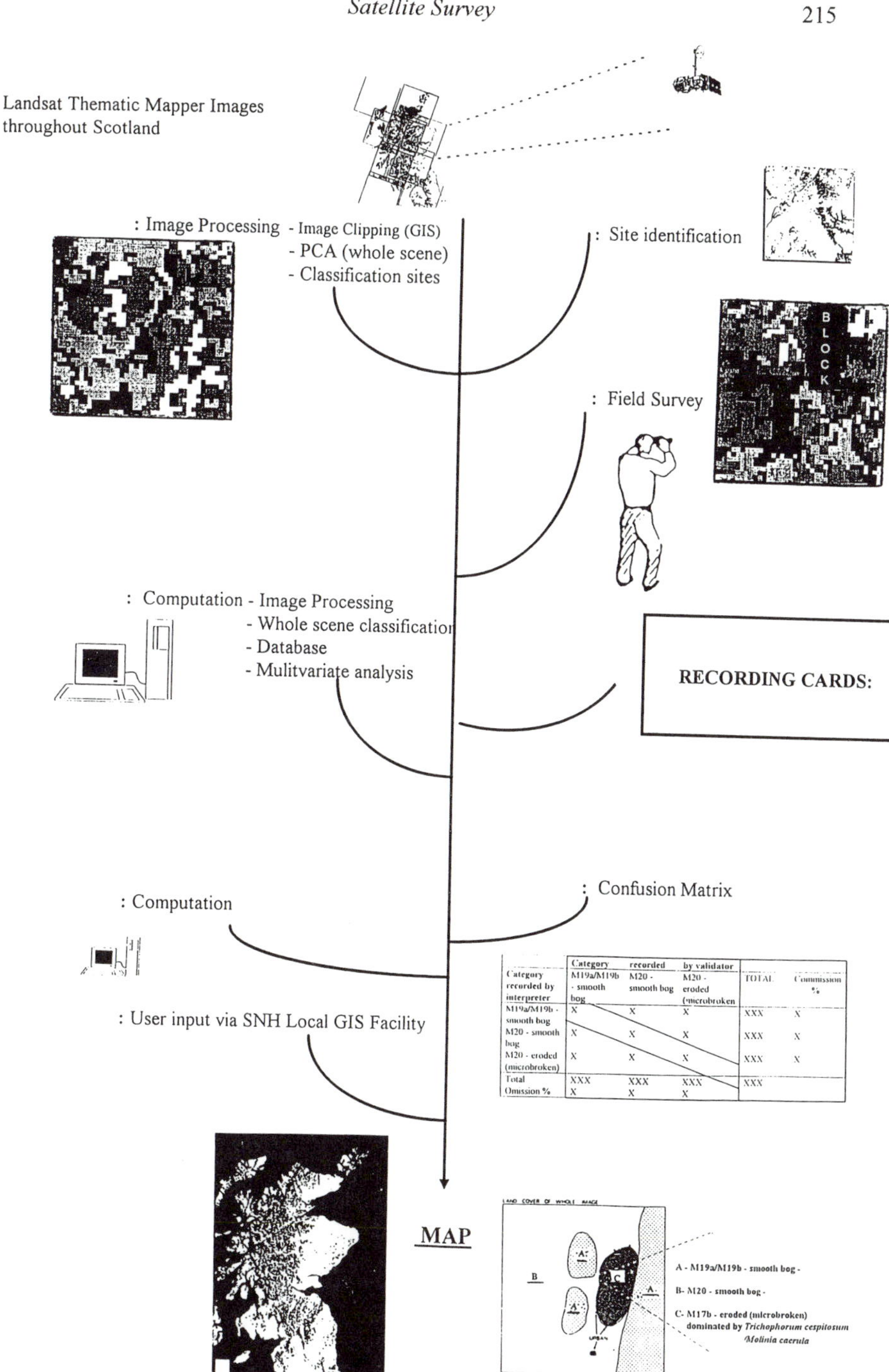

Category recorded by interpreter	Category recorded by validator: M19a/M19b - smooth bog	M20 - smooth bog	M20 - eroded (microbroken	TOTAL	Commission %
M19a/M19b - smooth bog	X	X	X	XXX	X
M20 - smooth bog	X	X	X	XXX	X
M20 - eroded (microbroken)	X	X	X	XXX	X
Total	XXX	XXX	XXX	XXX	
Omission %	X	X	X		

Fig. 22.5. Scottish Blanket Peatland Inventory methodology.

Table 22.4. Image spectral numbers linked with vegetation types which have been classified according to Rodwell (1991).

Spectral Classes No.	Major Vegetation Types Associated with each spectral class - type descriptions
39	Mosaic of: *Calluna vulgaris-Eriophorum vaginatum-Erica tetralix* (M19a) & *Empetrum nigrum* ssp. *nigrum* sub-communities (M19b)
19	*Calluna vulgaris-Eriophorum vaginatum - Erica tetralix* (M19a) sub community with small patches: M19b (see above) & M20 (see below)
37	*Calluna vulgaris-Eriophorum vaginatum - Empetrum nigrum* spp. *nigrum* sub-community (M19b)
10	Eriophorum vaginatum blanket and raised mire (M20)
36	*Eriophorum vaginatum* blanket and raised mire (M20) with small patches: (M19b)
26	*Eriophorum vaginatum* blanket and raised mire (M20) with small patches: *Sphagnum cuspidatum/recurvum* bog pool community (M2) & *Erica tetralix-Sphagnum papillosum* (M18) community

Conclusion

New technologies such as global positioning systems, Geographic Information Systems, and routine image processing software are revolutionising vegetation mapping and modelling, and aiding in the interpretation of remotely sensed imagery. These are currently contributing to a revival of applications in vegetation pattern and process over regional and national scales, and are facilitating the determination of plant community and ecosystem dynamics, and hence such techniques can aid in contemporary and strategic planning in the field of nature conservation.

This chapter has highlighted how particular spatial datasets and multidisciplinary methodologies/techniques can contribute towards SNH Inventory framework and hence allow SNH to fulfil its role in identifying key areas of importance in the natural heritage of Scotland, at different spatial scales. For blanket peatland habitat evaluation and for the development of a comprehensive inventory the SBPI technique has, so far, become a useful and cost-effective method for mapping and quantifying, at a coarse scale, large expanses of blanket peatland within Scotland. Here this heuristic methodology particularly emphasises these ecological characteristics in the field which SNH consider to be of prime importance in evaluating the scientific and conservation interest for sites. By applying this style of mapping, SNH can therefore identify and target potential areas of natural heritage 'value' within the blanket peatland habitat and within large designated or proposed designated sites. This can subsequently contribute to the evaluation of gaps in habitat designation, the wider countryside, and target ground-based survey at the small scale.

Chapter twenty-three:

Surveying and Monitoring of Mires in Switzerland

Mr Andreas Grünig
Swiss Federal Institute for Forest, Snow and Landscape Research (WSL/FNP), Switzerland.

The legal basis for mire conservation in Switzerland

Since the 1987 Referendum, when the Swiss people voted in favour of the Rothenthurm amendment to the Federal Constitution, Switzerland has probably been the only country in the world where the integral conservation of both mire habitats and mire landscapes is an explicit political goal (Kohli, 1994). This amendment lays down strict terms for the protection and restoration of the original conditions of mire sites of national importance:
Rothenthurm amendment to the Federal Constitution; Art. 24, item 6, Para. 5:
Mires and mire landscapes of particular beauty and national importance are protected areas. The construction of any kind of building or installation whatsoever, and any operations changing soil structure are strictly prohibited. Excepted are operations and installations necessary for the maintenance of the near natural landscape and existing agricultural use.

Inventories of Swiss mires and mire landscapes - the scientific basis for mire conservation in Switzerland

In the period 1978-1991, three individual surveys were made of the bogs (Fig. 23.1), fens (Fig. 23.2) and mire landscapes of Switzerland (Fig. 23.3) in order to meet the need for interpretation and definition of sites (Grünig *et al.*, 1986; Grünig, 1994; Broggi, 1990; Hintermann, 1992; 1994). From these surveys, we know that the total mire landscape area comprises about 80,000 ha or 2% of the country's surface and that most of the 20,000 ha (corresponding to c. 0.5% of the country's surface) of the mire habitats of national importance are located inside the perimeters of the mire landscapes.

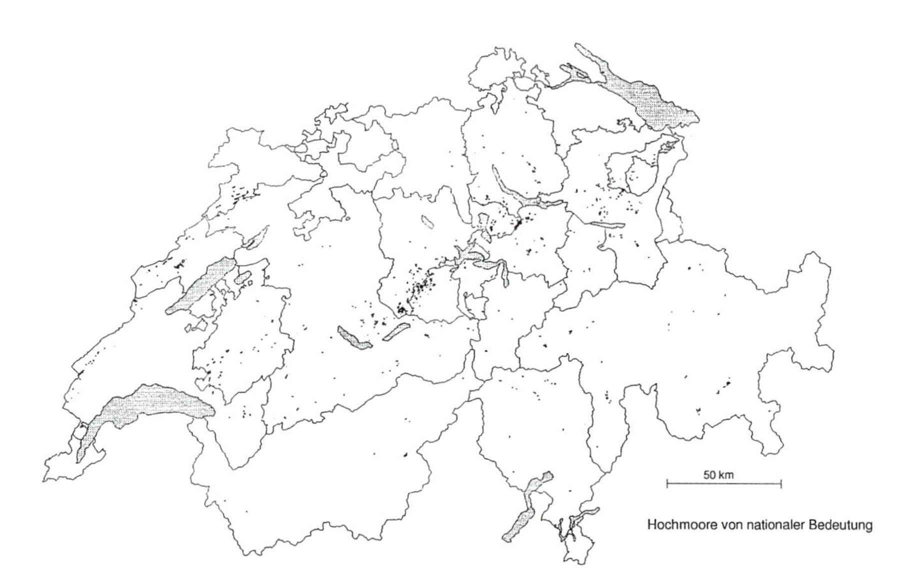

Fig. 23.1. Distribution of the raised bogs and transitional mires of national importance, according to the Federal Inventory (DFI, 1991).

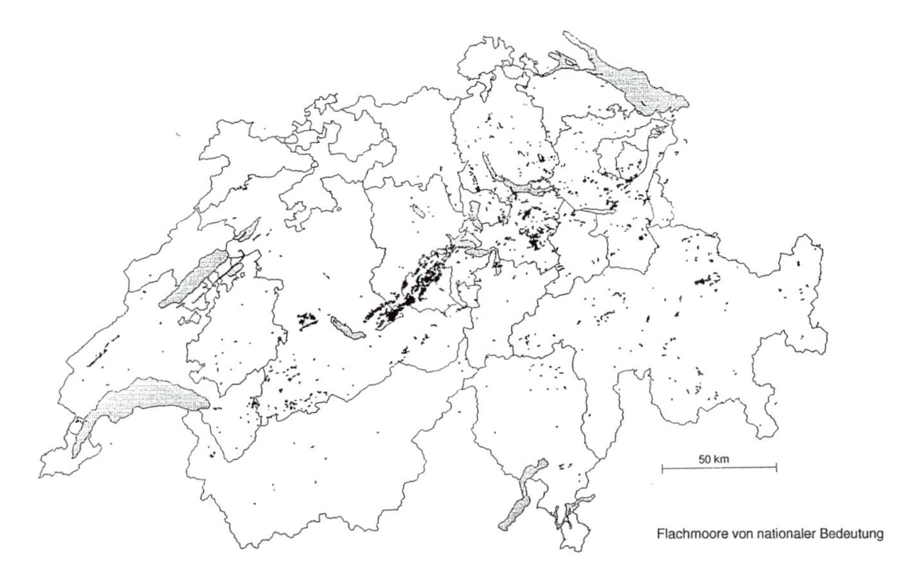

Fig. 23.2. Distribution of the fens of national importance, according to the national inventory (DFI, 1990).

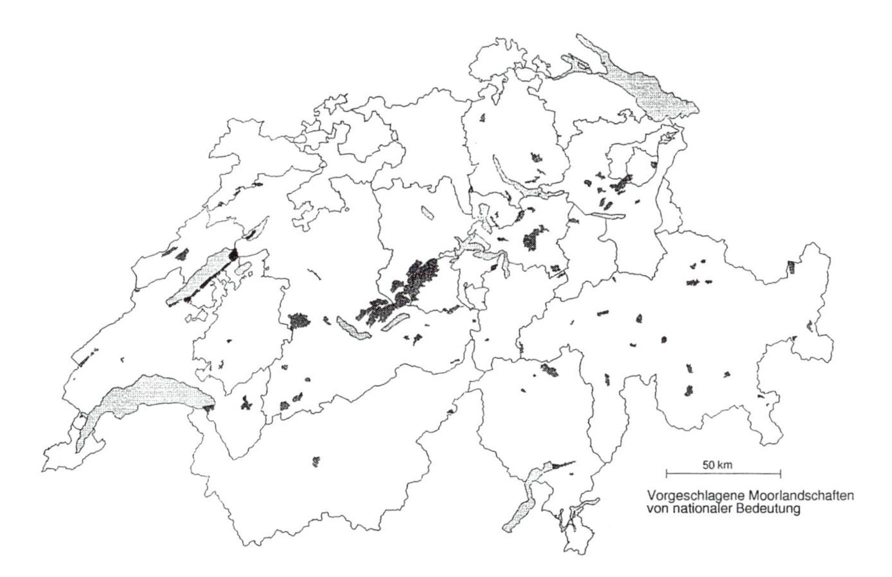

Fig. 23.3. Distribution and size of the proposed 91 mire landscapes of particular beauty and national importance (modified from Hintermann, 1992).

The first systematic survey of mires started in 1978, when Swiss non-governmental organisations initiated the Inventory of Raised Bogs and Transitional Mires with a time schedule of 8 months and a budget of SFr 50,000 (see Table 23.1).

Today, after almost 20 years of continuous work on mire survey and conservation and at a cost of several million Swiss Francs, the Federal Council has passed two Federal Decrees on Protection of Mire Habitats. They stipulate the integral conservation of 500 bog sites from the Federal Inventory covering a total area of 1,500 ha and a first series of 700 sites from the National Fen Inventory covering about 10,000 ha. With the enactment of the second and third series of the fen inventory (which will comprise more than 440 sites covering 9,200 ha), the targets in the field of mire habitat conservation will largely be met, at least as far as paperwork goes. Legislation on mire landscape protection will go on for some time. The corresponding scientific inventory contains 91 sites in total. As a result of local objections, a few of them will be deleted completely from the Federal Inventory, and most of the remaining sites will be reduced in size. However, Table 23.1 shows clearly, that:

1. the legislation process will be completed in 1996 (or 1997?);

2. the designation of fen sites and mire landscapes of national importance might also be concluded either this year or in 1997.

Table 23.1. Development of mire protection in Switzerland since 1978.

Year	Raised bogs	Fenlands	Mire landscapes
1978	Start of Survey		
1979			
1980			
1981			
1982			
1983*			
1984	End of Survey		
1985			
1986		Start of survey	
1987+,‡			
1988§			Start of survey
1989	Consultation Phase		
1990	Negotiations	End of survey	
1991	Enactment 1/2/91	Consultation phase	End of survey
1992		Negotiations	Consultation phase
1993		Negotiations	Negotiations
1994		Partial enactment 1/10/94, 1st part	Negotiations
1995¶		Partial enactment, 2nd part	Negotiations
1996		Negotiations	Enactment - 1/7/96
1997		Partial enactment, 3rd part	

* Text of the Rothenthurm Referendum submitted on 16/9/1983 with 160,293 signatures of Swiss people.
+ Bill for alteration of the Federal Act on Wildlife, Countryside and National Heritage Protection of 19/6/1987 enforcing the protection of habitats (e.g. mires, flood plains, semi-arid grasslands).
‡ Rothenthurm Referendum accepted by the Swiss population on 6/12/1987. 58% of the votes were in favour of a better mire conservation.
§ Enactment of the new Federal Act on Wildlife, Countryside and National Heritage Protection on 1/2/1988.
¶ Bill for alteration of the Federal Act on Wildlife, Countryside and National Heritage Protection of 24/3/1995 enforcing the protection of mire landscapes.

Implementation of the laws

The objectives for the protection of mire habitats are clearly stated in Art. 4 of the Federal Decrees on Protection of Mires (bogs and fens) of National Importance: '*The mires of outstanding beauty and national importance are to be preserved in their entirety. In mire areas already suffering disturbance, regeneration operations should be implemented where feasible. A particular objective is the protection and promotion of the endemic flora and fauna together with their ecological basis and the maintenance of the geomorphological peculiarity of the mires.'*

Due to the strict federal organisation of the country, it is now up to the 26 cantons to implement compulsory mire conservation measures and to devise ways which spell out the appropriate management methods to be adopted by landowners and land users. It will be interesting to see how far and how rapidly mire conservation will overcome the steps of Lorenz' epigrammatic sentence:

Said does not mean heard
Heard does not mean understood
Understood does not mean accepted
Accepted does not mean implemented
Implemented does not mean sustained
Konrad Lorenz (1903-1989), Ethnologist and Nobel Prize Winner

The Federal Government assists the cantons with annual subsidies of more than 20 million Swiss Francs. However, considering that the mire habitats are still dwindling in number and size, the question is whether these efforts are the most appropriate to meet the objectives of mire conservation and legislation. This question cannot be avoided given the high level of public and private spending involved.

Ecological monitoring - the art of turning vague data into precise statements

To answer the questions raised above, the Federal Office of Environment, Forests and Landscape commissioned in 1993, amongst others, the Advisory Service for Mire Conservation of the Swiss Federal Institute of Forest, Snow and Landscape Research to develop a sensitive nation-wide and long-term mire monitoring concept which should reveal, as early as possible, discrepancies between targets and real developments. It had to be a practicable instrument to provide those responsible with scientific results in order to assess and readapt, if necessary, their earlier policies.

The basic problem of any nature conservation monitoring scheme can be summed up with the question: 'How can you get precise information from vague estimates?' The following example by Regez and Tobler (1977) clearly illustrates that qualitative data can contain much more precise information than one would expect at first glance.

Given the task of locating all the capitals of the 26 Swiss cantons on a map, let us consider what kind of information is needed to solve the problem. If we had any quantitative data, such as precise co-ordinates, the capitals could be located without any problems. However, in this case we limit ourselves to qualitative data instead of precise co-ordinates. We simply consult four persons, each representing one of the four linguistic regions of the country, e.g. a German speaking inhabitant from Basle, one from Geneva where French is spoken, a third from Bellinzona representing the Italian speaking part of the country and a fourth from Chur belonging to the Romansh-speaking population of that canton. Each has to answer the following question: *'How far is it from your town to each of the other 25 cantonal capitals. Please do not indicate the distances in kilometres, but simply make a table of the capitals ranked according to their distance from your respective home-town.'* The person from Geneva will start with Lausanne and end up with Chur. ... Applying a proximity analysis we get the following map (Fig. 23.4).

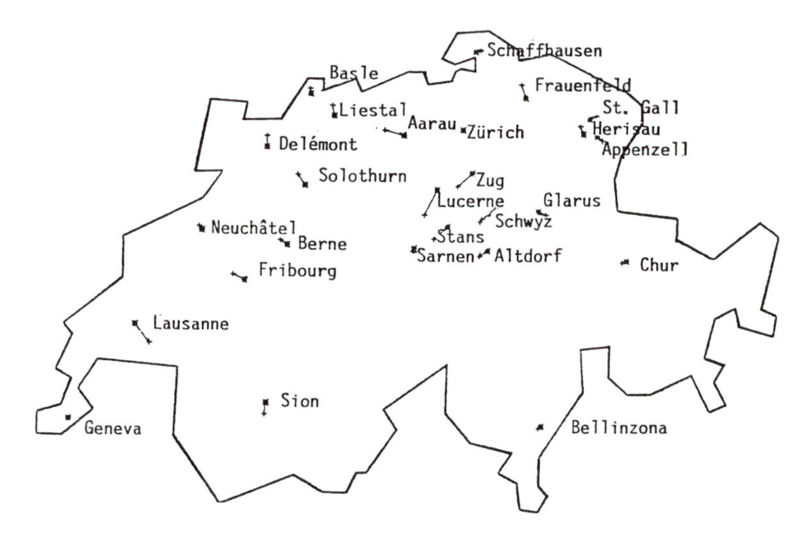

Fig. 23.4. Location of the 26 Capitals of the Swiss cantons by means of 25 distance ranks by four people. The real location of each cantonal capital is marked by a cross, whereas the asterisk indicates the locations calculated by the proximity analysis (from Regez and Tobler, 1977).

If we repeat our experiment with 26 inhabitants coming from the 26 capitals the resultant map is almost correct (Fig. 23.5). So the question arises, which would be the optimal or minimal effort - e.g. number of persons to be asked - to achieve an acceptable result in statistical terms. This example shows that it is not impossible to get accurate descriptions from vague data if they are processed adequately. However, the description of protected sites across a whole country is a more complex task than the cited example.

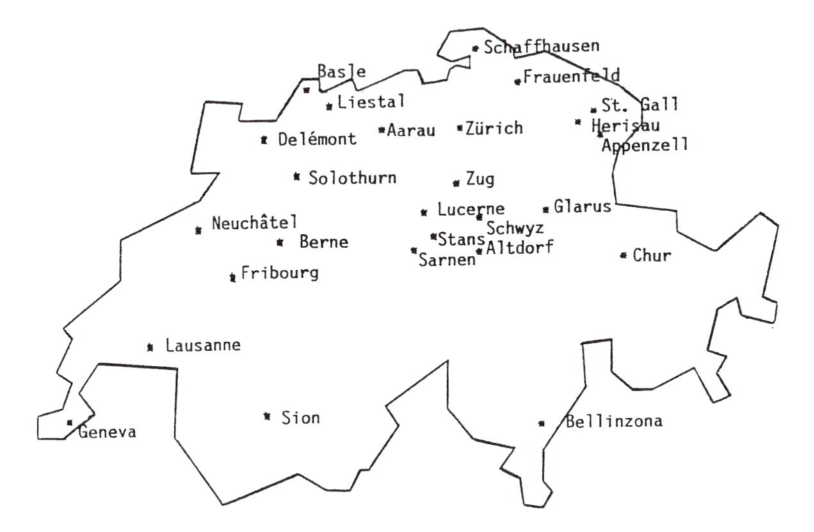

Fig. 23.5. Location of the 26 Capitals of the Swiss cantons by means of 25 distance ranks by 26 people. The real location of each cantonal capital is marked by a cross, whereas the asterisk indicates the locations calculated by the proximity analysis (from Regez and Tobler, 1977).

A monitoring scheme to detect changes in ecological systems must be particularly sophisticated with respect to:

1. the definition of aims and targets;
2. the selection of representative sites to allow inter-regional comparisons;
3. the selection of the variables which should be optimally sensitive;
4. the choice of the methods which should be practical, cost-effective and robust for both data gathering and data analysis;
5. criteria for significance;
6. flexibility for modifications.

This implies that when we design a monitoring scheme we must not see it only from the top down - e.g. mire quality targets supported by protection strategies and management objectives, which in turn imply monitoring objectives, which finally suggest monitoring activities - but also from the bottom up: which means starting from the description of a mire in the field e.g. by means of recording vegetation data on permanent plots, and continuing with surveying management activities, data storage, and statistical data analysis. Finally, to make sure that those in control receive the information they need, reports must be written and submitted to the federal government (being the employer) and cantonal authorities, who in turn make management decisions. These decisions will effect mire habitats and also future monitoring activities.

Facing the tremendous complexity of (geo-)biological systems (e.g. mires), we all recognise that peatland monitoring is a pioneering and presently inaccurate science; some would say an art. Therefore, monitoring of mire sites in Switzerland will remain - at least for the foreseeable future - the art of turning vague data into precise statements.

Monitoring the Swiss mire habitats

In a preliminary project, several methods were evaluated and tested for monitoring purposes, in particular for the observation of changes in the vegetation of mire habitats (Fig. 23.6).

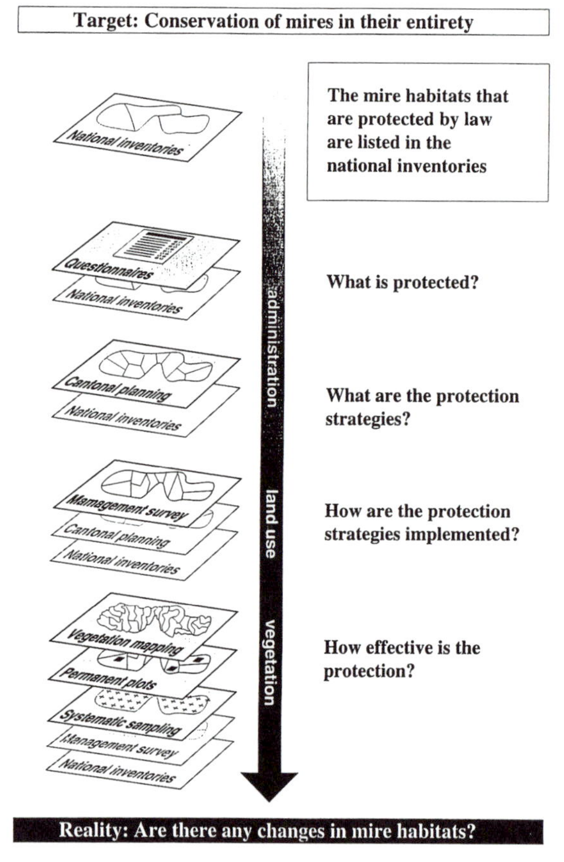

Fig. 23.6. Diagram of the proposed Swiss mire monitoring programme. Beginning with a screening of the administrative part of mire conservation, the programme finally focuses on possible changes in the composition of both vegetation types and species diversity. On the way from top to bottom, the precision of data collection increases whereas the number of monitored sites and surfaces (permanent plots) decreases.

As stated before, the mire monitoring concept should not be confined to the observation of the development of the mire vegetation alone. It should extend to all the planning and implementation procedures of the authorities (administration) and land users. The target of preserving mire habitats of national importance in their entirety, and questions about the effectiveness of protection measures, imply a quantitative and a qualitative aspect of the desired, as well as the undesired, changes in the mire vegetation. An effective monitoring scheme must therefore be designed to detect both types of change as early, and as accurately, as possible.

Does quantity matter? Is monitoring boundaries worthwhile?

Given that the overall perimeter length of protected Swiss mire habitats is about 6,000 km, boundary monitoring is a valuable exercise. If we assume that these boundaries can be fixed with a given precision of ±1 m in the field, quite a precise estimation, we will get an undefined area of about 1200 ha, which is more than 5% of the protected mire habitat surface!!

Further requirements are:

Establishing statistical design criteria

The 'population' (e.g. mire vegetation or parts of it) to be sampled has to be characterised in statistical terms.

Designing monitoring schemes

The monitoring programme must also be designed to ensure efficient answers to questions such as: Where to sample? What to measure? How often to sample?

Minimising expenditure

Cost effectiveness is essential to maintain a long-term monitoring scheme.

Consistency

The mire monitoring system must be consistent with the information objectives. This consistency can be achieved through complete designs with thorough documentation.

Reproducibility and accuracy (resolution)

Boundaries of protected mire habitats and vegetation units have to be delimited and identified with optimal precision.

Sensitivity and ease of rating

A set of parameters has to be established which enables the easy qualitative rating of vegetation changes from the point of view of mire conservation.

There is insufficient funding to conduct systematic monitoring of all 1600 mire habitats of national importance. Therefore, the monitoring programme has to deal with a statistically representative sample of mire sites. The methods of mire site selection and description of vegetation in the field are still being debated by the employer and the Advisory Centre for Mire Conservation.

To solve the issue, we are currently testing the advantages and disadvantages of:

1. Vegetation mapping by means of large scale CIR aerial photographs. The main question is which is the most appropriate resolution or minimum area of the individual vegetation unit that can be recognised and monitored.
2. Using permanent plots. The main problems with permanent plots are the stratification of the mire sites, the subsequent distribution of the plots and the choice of their optimal form and size.
3. Systematic sampling. This implies locating sampling points at regular intervals. Using this method the size of the sampling interval is extremely important, and also the impact of the size of the quadrats can play an important role.

To avoid the consequences of too radical abstraction (Fig. 23.7) and to provide an objective picture of possible changes in mire vegetation on a national level, a combination of several methods is likely to produce the best results in terms of accuracy and cost. At a future meeting, it should be possible to report with which degree of abstraction the Swiss mire monitoring programme has been realised.

Fig. 23.7. Einstein simplified: Consequences of too radical abstraction (Cartoon by Sidney Harris from 'Cartoons on Science').

Acknowledgements

I would like to express my sincere thanks to Peter Longatti and John Innes (WSL/FNP) for their helpful and critical remarks on the manuscript and the linguistic usage. Peter Longatti also prepared some of the figures. Assistance from Margrit von Euw with the manuscript was also much appreciated.

Part Seven

Commercial Uses of Bogs

An Introduction

Dr Rob Stoneman
Scottish Wildlife Trust, UK.

A tradition of hand-cutting for fuel, a desire to change 'unproductive' bogland to pastoral or arable land, widespread bog afforestation and peat extraction for horticulture has brought, for example, the UK's lowland bog resource to edge of annihilation (see Stoneman, Chapter 45, this volume). So why, you may ask, did the Scottish Wildlife Trust, arch peatland conservationists, invite on to the podium at the Peatlands Convention, representatives of the foresters and the peat extractors who have destroyed such beautiful habitat? Of course, the story is more complicated. As economic and socio-environmental concerns alter, in the UK at least, agricultural reclamation and forestry are *not,* at the present day, large threats to peatlands. Peat extraction for horticulture remains as a major concern as the DoE (1994c) appears to have given the green light to a further 1000 ha of lowland bog destruction. In other countries, the threats are different (see Joosten, Chapter 46, this volume for example).

Furthermore, despite a gradual shift of land ownership and/or control from peatland 'exploiters' to peatland conservationists, the vast majority of peatland still lies with those who have the ability, and sometimes the desire, to continue the degradation of our (semi-) natural peatland resource. Thankfully, as the debate continues, and the conservation significance of peatlands is more and more realised, these very people have taken the initiative to conserve at least some of the resource under their control.

This makes sense, as the 'exploiters' not only control the land but have the expertise to conserve it. The drains at Fenns and Whixhall Mosses were blocked by the operators who originally dug them, albeit under the auspices of English Nature rather than the peat cutting company. All three chapters in this section show that the term 'exploiters' is rather misleading; rather, these sectors can play a positive and constructive role in the conservation of peatlands.

This role falls into two areas. Firstly, their roles relate to the post-'industrial' use of the sites. Whilst extracted bogs can never be restored to their former glory, valuable wildlife habitat can be created. If the final phases of extraction are planned in such a way as to help this process, the cost of works are reduced dramatically. Further, the extractor's machinery can be used during the rehabilitation phase, again reducing costs and utilising extractor's expertise. The scale of these types of operations are significant. Bord na Móna control 83,500 ha of peatlands, of which 30% may be rehabilitated towards wetland (McNally) - some pretty big wetlands. Similarly, the impact of rehabilitating bogs from the effects of forestry could dramatically expand the raised bog resource in Scotland (see Brooks and Stoneman, Chapter 33, this volume). Anderson details the role of the Forestry Commission in this area.

The second area in which 'exploiters' can be involved is straight conservation. In Ireland, McNally notes the large areas of 'intact' bog which have been handed over, or sold, to the nation for conservation: a process paralleled in England (Temple-Heald and Shaw).

Particularly productive have been the development of partnerships between conservationists and these sectors. Agreements such as the English Nature - Levingtons 'deal' can be criticised (is it appropriate for peat extraction to continue on highly important sites such as Wedholme Flow?) but they have paved the way for large areas of land to come under conservation management.

Developing these partnerships is an on-going process. Describing peat extraction on lowland bogs as sustainable (Temple-Heald and Shaw) is clearly at odds with the views of the Scottish Wildlife Trust for example (see Stoneman, Chapter 45, this volume). It is particularly useful then, that the Edinburgh Declaration, which arose from the Peatlands Convention, has led to the creation of a forum between conservationists and extractors. Clearly, the inclusion of all interests in this debate can only be for the good of peatlands.

Chapter twenty-four

Peat Production and Conservation

Mr Nick Temple-Heald and Mr Alan Shaw
UK Peat Producers Association.

Introduction

As a consequence of the 'anti-peat' publicity campaign in the UK, which started in 1989, the role played by peat producers in balancing conservation and peat extraction has become increasingly important. This chapter reviews the positive conservation progress made by the peat production industry, within the context of government policy and the needs of the horticulture industry.

Before considering conservation progress amongst peat producers, it is useful to understand the influence of government policy:

'*It is government policy to maintain and encourage a competitive UK horticultural industry, while the general gardening public and the landscaping industry require a range of suitable, technically sound products at competitive prices. The government believes that there continue to be market demands for peat which should, in part, continue to be met by peat extraction from sites in Great Britain*' (DoE, 1995d).

Clearly, with an annual output of £1.4 billion (CSO, 1995), horticultural output is dependent on good quality growing media supplied from indigenous peat sources, which must continue if a healthy and competitive industry is to be maintained.

In volume terms, the UK market demand for peat stands at some 2.5 million m^3 annually (about 1% of global peat consumption), of which approximately 800,000 to one million m^3 is imported (Robertson, 1993).

It is the task of the peat producers to balance the demonstrable need for peat against the legitimate requirements of conservation. Having verified the need for peat, the other side of the equation - the need for peatland conservation in the UK - requires examination and definition.

Assessing the need and means for peatland conservation

Firstly, it is desirable to establish which elements are *not* relevant to the need for peatland conservation. This reflects the early days of the 'anti-peat' campaign in the UK, when environmental accusations against peat circulated which did not, and do not, stand up to analysis.

Firstly, horticultural peat extraction cannot seriously be attacked on the basis of being a significant greenhouse gas contributor. The total greenhouse gas contribution of UK peat extraction has been estimated at just 100,000 tonnes of carbon. This constitutes only 0.06% of the total UK emissions, with the clear implication that peat extraction cannot rationally be challenged on the basis of CO_2 emissions.

Secondly, it has been argued that peat should be eliminated in favour of alternatives based on waste materials from other industries, or municipal waste, in order to reduce landfill requirements. However all 2.5 million m^3 of the UK's peat consumption is the equivalent of less than 2% of the its compostable waste landfill requirements - as such the impact of such actions would be irrelevant.

Thirdly, the implication was made that peat extraction threatened the last vestiges of peatland of conservation value, and that soon it would all be consumed. On a global scale, this is manifestly untrue. Peatland is neither rare nor threatened. There are some 400 million ha of peatland (Bather and Miller, 1991) from which mankind extracts about 200 million m^3 $year^{-1}$. The earth generates around 600 million m^3 of peat $year^{-1}$.

In the UK, however, land use and species conservation is an important and legitimate concern, and peat extraction must be rigorously examined from this point of view.

Again, peatland is hardly rare in the UK - some 1.6 million ha exist of which about 70,000 ha are of the raised mire type (it is acknowledged that there exists considerable debate over the exact definition and terminology of these peatland types, but even under the most restrictive of interpretations, there are at least 70,000 ha in Great Britain). However most of these sites have been extensively manipulated by man for centuries in the interests of agriculture, forestry and, a long way third, for peat extraction. The result of all this activity has been to reduce the area of raised mire that can be described as 'largely intact' in Great Britain to, at the worst estimate, 8,800 ha (DoE, 1995d). This land is safe from any threat of peat extraction; in November 1990, the Peat Producers Association (PPA) agreed not to seek to extract such areas, and recent guidelines to planners (DoE, 1995d) formally reinforces this protection. Thus some 60,000 ha of raised mire exists which is considered to be of lesser conservation value. The entire horticultural peat industry in Great Britain utilises a mere 5,000 ha in total, and will only require a further 1,000 ha over the next quarter of a century at current trends i.e. less than 10% of a land type classified as 'of little or no conservation value'.

The definition of sustainability most commonly used is: '*the provision of resources to satisfy the legitimate needs of today's generation without jeopardising the needs of future generations*'. Even the most pessimistic view of the available data shows peat to be the UK's most sustainable mineral. To put this into further perspective, one company alone has already donated over 1,500 ha of vegetated peatland to immediate nature conservation, while other companies have made numerous additional donations and sales.

Alternatives to peat

Turning to peat alternatives, the consumer market for peat-free growing media has reduced over the past two years from a peak of 5% in 1993 to about 3% today. This is due to the failure of all potential products to fulfil the four key criteria of efficacy, reliability, safety and value for money on a truly commercially viable basis.

There is a flourishing market for bark based products intended for mulching and soil conditioning - some 75% of the market - and the majority of these products are marketed by peat producers.

Conclusion

The PPA are fully committed to the balanced management of peatland resources. The PPA supported and encouraged the protection for posterity of peatland areas of significant conservation value; these areas are now under no threat of extraction - either now or in the future.

PPA members supported and encouraged the development of planning conditions which place conservation provision as a priority consideration where areas are worked. On many of the larger sites, every square metre has a guaranteed future in conservation because they have been donated to English Nature or are subject to legally binding agreements with County Councils.

PPA members are committing substantial resources to after-use planning, regeneration research and archaeological projects, and are successfully pioneering techniques of protecting conservation areas from the effects of adjacent working.

PPA members have undertaken never to seek to extract peat without first seeking to work in association with English Nature, or other relevant statutory bodies, to ensure that legitimate conservation interests are not threatened.

We believe that this is the only sustainable way forward in the 1990s.

Chapter twenty-five:

Forestry and Peatlands

Mr A. Russell Anderson
Forestry Commission, UK.

Introduction

Any peat deposit more than 45 cm deep has been classified by foresters as 'deep peat' because most of the tree roots remain in the peat and the mineral substrate has little direct influence. Where peat is less than 45 cm thick, the mineral soil below can influence tree growth, these soils (peaty gleys, peaty podzols and peaty ironpan) are termed peat mineral soils, or shallow peats. This chapter is only concerned with deep peat as defined above. It is accepted that for inventory of deep peat areas a >1 m thickness definition is more useful, since geological survey maps of Great Britain have demarcated areas of peat over 1 m deep, and this information has been used initially in the National Peatland Resource Inventory. In the UK, forestry is an important land use on peatland and peat is an important soil type for forestry (Table 25.1). This chapter considers trees growing naturally on peat, and the development of peatland silviculture as a background to the environmental issues surrounding peatland forestry.

Table 25.1. The area of forests on deep peat show the importance of forestry as a land use on peatlands and the importance of deep peat as a soil type for forests.

	Area of afforested peat (> 45 cm thick)		
	(km^2)*	% of total peat area*	% of total forest area+
Scotland	1629	9	14
England	176	6	2
Wales	91	13	4
Great Britain	1896	9	8

* Figures given by Cannell, Dewar and Pyatt, 1993.
+Data on total forest areas in Forestry Commission, 1994.

Trees on peatlands - palaeoecological evidence

It has long been known that trees grew on deep peat during at least two periods since the last Ice Age (Fig. 25.1). Birch and pine stumps in blanket bog in the Eastern Highlands and Southern Uplands were first studied by Francis J. Lewis (1905; 1906a,b; 1907; 1911). His interpretation was that the forests had grown during interglacial periods and the peat had accumulated during glacial periods. The differing view of Samuelsson (1910), that the trees grew during the drier Boreal (9500-7500 BP) and Sub-boreal (5000-2500 BP) periods came to be widely accepted (e.g. Tansley, 1939). This fitted in with the climatic scheme suggested by Blytt (1882) and Sernander (1908) for Norway and Sweden. More recently, however, a re-examination using pollen analysis and radiocarbon dating (Birks, 1975) has placed doubt on this interpretation of the pine stumps. Hilary Birks found that pine remains from the sites studied by Lewis were most scattered over the period 7500-5000 BP, and concluded that they did not support the Blytt and Sernander climatic scheme.

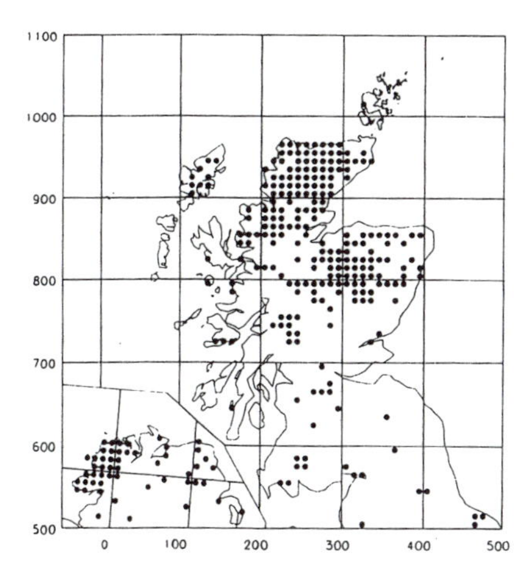

Fig. 25.1. The distribution of pine macrofossils in peat (Bennett, 1994). These show that trees have grown naturally on peats at certain times in the past.

It has been recognised by Birks (1975) and others that conditions suitable for pine growth on peat are also suitable for decomposition of stumps, so that a layer of stumps would only have been preserved when a sudden change from a dried out bog surface to wetter, anaerobic, surface conditions had occurred, rather than throughout the period of forest growth. Radiocarbon dating of 24 pine stumps preserved in blanket peat at Rannoch Moor (Bridge *et al.*, 1990) revealed two close groupings of dates (6220-6040 BP and 5200-5050 BP), each accounting for about a quarter of the trees dated, the remaining trees dating from before, between and after the groupings. These date groupings did not coincide with the main periods of pine growth indicated by the pollen record in peat cores from the sites, but fell within the period of low pine pollen percentages (6400-4970 BP). This led the authors to postulate that peaks in pine pollen represent times when conditions were favourable for pine growth whereas preserved stumps are those of trees which died during periods of rapid peat accumulation, i.e. in an environment unfavourable to their growth. This was supported by tree rings on 131 stumps from one of the Rannoch Moor sites, which showed that the trees whose stumps had been preserved had mostly been growing in difficult conditions. However the authors could not find further supporting evidence when they examined peat accumulation rates in dated cores from the sites; periods of rapid accumulation did not coincide with the dated stumps. Pearsall (1950) mentioned that birch and pine remains in deep blanket peat were found particularly where the topography provided better local drainage, such as along stream lines. Could damming of streams by the European beaver (*Castor fiber*) have been responsible for sudden rewetting of peat near streams and consequent preservation of stumps? It is surprising that there is not more evidence if that was the case, but perhaps it has never been sought.

Studies in Ireland (e.g. Doyle *et al.*, 1989) have shown that, between 4300 and 3200 BP, the Midlands raised bogs supported pine forests for short periods (up to 500 years), but that these forests existed at different times on the different bogs. In two out of the three bogs studied, fire destroyed the forest and in one of those it also initiated the growth of the forest. Stumps of individual pine and willow trees, as well as what appear to have been extensive pine forests, growing on blanket bog (Wilkins, 1984) have been found on the Hebridean island of Lewis, where previous examination of pollen stratigraphy had shown such low pine pollen percentages that this pollen was assumed to have come from the mainland. Likewise, pine stumps in Caithness and Sutherland, in the far north of Scotland, have shown that pine forests on blanket bog existed at least for a short period about 4000 years ago, right to the north coast, an area where conventional pollen evidence of them is lacking (Gear and Huntley, 1991). The East Anglian fens are also said to have been invaded by trees, especially pines, during the Sub-boreal period (Tansley, 1949).

The reasons why forests started and stopped growing on peatlands are complex, but it is certain that they can grow on them naturally and it is distinctly possible that bogs in Britain have supported trees until relatively recent times, but have only occasionally preserved their remains.

Present day tree colonisation

Today, self-seeded trees on peat are not unusual. Birks (1975) mentioned the pine-grown bogs of Scandinavia and listed some examples from Britain.

Ratcliffe (1964) considered the treeless nature of most Scottish peatland below the tree limit to be unnatural. He suggested that regular burning and grazing were maintaining this state, which had been caused by forest clearance, and that before widespread human disturbance, the edges of most bogs and the surfaces of the drier ones were at least partly wooded. The growth of pine trees on many peat-filled hollows within the native pinewood of Abernethy Forest provided evidence for this. In the case of raised bogs, the height to which the bog surface can grow above the level of its base is limited by the excess of rainfall over evaporation as well as by its basal area and hydraulic conductivity (Ingram, 1982). If a climate change were to occur, such that the excess of rainfall over evaporation was reduced, the theoretical maximum height of the groundwater mound would also be reduced, and a raised bog which was already close to its maximum height would suffer a drying out of its surface. Thus a raised bog can reach a state of climatic climax after which even small changes in climate could cause sufficient surface drying for natural tree establishment.

Not all examples of peatlands at present colonised by trees indicate **natural** conditions suitable for tree growth. In some cases, particularly on raised bogs, the surfaces have been drained by man, usually to dry them for sheep grazing, or to encourage heather growth for grouse, and it is this drainage which has allowed woodland or scrub to become established where grazing pressure has allowed. Thus man, in historic times, has prevented regeneration of trees on some peatlands cleared by his prehistoric ancestors, but has unintentionally aided tree colonisation of others. In this section and the last, we have concentrated on peatlands where trees are growing, or have grown in the past, but it is important to remember that the vast majority of Britain's deep peat area is treeless and some of it may have always been so, except during the early stages of accumulation.

The development of tree planting on peatlands in Britain

It is easier to find record of attempts to plant trees on peat after the formation of the Forestry Commission in 1919 than before. Earlier attempts were made by private estates and records were either not kept or not published. However a few accounts did find their way into print. In *A History of Peebleshire*, William Chambers (1864) mentions an attempt in 1730 to drain and plant trees on Blair Bog, 'a bleak heathy waste', 270 m a.s.l., 'destitute of shelter' and 'as unpromising as could possibly be imagined'. Archibald, third Duke of Argyll, bought the land, as part of the Whim Estate, near Lamancha 'for the sake of its shootings and also to amuse himself by an attempt at improvement'. He had difficulty in maintaining the deep open drains, because 'the soft moss crumbled down and filled up the cuttings' and many of the trees, which were planted in belts and clumps on the moss, perished. In the 1790s turf planting was used on thin flushed peat in Co. Tyrone (McEvoy, 1802). Cutting out and overturning a square of turf served 'to increase the depth of soil ... and also to drain and render it wholesome for the reception of plants'. A wedge of peaty land beside Loch Tulla, Argyllshire, was planted in 1858 and this forest was described 70 years later by J.A.B Macdonald (1928). Sitka spruce was first planted on peat at Durris, Kincardineshire, in the 1880s (Crozier, 1910). Sir John Stirling-Maxwell began planting up the blanket bog of his estate at Corrour in 1892, and by 1907, he appreciated the need to drain (Stirling-Maxwell, 1907). He advocated the use of the 'Belgian system' of digging a ditch, spreading out the upturned turfs, and planting Sitka spruce (*Picea sitchensis*) or Scots pine (*Pinus sylvestris*) trees on top (Fig. 25.2). The Belgian system became standard forestry practice and was used until spaced furrow ploughing took over in the 1950s, but another of Stirling-Maxwell's innovations for peatland planting was not so widely accepted. Turf nurseries (Stirling-Maxwell, 1936) were made by overturning blocks of peat in continuous rows, leaving just enough ground between the trenches to carry the line of upturned turves (Fig. 25.3). The seedlings were planted on these turves and, after two growing seasons, hand barrows were used to lay out the turves at the required spacing on the surrounding land. The advantages were that the turves were much lighter to handle after two summers above ground, and that tree growth was not interrupted because the roots were still within the turf. Unlike the Belgian system, there was no drainage of the land being planted up, except within the turf nursery itself, so it was not suitable for planting extensive areas of peat. Many other accounts of planting on peat must exist, but it is not necessary to read them all to realise that it was the better quality peats on which the early planters were successful, not the nutritionally poor ones (Zehetmayr, 1954).

'My object has been to commend this fascinating line of study to all those who desire to see our bare glens re-peopled with trees and busy with the healthiest of industries.' Thus Stirling-Maxwell (1907) ended his article on the planting of high moorlands and from its earliest days the Forestry Commission research branch experimented with methods of afforesting peatland. In 1920 it devised a classification of peatland types, based on vegetation, for Inverliever Forest, Argyll, crown land where tree planting had begun in 1909. It contained six classes, of which two were doubtfully plantable and two were unplantable (Zehetmayr, 1954). Research on peatland site types, ground preparation methods, fertiliser needs, planting methods, species choice, species mixtures and latitudinal planting limits, were described by Zehetmayr (1954). His summary included planting prescriptions for different peatland site types. By then, only *Sphagnum*-dominated sites were considered unplantable.

During the 1960s, tree establishment experiments were set up on the nutritionally poor *Sphagnum*-dominated raised bogs, such as Lochar Moss and Flanders Moss and on very wet areas of deep blanket peat. The advent of low ground pressure tractors capable of ploughing such sites, together with an improved understanding of fertiliser demands, and the use of lodgepole pine (*Pinus contorta*), had allowed this group of peat types to be afforested successfully.

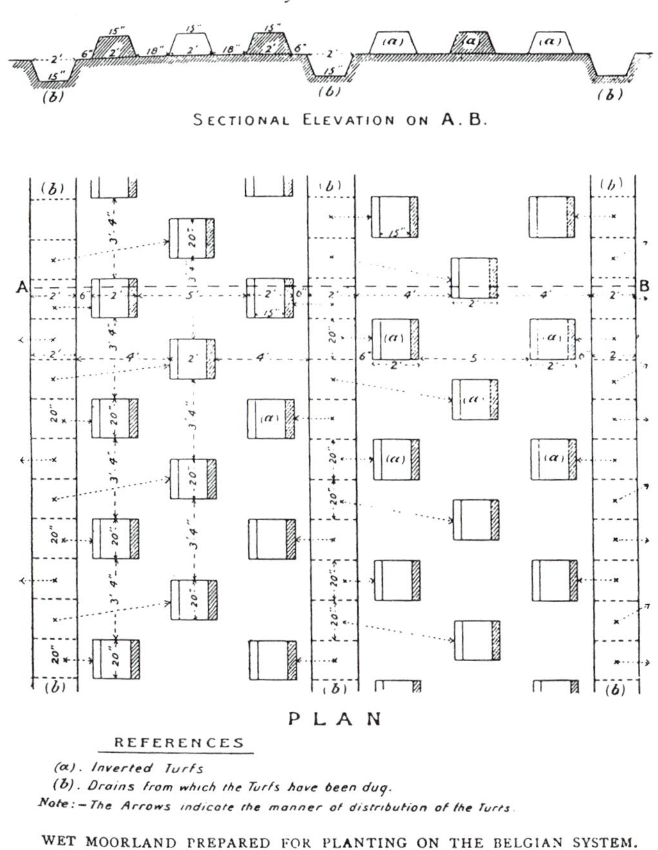

Fig. 25.2. The 'Belgian system' of turf planting used to plant trees on blanket peat by Sir John Stirling-Maxwell at Corrour in 1907. The method became normal practice on all wet soils until spaced furrow ploughing took over in the 1950s.

Summarising recent research results up to 1988, Pyatt (1990) concluded that they provided grounds for optimism over the future of forests on peatlands. The tree growth rate varied with the species and seed origin used, the site, and the fertiliser regime. At one extreme, fast growth of lodgepole pine and Sitka spruce had been achieved at Lochar Moss, a very infertile *Sphagnum* bog near Dumfries, although the spruce had needed repeated N, P and K fertiliser applications. More modest, but acceptable, growth rates were achieved in the colder climate of the north of Scotland. Growth rates and fertiliser requirements of first rotation plantations of lodgepole pine and Sitka spruce were now known, and the use of self-thinning mixtures of these two species, with the spruce forming the majority of the final crop, was recommended. The earlier practice of using the fastest growing, south coastal, provenances of lodgepole pine (i.e. from the coast range of Northern California, Oregon and Washington), had proved to be a mistake because these trees seldom grow straight enough to provide valuable saw-milling timber and are very susceptible to damage by snow. However, provenances which combine fast growth and straighter stems were identified in a provenance experiment on Lochar Moss. Observations of peat cracking and large buttress roots crossing plough furrows in a 40-year-old Sitka spruce stand at Rumster, Caithness and in 60-year-old stands of both lodgepole pine and Sitka spruce on the Lon Mor, Inverness-shire, suggested that trees on pseudofibrous peats would be more windfirm than on gleyed mineral soils. Young second rotation Sitka spruce plantations on the Lon Mor and at Rimsdale, Sutherland were growing very fast, even without fertiliser, indicating that second rotation forests on peatlands could be expected to give higher yields with less fertiliser. There was no doubt that the development on peatland forestry in Britain for wood production was a silvicultural success story.

Fig. 25.3. A turf nursery on blanket peat at Corrour. Seedlings were planted on the closely spaced turves and after two summers the turves were spread out over the surrounding area using hand barrows. The system was not widely used.

Growing awareness of the need for peatland conservation

The Flow Country controversy of the 1980s brought to the fore the issue of the afforestation of blanket bog in the far north of Scotland. The forestry industry, which was only able to obtain land with little agricultural value (Zehetmayr, 1953), had, since the 1950s, been planting trees on the moorlands of Caithness and Sutherland. By the 1980s, the rate of afforestation in the Flow Country, a large expanse of deep blanket bog in these counties, was worrying the Nature Conservancy Council (NCC) and conservation bodies such as the Royal Society for the Protection of Birds (RSPB). They saw that the rate of loss of blanket bog habitat in the area during the 1980-85 period presented a risk to internationally important populations of breeding birds, especially greenshank (*Tringa nebularia*), dunlin (*Calidris alpina*) and golden plover (*Pluvialis apricaria*) (RSPB, 1985). The NCC, who had begun a programme of detailed survey of the Flow Country in the late 1970s, pointed out the international rarity of blanket bogs (Lindsay *et al*., 1988). In attempting to conserve examples of the various types within the complex, the NCC joined with the RSPB in calling for an embargo on further forestry planting in the Flow Country. It was not until the publication of a land use strategy report by the Caithness and Sutherland Working Party, convened by Highland Regional Council (Caithness and Sutherland HRC Working Party, 1989), that the issue, which had received much attention in the media, was resolved. The report contained an Indicative Forestry Strategy for the two counties, a planning document which mapped the region into four forestry policy areas (unsuitable, undesirable, possible or preferable) and which has been used as an early model for strategies in the other Scottish regions.

Concern over the loss of peatland habitats, particularly raised bogs, to agriculture, peat extraction, open-cast mining and forestry has increased in the 1980s and early 1990s. Plantlife's Commission of Enquiry into Peat and Peatlands (Plantlife, 1992) aired the views of many different people and organisations with an interest in peatlands. A case was made by the RSPB and Plantlife for conservation of lowland raised bogs, with a list of suggested action points including an end to tree planting on raised bogs and adoption of raised bog rehabilitation policies (RSPB and Plantlife, 1993). The Scottish Raised Bog Conservation Project is another important example of the growing interest in conservation of peatland habitats, which has raised awareness among foresters and encouraged discussion of the related issues. The EC Habitats Directive (92/43/EEC) has identified active raised bog, active blanket bog, and bog woodland as priority habitat types for conservation, within a European context. Those recognised as needing conservation action, but with lower priority, include degraded raised bog and blanket bogs still capable of natural regeneration, and transition mires or quaking bogs. Special Areas of Conservation for the priority types have recently been proposed by the UK government. Measures to bring habitats into 'favourable conservation status', as required by the Directive, have still to be detailed. Most recently, as part of the UK Action Plan for Biodiversity, habitat action plans for certain habitats, including active lowland raised bogs and active blanket bogs, have been drafted by a group of voluntary conservation organisations (Biodiversity Challenge Group, 1995) and the Department of the Environment is consulting widely with a view to adopting agreed versions of these plans as government policy.

Possible contribution to an enhanced greenhouse effect

There are concerns that afforestation of deep peat soils may affect atmospheric concentrations of carbon dioxide, methane and nitrous oxide, some of the so-called 'greenhouse gases' which drive global climatic change. There is no doubt that afforestation of peatland results in major changes to the peat, through drying with its associated shrinkage, cracking and aeration, caused by the initial drainage and then by tree growth (Pyatt and John, 1989; Anderson *et al.*, 1995). It is also clear that decomposition of peat in the upper layer which becomes aerated after afforestation, can lead to a release of carbon dioxide into the atmosphere. The extent to which this is counter-balanced by the trees taking in carbon dioxide for photosynthesis, and by a reduction in the emission of methane to the atmosphere, is not yet clear.

The UK government has committed itself, as a result of the Climate Change Convention at the Rio Earth Summit, to securing an annual increase in the net amount of carbon collected by sinks in the UK and to reducing emissions of greenhouse gases (UK Government, 1994). At present, guidance on forestry practice on peatland, to mitigate the greenhouse effect, cannot be definitive.

The future of forests on peatlands

The UK Government recognises the value of the various peatland habitats for nature conservation, as components of the landscape, as carbon sinks, and as sources of archaeological, palaeoecological and palaeoclimatic information. These values are carefully considered, together with advice from other interests, when the FC makes decisions affecting forests on peat, including new planting and replanting of existing woodland.

UK Government policy now aims to encourage more new planting on better quality agricultural land, partly to avoid conflicts with conservation of valuable semi-natural habitats. Very little planting is currently being done on deep peat, but there is some planting on shallow peat on agricultural land, and some planting of native woodland on shallow peat in the north of Scotland. The FC, through the application of its own guidelines (e.g. Forestry Commission, 1990; 1993; 1995) and the consultations which occur over both private and Forest Enterprise planting, seek to avoid planting areas which are recognised as having high conservation or other environmental value as open habitats. Active raised bogs, for example, are highlighted as important habitats in the consultation process, indeed many of them are Sites of Special Scientific Interest (SSSIs). The FC makes compliance with the guidelines a condition for the award of grants under the Woodland Grant Scheme. Forest Enterprise planting does not receive grants but it also has to meet the FC Guidelines standards. The Forest Nature Conservation Guidelines recommend that 10-20% of the total area of a woodland should normally be open ground, and that this should be organised so that it yields the maximum environmental dividends. Small areas of bog and other valuable wetland habitat are normally selected for retention as open ground on the advice of Forestry Authority staff and consultees.

When parts of existing forests are felled, there is a requirement to redesign open ground areas to achieve greater environmental benefits. This gives opportunities for some formerly planted peatland areas to be left unplanted, if they are potentially valuable for nature conservation. The Nature Conservation Guidelines also encourage forest managers to create transition zones between open and wooden ground to produce a more structurally diverse edge. On peatland sites, there may be potential to mimic the transition found on bogs in natural forests, with closed canopy forest around the outside graduating to a more open woodland of smaller trees, then to open ground with a few stunted trees, or to treeless bog in the central, wetter area. This type of habitat is rare in the UK, but exists in remnants of semi-natural woodland and is an integral part of the boreal forests of Scandinavia, Russia and North America. The FC encourages the creation of some new native woodlands on shallow peats, dominated by downy birch, alder or willows, for example at the margins of extensive blanket mire. However, this encouragement is not extended to agricultural districts where marshes, fens and bogs are small and fragmented and are often more valuable as open habitats (Forestry Authority, 1995).

Government policy presumes against the conversion of forests to other land uses except where there are strong public benefits from not replanting. Large scale conversion of forested land to open peatland habitats could not therefore be justified without strong evidence of the benefits. Research would be required to evaluate whether, and at what cost, raised bog or blanket bog habitats could be restored after deforestation; the FC is willing to undertake this research in partnership with others. The FC will continue to balance the various environmental and economic benefits and disadvantages determining the future of existing and new woodlands on peatland and to take account of new understanding of the issues as it becomes available.

Current peatland conservation projects in forests

Despite uncertainty over the effectiveness and benefits of bog restoration, a number of local projects have been undertaken in recent years, notably in Forest Enterprise forests, in partnership with wildlife trusts, other non-governmental organisations, national parks and the nature conservation agencies. These include some deforestation on a moderate scale, as well as restoration of water levels.

The Border Mires are described elsewhere in these proceedings (Burlton, Chapter 29, this volume), so needs only the briefest outline here. A consortium comprising the Kielder District of Forest Enterprise, English Nature, the Northumberland Wildlife Trust and Northumberland National Park, jointly funds and manages a large number of mires within Kielder Forest, with habitat conservation as the main objective. Trees are being removed from their margins and drainage ditches dammed up in an effort to conserve the habitat. Monitoring of vegetation composition and water levels is being done to gauge success.

At Dolgellau Forest District, in Wales, spruce forest has been cleared from areas of blanket bog at Penaran and these are being managed as open space with the main objective being to provide suitable habitat for black grouse (*Lyrurus tetrix*). The project has involved collaboration between Forest Enterprise and the RSPB. At Gamlan, in the same district, a raised bog is being rewetted by damming ditches after the removal of trees. In Borders Forest District, Forest Enterprise has attempted the restoration of a raised bog, Stell Knowe, after harvesting the first rotation forest. They removed debris, consisting of branches and tops, which is normally left on site after felling a mature tree crop, and have blocked ditches to rewet the bog. It has now been written into the District's conservation plan and forest design plan, ensuring it will be protected in future.

The suggestion by Avery and Leslie (1990) that the plantations of Caithness and Sutherland might provide a route for birds which use wooded bog in Scandinavia to colonise forest bogs in Britain has prompted a project in the far north of Scotland. The edges of some lodgepole pine plantations on blanket peat in Caithness have been 'feathered' by cutting out trees to graduate the change from open bog to forest. The objective is to make forest edges and the adjacent open ground more attractive for birds such as redwing (*Turdus iliacus*), greenshank (*Tringa nebularia*), wood sandpiper (*Tringa glareola*), brambling (*Fringilla montifringilla*) and bluethroat (*Luscinia svecica*), by simulating the natural gradation found in boreal forests in Scandinavia. The project's success is being monitored by surveying bird populations in these and in untreated control, areas.

Acknowledgements

I thank Gordon Patterson and Richard Toleman for their substantial input to this chapter, Graham Pyatt, Jim Dewar, Simon Hodge, Wilma Harper and James Simpson for their helpful comments, and all the Forest Enterprise staff who enthusiastically described local conservation initiatives, including some not mentioned here.

Chapter twenty-six:

Peatlands, Power and Post-industrial Use

Mr Gerry McNally
Bord na Móna, Ireland.

Introduction

Extent of peatlands

Peatlands once covered 1.18 million ha of the Republic of Ireland - some 17% of its land surface. Only two other countries, Canada and Finland, have a greater proportional coverage by peat.

Within this 1.18 million ha three broad categories of peatland are recognised: raised bog, blanket bog and fen.

- **Raised bogs** once covered 311,000 ha of the Midland counties of the country. They are complex structures of organic debris attaining thicknesses of 9-12 m in the undrained state.
- **Blanket peatlands** extended to 772,000 ha mainly along the Western coastal countries and the mountainous regions of the South and East. The term blanket was first used by Tansley (1949) to describe this type of peat terrain which conforms to the underlying topography. Peat depth usually range from 1 to 6 m.
- **Fen peats** are of lesser importance covering approximately 100,000 ha. They occur throughout the country contiguous to raised bogs and in river flood plains. Unlike the other peat types they have largely been reclaimed for agricultural purposes (especially grassland) and undisturbed fens are now very rare in Ireland.

History of peat use in Ireland

Peat has been used as a source of fuel in Ireland since prehistoric times. The respective rights of bog owner and turf cutter were defined by tribal customs long before these were codified and there is documentary evidence of the use of peat as a household fuel from the eighth century onwards. In medieval warfare the peat ricks of the vanquished shared the fate of their other perishable possessions - destruction by fire.

Peat fuel was used not merely by the rural populations but also by townspeople and as such was subject to tolls and tithes, usually paid in kind. Even whilst wood was in plentiful supply, peat was the preferred fuel because it was easier to harvest; but, with the clearance of Ireland's forests in the seventeenth century - peat became the only fuel available to the vast population which reached a peak of 8.2 million people by 1840. Peat usage at this time has been estimated at 6-7 million tonnes. Even the bog wood unearthed during peat digging was used widely in house building, roofing and in the making of domestic vessels.

Historical attitude to bogs

The earliest printed records of Irish bogs, written for the edification of Cromwellian settlers, claimed that the occurrence of bogs was due to the recklessness of the Irish who, due to ignorance or laziness, let daily more and more good land become covered with bog.

Coming forward to the mid thirties of this century; a time of economic depression and virtual isolation of the Irish nation, Todd Andrews - the founder of Bord na Móna remembers: 'The works were only accessible to a car driving along the tow-path of the canal and that was a perilous journey. In the nearby village of Pollough there was evident squalor and poverty on a scale much worse than I have ever seen in the Dublin slums of my youth. Pollough was so isolated as to be virtually an island disconnected from the outside world.'

Industrialisation of peatlands

It was against this background that Bord na Móna was established in 1946 with a mandate 'to produce and market turf and turf products to foster the production and use of turf products and to acquire bogs.'

From humble beginnings in 1946, Bord na Móna has grown to be one of the leaders in world peat technology. Today the company:

- owns 83,500 ha of peatland
- produces 5 million tonnes of peat annually
- produces 400,000 tonnes of briquettes
- produces 1.5 million m^3 of horticultural peat
- employs 2,000 people

- generates 14% of Irelands' electricity at five peat-burning stations.

New peat fired power station

The primary objectives of Irish Energy policy are to supply a choice of fuels at competitive prices, to consume this energy in an efficient manner, and to produce as much energy from indigenous sources as economically possible.

The present and future energy sources for electricity are shown in Table 26.1. As can be seen from the table, the contribution of peat falls off significantly as electricity energy demand increases. Demand is predicted to double over this time period.

Bord na Móna has an uncommitted peat resource on its already developed peatlands of 67 million tonnes. This reserve is in peatland areas already drained and in peat production. The Board has therefore proposed the building of a 123 MW fluidised bed peat-fired power station. This station will be a base-load plant producing 7,500 ha $year^{-1}$ consuming 1 million tonnes of peat, with an operational life of 35 years and a net efficiency of 36.7%. This new station will ensure that peat will contribute more than the 4% indicated in Table 26.1.

Table 26.1. The energy sources for electricity generation in Ireland (past and predicted).

	Energy Sources for Electricity Generation (%)					
Year	Coal	Gas	Oil	Peat	Hydro or Wind	Imported Power
1991	40	25	16	14	5	-
2001	29	27	27	7	4	6
2011	39	35	13	4	4	5

[shaded] Indigenous Energy Source

This proposal has been accepted by the EU commission and the Irish Government and is at present in preparatory stages prior to public tender. The key benefits from this proposal are that a significant industry will be established in a socio-economically disadvantaged area utilising an indigenous energy source from already developed peatlands. Much needed employment in an industrially declining area will be created.

Peat extraction policy

The peat extraction policy of Bord na Móna is to remove as much peat as economically possible from all its developed peatland. This primarily arises because developed peatlands with an infrastructure have a value of £25,000 ha^{-1} m^{-1} deep. The capital costs associated with peatland development - initial drainage, provision of rail infrastructure, purchase of specialised machinery - are all incurred in the early start up years. Therefore, to recoup this investment, it is necessary to remove as much peat as economically possible; this does not, however, equate with total peat removal. None of the presently available future uses for cutover and cutaway peatland can justify leaving behind extractable peat resources.

Industrial removal of peat using the milled peat technique differs considerably from the traditional digging system of peat removal. Under milled peat the total bog area is drained and then lowered by horizontal peat removal at the rate of 1 m per 10 years. As there are usually 4.5-5 m of extractable peat in each bog, it takes 45-50 years to remove the peat resource.

Conservation policy

Adopting such a policy on peat extraction obviously leads to public concerns with regard to the destruction of unique ecosystems, the destruction of wildlife habitats, the possibility of creating wastelands, and the addition of greenhouse gases.

Bord na Móna have always recognised the uniqueness of the peatland ecosystem, and the need to conserve representative peatlands for future generations. As its contribution to European Conservation year in 1970, Bord na Móna handed over Pollardstown Fen and Raheenmore Bog for conservation. As awareness of the uniqueness of peatland grew the Board handed over Mongan Bog, Clara Bog, Redwood Bog and All Saints Bog for conservation, during the late seventies/early eighties. When lists of bogs worthy of conservation were drawn up, the Board handed over any such bog in its possession to the Office of Public Works. Over the years the board desisted from acquiring any bog area deemed worthy of conservation. Thus the Board has co-operated over the last 25 years with the Office of Public Works and has contributed significantly to the national target for conservation of 10,000 ha raised bog and 40,000 ha of blanket bog.

Before reverting to the other issues, it is necessary to discuss the outcome of the extraction policy, namely cutaway bog.

Cutaway bog

Cutaway bog is the term used to describe areas of peatland from which it is no longer possible to extract peat economically. Cutaway bogs contain various peat types and depth which overly complex subsoil with differing drainage potentials. While all these factors are important perhaps the key issue is the drainage potential. Drainage water is removed from peatland either by gravity into the surrounding arterial drainage system or by a stepped series of pumping to eventually reach the external drainage system. As a broad general statement the drainage water from peatlands on the eastern part of the country are removed by gravity while those straddling the Shannon are pumped to facilitate peat removal.

The future use of those areas which are gravity drained will be a combination of forestry and grassland. Those areas at present pumped offer the opportunity to create quality wetlands which can become significant wildlife habitats.

Matching the technical requirements of each use to the conditions prevailing, the likely future use of the industrial cutaways in Ireland are:

1. 40-50% coniferous forestry
2. 20-30% wetlands, natural landscape
3. 20-30% grassland
4. 10-20% hardwood forestry wetlands.

While the forestry and grassland uses of cutaway bog have been long recognised, the potential for wetland creation and its wildlife use have not been fully appreciated to date. However, relevant works which have been ongoing for the past five years on a 100 ha site at Turraun bog and these are described below.

Site description

The site area covers approximately 20% of a 500 ha bog complex. The remaining 80% is still under industrial peat extraction today. The ordinance data for the site show that the elevation drops from OD 55 m to OD 47 m i.e. a spread of 8 m from highest to lowest point. The highest point is associated with a gravel moraine which always remained above the intact bog surface, and thus was never covered with peat. Peat did however reach as far as the 52 m contour. A total of 75% of the site is below the 48 m contour and this seems to mark the boundary of the old post-glacial freshwater lake. The soils below the 48 m contour are all of alluvial origin comprising, in the main, silty clays. These in turn are overlain by a calcareous marl deposit which resulted from limnic deposition from the stoneworts which grew in the post-glacial lakes. This marl contains the intact shells of water snails *Lymnaea peregra* and *Valvata piscinalis.*

Above the 48 m contour the site would be described as dry and below it as wet. The peat depth remaining after extraction varies from zero to 1.5 m, with an average of less than 0.5 m.

All post-industrial conditions likely to be encountered were thus present at this site (i.e. shallow and deep peat overlying well drained weathered drift soils, and shallow and deep peat overlying poorly drained silty clays and marls).

Baseline ecological study

A baseline ecological study showed that the colonisation of the cutaway was influenced by the above mentioned soil factors, leading to a mosaic pattern of plant growth.

Initial colonising species

In the latter years of industrial peat extraction the drainage ditches, which remain relatively free of disturbance, act as pathways along which the pioneer plant species become established waiting to spread across the production fields once ground disturbance ceases. The two most common pioneer species are marsh arrowgrass, *Triglochin palustris,* and the marsh horsetail, *Equisetum palustre,* the spread of which is facilitated by vegetative propagation. The acidity and degree of saturation strongly influence the subsequent colonisation progression. Dry alkaline areas are invaded by grassland species often dominated by *Holcus lanatus.* The wetter alkaline areas favour the spread of *Phragmites australis* as the dominant vegetation with *Mentha aquatica* and *Hydrocotyle vulgaris,* as subordinate species. The vegetation close to the mineral spring areas tends to be dominated by *Typha latifolia.*

In areas with significant peat depths 0.5 m+ and high moisture content combined with low pH, the dominant vegetation is *Juncus effusus.* The vigorous growth and tussock like form of *Juncus* makes for a dense vegetation with little species diversity.

Successional stages

The grassland sites are usually invaded by tree species principally birch, *Betula pubescens,* and crack willow, *Salix cinerea,.* Trees of lesser importance are *Pinus silvestris* and *Pinus contorta.*

Over the period of the study, successional stages were most pronounced on the drier shallow alkaline peat areas where arrowgrass, *Triglochin palustris,* was replaced by grassland, giving way to scrub and later still to woodland. In shallow wetter alkaline areas, *Phragmites australis* became dominant whilst around springs, *Typha latifolia* predominated. The *Juncus* dominated areas were slowly being invaded by *Salix* and *Betula* and will most likely proceed to woodland.

During the 17 year colonisation phase the industrial drainage system remained intact and thus all the spring-derived water left the site. All available evidence indicated that if the site was allowed to develop unaided that it would eventually progress to a birch/pine woodland with consequent loss of habitat diversity.

Creation of open water body

The availability of several mineral-rich springs indicated that the site had potential as a wetland so, in the autumn of 1991, it was decided to block up the drainage channels and create an open body of water and in order to maximise the potential of the site as a habitat for winter migratory bird species. In the course of doing this, a significant area was stripped of its peat cover exposing marl substrate (which was then used to create islands).

The water level rose rapidly within the site and, within weeks, 40 ha were covered by depths ranging from a few centimetres to 1.5 m with the majority having 1 m depth of water.

Wild bird usage

This water build up coincided with the arrival (to Ireland) of the winter migratory bird species. They were immediately attracted to the site. Subsequent bird counts showed that up to 300 whooper swans and several hundred duck species principally wigeon, teal, mallard were occupying the site. These counts have been repeated for 3 years and similar numbers have occupied the site each year.

A significant population of resident bird species principally mallard, coot, tufted duck, little grebe, crested grebe, lapwing, snipe, mute swans are also occupying the site and are breeding successfully. In total, 96 bird species have been seen on the site of which 30 are now confirmed breeding. The site is about to be declared a national reserve for fauna.

All the above confirms that, with proper management, high quality wetland habitats can be created after peat extraction.

Part Eight

Peat Free Horticulture

An Introduction

Dr Rob Stoneman
Scottish Wildlife Trust, UK.

In the 1980s, a major concern of peatland conservationists was the afforestation of the Flow Country. With the closure of concessions for such irresponsible forestry in 1986, attention turned to the plight of lowland raised bogs. The destruction of Thorne and Hatfield Moors were of particular concern as the sites represented the last raised bogs of eastern England. The ensuing debate (confrontation!) led to the formation of the Peatlands Campaign Consortium and a wider realisation that raised bogs were under particular threat. The Peat Producers were targeted.

Peat extractors have often pointed out that the total area of peatland exploited for horticulture is small - a mere 5,400 ha or just 0.5% of Britain's total peatland area (Robertson, 1993) and that the industry requires only 1,000 ha more over the next 20 years (DoE, 1994b).

This misses the point. Whilst peat extraction presently affects only 1,100 ha of Scotland's original resource of 27,000 ha of raised bog, it remains as the most significant direct threat to the remaining raised bogs. This is because afforestation and 'reclamation' for agriculture have virtually ceased. Given that there is only 2,400 ha of 'near-natural' bog on uncut surfaces remaining, the prospect of a further 1,000 ha of peat extraction in Scotland is *potentially* horrific.

The threat is all the more lamentable given that this debate could lead to the conservation of bogs and the development of a sustainable horticultural products industry. As the following two chapters show, peat alternatives are viable, they do work and they could replace peat. Indeed, given the less exacting requirements of amateur gardening as compared to the horticultural industry, 58% of peat usage could be switched to alternatives immediately. For the professional horticulturalists, more time and research is required to switch from peat-based products entirely.

This is not a conservation versus jobs conflict either, since it is the peat producers which have the expertise and facilities to produce the alternatives. Indeed, some of the best peat alternatives (e.g. Levingtons Peat-free mix) are produced by peat producers.

We have a choice: We can muddle through slowly developing peat alternatives with limited will and funding, whilst at the same time destroying more bog-land (as has been proposed). Alternatively, we could stop all future peat destruction now and turn instead to the development of a sustainable industry producing good quality peat alternatives for home markets and export.

Britain could lead the world in this endeavour and help the conservation of, not only of our bogs, but of those threatened across the globe.

Chapter twenty-seven:

Gardening without Peat

Mr Nigel R. Doar
Scottish Wildlife Trust, UK.

Peatlands, peat and the British back yard

Over the last 30 years, peat has become a seemingly essential part of any gardener's equipment. It is dug into flower-beds to improve soil structure, spread on to lawns as a top-dressing, applied to the roots of acid-loving plants, or used in containers to grow seedlings and house plants.

Every year, amateur gardeners in the UK use about 1.5 million m^3 of horticultural peat (DoE, 1995d), which is about 58% of all the horticultural peat used in this country (see Fig. 27.1). Nearly all of it originates in the lowland raised peat bogs of the UK (≈60%) and the Republic of Ireland (≈33%). Such peat bogs are every bit as valuable as, and considerably more endangered than, the Flow Country's blanket bogs which attracted so much attention in the 1980s.

By the start of the UK Peatlands Campaign, in 1990, less than 2% of England's original raised bogs remained in a near-natural condition (DoE, 1994b), and raised bogs were believed to be in danger of extinction within the European Community. In England, after centuries of drainage, ploughing, tree-planting and neglect, commercial peat extraction for horticulture was threatening to finish off what little natural raised bog remained.

Even in Scotland, where considerably more raised bog remains in a near-natural state, commercial peat extraction for horticulture is currently damaging more raised bog than any other operation which is subject to planning control (SWT, 1995b). With amateur gardening creating more than half of the UK's total demand for horticultural peat, there is clearly a very large role to be played by amateur gardeners in protecting our remaining raised peatland resource.

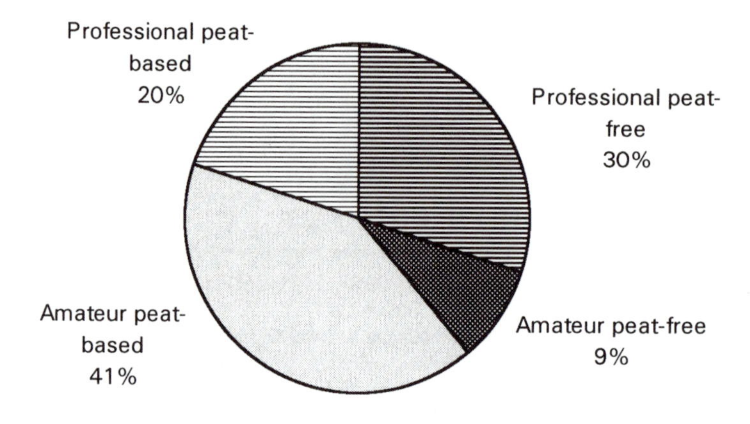

Total substrate used	**3.6 million m^3**	**100%**
Total peat used	2.6 million m^3	71%
Total peat-free used	1.0 million m^3	29%
Total amateur use	1.8 million m^3	50%
Total professional use	1.8 million m^3	50%

Fig. 27.1. The use of peat-free and peat-based products by professionals and amateurs in the early 1990s. Data source: DoE (1995d).

Plants without peat

In 1991, horticulturist and landscape architect, Professor Chris Baines expressed a view held widely amongst professionals when he commented 'peat is *not* the best choice of material for most of the uses to which it is being put' (Baines, 1991). Alternatives to peat exist for every use to which it is currently put. Some alternatives are relatively new, while others have been used for centuries. Even in areas such as growing media, where most of the established high performance commercial products are based on peat, peat-free alternatives which can amply satisfy the needs of the amateur gardener are now readily available. When Friends of the Earth published their Peat Alternatives Manual, 47 companies were listed as manufacturers or suppliers of peat-free products (Pryce, 1991). The availability of peat-free products has continued to increase since, as has their quality - whilst prices have reduced. Some companies, such as B&Q, now pursue an equal pricing policy between peat-based and peat-free products (Knight *et al.*, 1995).

Soil improvers

More than half of all the products used as soil improvers by amateur gardeners in the UK are peat-based (see Fig. 27.2), even though there is considerable consensus that peat is not the best material for these purposes. Alternatives almost always perform better.

Mulches are spread on to the soil surface to suppress weeds, retain moisture, buffer extremes of temperature and to improve the soil's visual appearance. It is now generally accepted that mulching is a poor use of peat. Michael Pollock, technical advisor to the Royal Horticultural Society, expressed his views on the use of peat as a mulch thus: 'The use of peat for mulching is not justified in view of the many acceptable alternatives, including well rotted manure, composted plant remains and woodchips, shavings or different grades and forms of bark'. (Bennington and Steel, 1994).

Similarly, Professor Chris Baines remarked 'For mulching of the soil surface, granulated peat offers a fine seed-bed for incoming wind-blown weed seeds, it dries out and either blows away or creates surface rooting and drought stress problems. A much coarser material is needed, and the renewable waste product, coarse-chopped tree bark is far more effective.' (Baines, 1991).

In 1992, the Consumers Association did not even consider peat-based products in a trial of mulching materials - such was the consensus that this is an inappropriate use of peat. The results of their trial showed that cocoa shells (a by-product of the chocolate industry), spread to a depth of 5 cm over weed-free and well watered soil was the most effective commercial mulch tested. (Consumers Association, 1992).

Most professionals now avoid using peat as a mulch, but despite the trials, the consensus of experts and the efforts of peatland conservationists, a small minority of amateur gardeners still persist in spreading peat on the surface of their flower beds.

When **planting trees and shrubs**, there is often a perceived need to add some organic matter round the roots to encourage establishment and early growth. As with mulching, a good deal of very sound, and widely accepted evidence indicates that organic matter is not normally required, and that, where organic matter is needed, there is no reason to believe that peat is better than other bulky organic substances.

Research reported by the Forestry Authority concluded that 'Bulky organic soil amendments added to the planting pit at the time of planting are unlikely to confer benefits. There appears to be little justification for their use' (Hodge, 1993). Yet there are still amateur gardeners who use peat for this purpose.

Perhaps the biggest of the outdoor, open-ground uses of peat is as a general **soil improver** - to break up heavy soils, increase water retention in light ones, to increase organic content and to add nutrients.

Whilst investigating open-ground alternatives to peat, small-scale trials run by Geoff Hamilton for the BBC's *Gardeners' World* television programme led him to conclude 'My own limited tests showed clearly what was already well known - that most bulky organic materials will improve soil condition' (Hamilton, 1991). Professor Chris Baines went one step further, saying that 'Peat is difficult to wet and difficult to integrate into existing soil. It is not nutrient-rich, and therefore adds nothing significant to soil fertility, although it does boost organic content and can improve crumb structure. However, coarse fibrous compost made from chopped straw is far more beneficial, especially in light soils, and a lime-based compost is more valuable in clays.' (Baines, 1991).

One of the most well-established yet under-valued soil improvers is common-or-garden compost - produced from the action of micro-organisms on dead vegetable matter when it is piled together in warm, damp, sheltered heaps with good ventilation. The composting process has been used by the Chinese since 7000 BC and the end product is no less useful today. Every year, the UK produces 20 million tonnes of household waste - around a third of which could be composted. It has been estimated that if only 2% of our compostable waste was recycled in this way, it would remove the need for horticultural peat at a stroke.

Some gardeners still tend their own compost heaps, or take part in Community Composting Initiatives (Crofts, 1992; Bennington, 1996). Increasingly, local authorities are taking a lead in compost production. A municipal composting operation run by Dundee District Council since March 1992, processes the green waste from around 5000 selected Dundee households, together with green waste from municipal parks and gardens. After extensive tests, it was concluded that their *Discovery* compost was 'an excellent product for general purposes and as an organic soil conditioner'. (Dundee District Council, 1994). The vast majority of professionals now avoid the use of peat as a soil improver as it is far surpassed by other materials. Again, many amateur gardeners continue to use peat in this way - despite the evidence that it is not the most appropriate material to use.

The final open-ground use to which peat is often put is **soil acidification**. Many popular garden plants such as azaleas, rhododendrons and heathers require an acid soil to thrive. Peat's low pH and low price have made it a popular material for providing this acidity.

Even so, pine needles and composted heather or bracken provide good, low pH alternatives to peat (Bennington and Steel, 1994). Colwyn Borough Council use ground composted bark for their acid-loving plants, whilst cocoa shells have been shown to provide stable acidic soils when dug in (Hamilton, 1991), and liquid soil acidifiers are also commercially available.

From a purely ecological viewpoint, it may be better for gardeners to work with the soils they have, rather than against them. If a garden doesn't have naturally acid soils, then perhaps more benefit could be gained from growing alternative, equally attractive plants better suited to the natural conditions. This may have been in Michael Pollock's mind when he commented 'The value of peat for soil acidification is generally overstated; sites with a high pH requiring unreasonable quantities of acidic peat for a useful lasting effect'. (Bennington and Steel, 1994).

Growing media

Growing media for container plants are considerably more exacting in their requirements than open ground soil improvers. They must hold ample water (but not too much) and they must provide sufficient nutrients (in the correct proportions for different plants and at the correct rates of release). Their structure must be open enough to allow air to reach the roots and firm enough to support the plants as they grow. To be commercially viable, they must also be available in quantity, they must be safe, sterile, consistent, stable when packaged and, for the amateur gardener in particular, they must look, feel and smell pleasant.

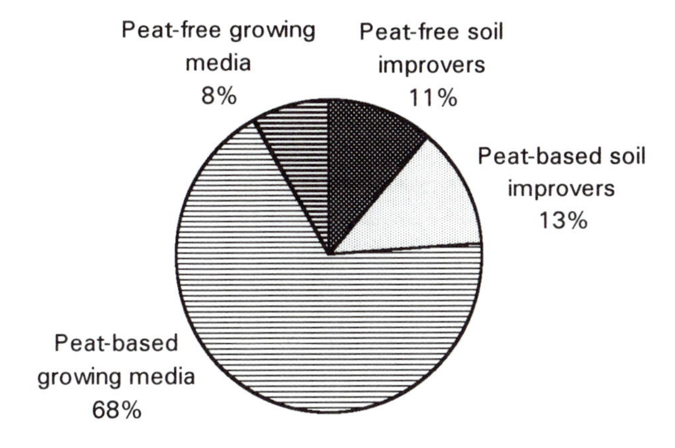

Total amateur substrate use	**1.8 million m^3**	**100%**
Amateur peat use	1.5 million m^3	81%
Amateur peat-free use	0.3 million m^3	19%
Amateur growing media use	1.4 million m^3	76%
Amateur soil improver use	0.4 million m^3	24%

Fig. 27.2. The use of peat-free and peat-based growing media and soil improvers by amateur gardeners in the early 1990s. Data source: DoE (1995d).

Because of these demanding requirements, the development of peat-free growing media has not been easy. In the early 1990s, many amateur gardeners tried the newly emerging peat-free growing media and were disappointed at their poor performance, unattractive appearance or unexpected difficulties in their use. This has led to many gardeners reverting even more steadfastly to the use of peat-based products (Knight *et al*., 1995). In the years since the start of the Peatlands Campaign, amateur gardeners have been confused by publicity surrounding growth trials which concluded that 'many,' 'most,' or 'nearly all' of the peat-free products tested failed to perform well. Only a steadfast few (representing 8% of all sales of growing media to amateurs - Fig. 27.2) saw through this meaningless smokescreen and pursued the peat-free path, by adopting the few peat-free products that undoubtedly do perform well, adapting their gardening habits to suit these new products (rather than expecting the peat-frees to behave exactly as peat used to) and keeping a close eye on a very rapidly changing market.

In the 1995 *Gardening from 'Which?'* Compost Directory, of 17 peat-free products listed, four were noted as being discontinued for 1995 and six had been reformulated since the tests were carried out. Two products were noted as giving satisfactory results and another two gave good results. One of these (Levingtons Peat Free Universal) was highlighted as a 'best buy' for potting (Consumers Association, 1995).

The most widely available and successful of the peat-free growing media are based either on coir (a recycled by-product of the coconut and coconut matting industries) or on various wood products (bark, wood fibre, woodchips etc.).

If treated like peat, even the best products may perform poorly, but if watered carefully and fed regularly, plants grown in either coir or timber-based growing media can grow better than in peat. In 1994, tests carried out on carefully sourced Lignocell coir product, concluded that 'Plant growth in our coir composts has always been at least as good as in peat and, as with Fuchsia, it can sometimes be better.' (Smith, 1994). That study also made some observations on the handling of coir-based growing media, which amateur gardeners would be well advised to take note of: 'Coir dries quickly on the surface and growers can easily be fooled into believing that the pot is dry. Watering at this stage merely washes out the nutrients, particularly nitrogen in the compost. So long as moisture content is judged instead by pot weight this can be avoided.'

If some can - everyone can

As demonstrated above, there are clear, real examples of peat-free products being available as effective replacements for peat, here and now, in every area where amateur gardeners currently use peat-based materials. There are documented examples of how people have switched successfully from peat-based to peat-free products - in both the amateur and professional sectors (Bennington and Steel, 1994).

If all Britain's hobby gardeners adopted 'environmental best practice' tomorrow, the demand for horticultural peat in the UK would be cut by 58%. Of course, this is unlikely to happen. Overall, nearly a third of growing media and soil improvers sold in the UK during the early 1990s were peat-free (see Fig. 27.1), but the proportion of peat-free products bought by amateur gardeners has remained steadfastly small, at about 19% (see Figs 27.1 and 27.2). Even in large garden centre chains where a relatively positive attitude has been taken to the marketing of peat-free products, the market share has remained rather small. In 1994, 24% of B&Q's 80 litre multipurpose compost range sold was peat-free

(Knight *et al.*, 1995). Similarly, only 19-20% of B&Q's 1994 grow-bag sales were peat-free. Why should this be?

Perhaps some indication was given by an exercise carried out jointly between B&Q and the Scottish Wildlife Trust over Easter weekend in April 1995. Over three days, trained volunteers and staff from the Trust were present in the garden centre at B&Q's Milton Road superstore in Edinburgh. They spoke to around 80% of the customers who bought any soil improvers or growing media that weekend, and recorded customers' comments. They also kept a record of the products sold. From the customers' responses and the sales pattern noted, it became clear that most customers bought peat-based products out of habit or because they were unaware of the alternatives - not because they were particularly wedded to peat (Doar, 1995). Without good advice and guidance at the point of sale, customers tended to buy the cheapest, most prominent product, or the one which they regularly used - almost irrespective of its environmental credentials or its horticultural suitability. When their attention was drawn to the issue, about 80% of customers either bought peat-free (where an equivalent product was available) or said they would have done so if an equivalent peat-free product had been in stock. (Several peat-free ranges sold out over the course of the weekend).

It seems that the public would rather do the right thing for the environment, given the choice. In this case they are not yet sure what the choice is, and are wary of change without ample support and guidance.

Conclusions

There is adequate evidence that peat-free materials able to satisfy the demands of amateur gardeners are technically feasible, widely available and that they can be competitively priced.

Equally, it is clear that the potential for wide-spread peat-free gardening that this indicates is not yet being realised. The continuing development of peat-free products, increasing their technical specifications and ease of use, is undoubtedly important. Research is also needed to accelerate the production of viable home-grown recycled alternatives to both peat and coir. In addition, it seems that no amount of technical research and development work will assist in moving the amateur gardening market away from peat use unless government, peat producers, environmental bodies, gardening organisations, the media and the gardening retail industry collaborate whole-heartedly in the marketing and promotion of peat-free products and in the education of the gardening public in their use. This must be an immediate priority if both Britain's gardeners and our raised peat bogs are to have a sustainable future.

Chapter twenty-eight:

Testing Alternatives to Peat as Horticultural Growing Media

Dr Chris A. Smith
The Scottish Agricultural College (SAC), UK.

Horticultural uses of peat

By the end of the 1980s peat had become the predominant organic substrate used in horticulture, including amateur gardening and landscaping. Its low price, ease of availability, sterility, high moisture retention and low nutrient content rendered it usable for a wide range of applications. These can be split into two broad groupings: 'growing media' - used in pots and other containers in isolation from the soil; 'soil improvers' - which are added to mineral soils to improve their fertility.

During the 1990s, the total horticultural market for growing media and soil improvers in the U.K. has been around 3.5 million m^3 per annum (Table 28.1). Between 1990 and 1993 the amount of non-peat substrates used annually in horticulture almost doubled to around 1 million m^3, primarily due to increased use as soil improvers in gardening and landscaping, the majority of which are now non-peat (DoE, 1994a). Landscapers readily acknowledged the environmental impact of peat extraction and switched to non-peat soil improvers, realising in addition that their more favourable properties, such as higher levels of both nutrients and readily decomposed organic matter made them intrinsically more suitable than peat.

Recent estimates put the proportion of non-peat substrates used by professional horticulture in growing media at less than 5% (DOE, 1994b), and of these the majority (e.g. bark, perlite, grit) are used as openers to modify the physical properties of peat. This chapter assesses the technical reasons for limited continuing use of non-peat growing media.

Table 28.1. Horticultural use of peat and alternative products in the early 1990s.

	Growing media		Soil improvers	
Market Sector	Total ('000 m^3)	Non-peat (%)	Total ('000 m^3)	Non-peat (%)
Amateur gardening[1]	1346	8.6	396	39
Local authorities[1]	21	27	197	96
Private sector landscaping[1]	26	4.2	530	94
Professional horticulture (plant production)[2]	942	<5	50	
Total	**2335**		**1173**	

1. Figures for 1993 (DoE, 1994a).
2. Figures for 1990 (Blackman, 1993).

Evolution of potting mixes

The history of potting mixes teaches a number of lessons. In the early 1900s, most growers used their own potting mix recipes which typically contained soil, conditioned with one or more of a range of organic and mineral amendments such as leaf mould, garden compost, ashes, grit, etc. Variability of performance both between batches and between nurseries was a common problem, as were soil-borne pests and disease.

In the 1930s, workers at the John Innes Institute developed a uniform and reliable growing media for their research with plants. John Innes Compost's ingredients were mixed to a strict recipe; sterilised mineral topsoil - 'loam' - was conditioned with uniform and freely available grit and peat, to which was added organic and inorganic fertilisers (Lawrence and Newell, 1939). These 'composts' became the industry standard as growers benefited from using the same reliable growing media.

However, John Innes mixes had drawbacks; they were labour intensive to produce (mainly due to preparation and sterilisation of the loam), storage over long periods resulted in phytotoxicity following mineralisation of organic nitrogen, they were heavy to handle and, as the industry expanded, reliable supplies of loam became difficult to obtain. Peat-based mixes, which came to prominence in the early 1970s, largely overcame these problems. They were light, moisture retentive, required no sterilisation, were cheap to prepare, and generally supplies of peat were reliable and inexpensive.

Deficiencies in non-peat substrates

The popularity and success of peat as a potting substrate in the past 25 years is due to its almost perfect suitability. Cultivation of container-grown plants places exacting demands on the properties of a substrate; most alternative substrates fall short of this ideal to a greater or lesser extent (Tables 28.2 and 28.3).

Table 28.2. Desirable properties for potting substrates.

Property		Peat	Coir	Perlite	Composted MSW[1]
Physical	Water retention	•••	•••	••	••
	Aeration	••(•)	•••	•••	•••
	Microbial stability	•••	••	•••	•
	Light weight	•••	•••	•••	••
Chemical	Absence of toxins	•••	•••	•••	••
	Low nutrient content	•••	••	•••	•
	Low pH	•••	••	••	•
	Buffering capacity	•••	•••	•	••
Biological	Pest, weed & disease content	•••	•••	•••	•(•)
General	Handling characteristics	•••	•••	••	•(•)
	Appearance	•••	•••	•	••
	Reliability of quality	•••	••	•••	•
	Reliability of supply	•••	••	•••	•(•)

Key: ••• Good; •• Moderate; • Poor; brackets indicate variability in a property;
[1] Municipal solid waste.

Table 28.3. Substrates for non-peat growing media.

Organic, viable	Organic, not proven	Inorganic
Coir	Bracken*	Perlite
Bark*	Municipal solid waste*	Vermiculite
Wood waste*	Animal manure*	Rockwool
Garden refuse*	Paper waste	Clay granules
Rape straw*	Leaf mould	Polystyrene
	Spent mushroom compost*	

* Materials which must be composted, or otherwise processed, before use.

The more common deficiencies include:

High pH and nutrient content

Whilst lime can be used to raise the pH of acid substrates it is practically difficult to reduce pH to acceptable levels for optimum plant growth. A high nutrient content, in excess of plant requirements, can both induce nutrient disorders and raise the salinity of the growing media thus impairing plant water uptake.

Phytotoxins

Both organic and inorganic substances may be present at phytotoxic levels in substrates; spruce bark, for example can contain phytotoxic levels of both manganese (Maher and Thompson, 1991) and monoterpenes (Aaron, 1982).

Poor physical properties

Few substrates have the high moisture retention of peat, and thus require more frequent irrigation (Smith, 1995).

Microbial instability

Susceptibility to microbial decomposition is a common failing of organic substrates. Where carbon-rich substrates are affected, microbial activity competes for, and immobilises, plant-available nitrogen to the detriment of the crop (Handreck, 1992). The decomposition of more nitrogen-rich substrates can lead to the release of plant-available nitrogen (by mineralisation); the rate of release is dependent on both temperature and moisture content and is rarely in step with plant demands. More problematic is the consequent accumulation, to potentially phytotoxic levels, of mineral nitrogen in stored media. Rapid decomposition can also lead to shrinkage and loss of substrate volume.

High processing costs

Aerobic composting is commonly used to minimise some deficiencies; organic matter can be stabilised thus reducing either nitrogen immobilisation or mineralisation potential, organic phytotoxins can be denatured, and physical characteristics can be improved.

New container substrates - case studies

Whilst many substrates are less than ideal as primary ingredients of growing media, their deficiencies are not always insurmountable. Research with perlite and coir has demonstrated how it can be possible to develop successful mixes and cultural strategies. However, even for these substrates, which exhibit only subtle differences to peat, a substantial research effort spread over a number of years has been required. To successfully develop growing media from less suitable materials like composted municipal waste (Table 28.2) a substantial research effort will be required.

Perlite

Perlite, a porous lightweight mineral substrate, is manufactured by high temperature expansion of crushed volcanic rock. It has many advantages as a potting substrate (Table 28.2). It is already familiar to commercial growers, being used as an 'opener' to improve the aeration of peat-based potting mixes, and as a substrate for hydroponic culture of edible crops. It is readily available, biologically sterile, lightweight and of consistent quality, characteristics which are essential to commercial growers. Perlite is available in a wide range of particle sizes and it is hence possible to produce different perlite-based mixes with physical characteristics to match the requirements of a variety of crops and cultural conditions. Its low nutrient content allows the grower to carefully control crop nutrition using liquid feeds and controlled-release fertilisers which are now industry standards (Smith and Hall, 1994; Smith *et al.*, 1995). The productivity and quality of both flowering and foliage pot plants (*Ficus elastica* 'Robusta' and pot *Chrysanthemum* 'Bright Golden Anne') grown in the greenhouse, was shown to be as good in perlite-based, as in peat-based, growing media (Smith and Hall, 1994). Subsequently, similarly favourable results have been demonstrated for a range of six non-ericaceous hardy ornamental shrubs, grown outdoors in perlite and peat-based potting mixes containing controlled-release fertiliser (Smith *et al.*, 1995).

Research has demonstrated that apparent shortcomings, such as a lack of buffering capacity, are not necessarily detrimental to the performance of perlite-based container media. The addition of limestone provides a check against acidification whilst increases in pH can be avoided by including a proportion of ammonium nitrogen in the feed or fertiliser, which induces the plant roots to release hydrogen ions. The consequent low pH of the rhizosphere also helps to maintain adequate trace element availability despite the alkalinity of the bulk substrate (Hall and Smith, 1994). However, despite their proven potential and productivity, perlite-based substrates are unlikely to achieve mass appeal with commercial growers or gardeners in the UK; perlite is visually unappealing, about twice as expensive as peat, has a reputation for being dusty and unpleasant to handle, and finally, being a non-renewable resource, has apparently poor environmental credentials.

Coir

Coir is the waste pith from the husk of the coconut which is discarded after fibre extraction for rope or matting manufacture. It had entered the U.K. growing media market before the peat issue gained widespread publicity. The substrate has many similarities to peat (Table 28.2), however, initial results were mixed as growers tended not to take into account some significant differences, such as its higher pH and content of soluble salts, including potassium. Other drawbacks reported in the horticulture press included; difficulties with handling/pot filling because of stringy fibres, a tendency to be nitrogen 'hungry', problems with rapid drying-out, and variability of quality.

Subsequent commercially funded research by SAC clarified the facts behind some of these assertions, and provided substantive data on which sensible and effective growing media mixes and crop management advice could be based (Smith, 1995). This work used a single, carefully sourced, supply of Sri Lankan 'Lignocell' coir, which is well 'aged' to reduce the content of readily decomposed compounds. It is also free from long fibres and has a relatively low mean content of soluble salts and potassium (96 $\mu S\ cm^{-1}$ and 79 $mg\ l^{-1}$ respectively in a 1:6 water extract).

Growers had reported that plants grown in coir became pale and nitrogen deficient more rapidly than those grown alongside in peat. The possibility of microbial immobilisation and more rapid leaching of nitrogen were investigated. Laboratory incubation studies demonstrated quite clearly that very little plant available nitrogen was immobilised by the coir; the losses, although greater than in peat, would be readily offset by normal liquid feeding. When peat and coir mixes were given the same addition of nutrients and the same amount of over watering it was shown that marginally less nitrogen was leached from coir than from peat. The explanation as to why coir appears to be 'nitrogen hungry' can actually be found in the apparent problem of rapid drying out. The water-holding capacity of coir is much the same as that of peat, around 80% (v/v), however, if plants are genuinely being under irrigated, there is a tendency for wilting to occur more quickly in coir than in peat. Investigations showed that although the total water-holding capacity of the two substrates is the same, the moisture in coir is more easily extracted by plants and therefore could be utilised more rapidly, thus requiring more frequent irrigation of the crop. Many people have observed, however, that the coir on the surface of a pot dries out rapidly, presumably by surface evaporation. This can fool growers into believing that the coir needs watering when in fact the bulk of the coir in the pot is perfectly moist. The consequence of watering at this stage is to over water and hence cause excessive leaching of nutrients. The solution is to use pot weight rather than the appearance of the surface to judge when irrigation is needed.

In growing trials of 'Lignocell' coir mixes formulated to account for coir's higher pH and nutrient content, the quality of bedding and pot plants was at least as good as in peat-based mixes, and for some subjects better. The liquid feeds used in the trials were the same for peat and coir mixes, with nutrient ratios typical of those currently used in peat-based growing systems. The research demonstrated that the only significant management change needed to work successfully with correctly formulated coir mixes is in scheduling of irrigation to maintain good moisture supply and avoid excessive leaching of nutrients - particularly nitrogen.

Conclusions

Increasing the use of non-peat growing media presents a number of challenges to both conservationists and the horticulture industry. The need to raise awareness of peatland conservation issues and change the attitudes of both professional growers and amateur gardeners is clear. However, this will be of little value unless cost-effective non-peat growing media can be developed. A number of factors constrain the pace of such developments: the inherent unsuitability of most non-peat substrates; the high costs of research and development needed to overcome the inherent deficiencies and variability of non-peat substrates; the lack of incentive for growing media manufacturers to develop and market non-peat mixes, given that the majority have the manufacture of peat-based mixes as their core business, and that there is little demand for non-peat products from, particularly professional, consumers; the low price of peat in the UK which is currently falling due to competition from imports; the high costs of essential processing (e.g. composting), particularly in low-volume production, which make non-peat mixes more expensive; a lack of familiarity in the UK with the technical requirements of successful large-scale composting; a lack of funding for independent research and technical advice for growers on non-peat media and their use.

Part Nine

Conservation Management, Rehabilitation and Monitoring

An Introduction

Dr Rob Stoneman
Scottish Wildlife Trust, UK.

The pursuit of favourable conservation status for Europe's rarest habitats and species is now the goal of the Union's member states. As I show (Chapter 45, this volume), for the raised bogs of Scotland, this is a considerable task given a highly degraded and degrading resource. Löfroth (Chapter 20, this volume) points to a similar degree of damage across all of Europe's peatlands.

At a site-level, the near loss of Glasson Moss National Nature Reserve in the dry summer of 1976 serves as an example (Lindsay, 1977). Dry weather and partial peat extraction from the moss (which had upset the bog's hydrology) led to a severe fire which began in August, crossed and re-crossed the site many times and was only finally extinguished by autumn rains in November. D. A. Ratcliffe (the Nature Conservancy Council's chief scientist) felt that the Moss was now possibly irretrievably damaged. Twenty years later, Glasson Moss is considered one of Britain's most intact raised bogs. The spectacular recovery of Glasson Moss came about because of positive conservation management.

Two different approaches have been taken. Some sites have only been partly damaged. For these, management usually requires a gentle tweaking of their hydrology for success. An example is given by Mr Burlton on the Border Mires. Understanding the ways bogs function is vital here. Herr Ginzler demonstrates how ground water mound modelling can be used. Such modelling has profound implications. In many cases, damage relates to regional ground water changes; raising waterlevels around sites normally involves complex and expensive schemes.

The second approach concerns sites which have been totally altered by human activities. Examples of rehabilitating such sites from peat extraction (Rochefort and Campeau, Chapter 31, this volume; Sliva, Maas and Pfadenhauer, Chapter 32, this volume) and from afforestation (Brooks and Stoneman, Chapter 33, this volume) are included here. In these cases, conditions have been so altered, workers have to carefully examine the conditions required by bog plants and attempt to engineer an appropriate environment. It is interesting to consider Professor Clymo's submission (Chapter 10, this volume) in this context where he demonstrates the remarkable ability of *Sphagnum* to alter conditions to suit itself. It is not surprising, therefore, that the largely experimental attempts at rehabilitation revolve around *Sphagnum* recolonisation.

Whilst fenland management is a traditional agricultural practice, conservation management and rehabilitation (of bogs especially) is a new and experimental discipline with only a fragmented 30-year history. Monitoring such work is vital and an integral part of a dynamic process to evaluate and exchange bog management information. Part of this dynamic is being fulfilled by the publication this year (1996) of *Conserving Bogs: A Management Handbook* by SWT, SNH and RSPB. However, good reporting of management and rehabilitation work is necessary to add to the information gathering, evaluation and exchange process.

Dr Johnson also points out the other uses of monitoring sites. These include international and national needs (e.g. statutory designations) and local needs (e.g. site management). Dr Johnson calls for the multi-purposes of monitoring to be fulfilled with minimum-level, standardised, information technology compatible methods. The collation and assessment of such information should be taken forward to a lead agency.

Achieving the elusive goal of favourable conservation status for our bogs will be difficult. It is a stark fact that most, if not all, bogs across western Europe are damaged and require positive conservation management and allied monitoring. The resources and will for this task are simply not available as yet. This Convention, however, did show that the expertise to do the job is in place and growing.

Chapter twenty-nine:

The Border Mires Approach

Mr Bill Burlton
Forest Enterprise, UK.

Introduction

Kielder Forest is situated in northern England adjacent to the Scottish border. The forest is owned by the nation and managed by Forest Enterprise - the management arm of the Forestry Commission. It is the largest man-made forest in northern Europe with some 50,000 ha of planted forest in one contiguous block and an additional 10,000 ha of open land, much of which is blanket mire. The plantations are largely made up of exotic conifers particularly Sitka spruce (*Picea sitchensis*) from west-north America.

The Border Mires are a collection of, mainly, blanket bogs with lenses of deeper peat in and around the southern part of Kielder forest.

The Border Mires - a brief description

Geologically the area comprises of a series of shales, thin bands of limestone and coal seams overlain by glacial drift which may be up to 30m thick. Over some of the area in the southern part of the forest the glacial drift is in turn covered by peat deposits of varying depths. At higher altitudes, these deposits are typical blanket peats formed directly over the drift by paludification, whilst at lower altitudes peat has formed in hollows by a succession from open water to fen, carr woodland, and finally *Sphagnum-Eriophorum* peat. These 'raised bog' units have subsequently spread out over the surrounding land. Peat depths measured on various mires range from 2 m to >10 m.

These extensive peat areas have been substantially afforested with the exception of small, isolated islands of wetter, deeper peat. Although a number of the remnant deep peat areas have been modified by drainage following forestry ploughing, the best examples still remain relatively unchanged.

The vegetation on much of the open mire areas comprise a carpet of bog mosses (*Sphagnum* spp.) together with other plants typical of ombrotrophic conditions. The main NVC type is M18a *Erica tetralix - Sphagnum papillosum* raised mire *Sphagnum magellanicum - Andromeda polifolia* sub-community.

Amongst the plants of interest on the Border Mires are the nationally scarce *Sphagnum imbricatum* and *Carex magellanica* and regionally scarce *Carex pauciflora, Carex limosa* and *Rhyncospora alba*.

Several nationally notable invertebrate species also occur (e.g. the beetle *Trechus rivularis*, a Red Data Book species).

The topography and variation of climate has allowed a wide range of ombrotrophic mire types to develop including watershed mires, spur mires, valley mires, valleyside mires, basin mires and saddle mires. This range of mire types is a very important characteristic of the Border Mires and adds greatly to their conservation interest and value.

History of afforestation

Prior to afforestation most of the land surrounding the mires was extensive upland grazing accessible to sheep and cattle and periodically burnt. Today - apart from a few mires on the forest edge which are bounded in part by rough semi-improved pasture - most of the mires are surrounded by the plantations of Kielder forest. Most of the plantations were established in the post World War II period (1945-1960). The creation of the Forestry Commission in 1919, following World War I, resulted from the wish to create a strategic reserve of timber should another war occur. In the immediate post World War II period, this policy was reaffirmed and there was a rapid expansion of afforestation throughout the country. Little knowledge was available at that time about the importance of the Border Mires, and in the 1940s and 1950s conservation was of a much lower priority than is the case today.

The main technique employed to establish plantations in the uplands was to cultivate the land with heavy ploughs and plant the trees on the upturned furrows. At the time of planting in the Border Mires area the ploughing equipment was not able to travel over the very wet, soft deep peat areas so they were left predominantly unplanted (although many of the mire edges and some complete mires were drained and planted).

The forest crops are felled on an average 45-year cycle, so many of the tree crops around the mires are now starting to be felled.

Conservation history

Interest in, and management of, the Border Mires has been ongoing since the late 1950's. Lunn (1979) drew attention to, and mapped, the '*Sphagnum* Mires'. Concern over the expansion of forestry in this area led to eight sites becoming reserves of Northumberland and Durham Wildlife Trust (now Northumberland

Wildlife Trust), in the mid 1960s. Management plans were written for these sites and early work by Trust volunteers ditch blocking using peat is recorded on Haining Head Moss.

The earliest site to receive statutory protection was Coom Rigg Moss in 1959. In the 1960s, a further seven sites were notified. The eight notified sites were included in *A Nature Conservation Review* (Ratcliffe, 1977) under the generic name Irthinghead Mires. Further recognition of the importance of the mires came when the eight Irthinghead Mires were designated as important international wetlands under the 1976 RAMSAR Convention.

Further notifications have occurred in the 1970s, '80s and '90s and Coom Rigg Moss and Grain Heads Moss were given National Nature Reserve status in 1983.

Border Mires Committee

In the 1980s, the Forestry Commission objectives changed to a multi-purpose role where conservation and recreation became an integral part of forest management.

Smith and Charman (1987) discussed the effect that the extensive plantations of Kielder Forest might be having on the Border Mires. Coming at a time when the Forestry Commission was adapting to its new multi-purpose role, this paper acted as a catalyst in encouraging the Forestry Commission at Kielder to take a proactive approach to managing its mires. In March 1987 the inaugural meeting of the Border Mires Committee was held and 20 sites covering some 1000 ha were adopted for management. The Committee is led by Forest Enterprise, and includes representatives from English Nature (N.E. and N.W.), Northumberland Wildlife Trust, Northumberland National Park, Newcastle University Agricultural and Environmental Science and Geography Departments and the Forestry Authority.

The Committee has been very proactive with a successful blend of academic and practical knowledge and a willingness to work in partnership. Over eight years, much has been achieved particularly with ditch blocking (over 2200 dams to date) and removal of small trees. Surveys of vegetation, invertebrates and hydrological monitoring have been instigated on several mires. Having a single land owner is no doubt a key factor in helping the partnership to achieve much work. Over the years, different partners have taken the lead on practical work or research depending upon which agency was able to raise funds in any particular financial year. It is accepted that each partner's ability to contribute to the mire management will vary from year to year, some years a partner may be limited in what they can offer but none-the-less they remain a valued member of the partnership. This approach has helped foster a very real sense of joint achievement which has been the key to completing so much work.

Key projects: 1991-1995

Several major projects have been undertaken in the past four years which have shaped current policy and direction on the management of the Border Mires.

1991/92 Border Mires reassessment (Holmes, 1992)

An evaluation of all the open mires sites in the forest in Northumberland resulting in a list of 48 sites (28 more than was first adopted in 1987), which were considered of high conservation value and worthy of managing under the auspices of the Border Mires Committee.

1991 Border Mires hydrological boundaries: a reconnaissance study (Newson and Rumsby, 1991)

A pilot study to look at ways of defining the natural extent of the mires pre-afforestation.

1992/93 Redefinition of hydrological boundaries (Lowe, 1993)

Drawing on experience from Newson and Rumsby (1991) and Lunn (1979), allied to 1947 aerial photos along with ground verification of peat depths, this survey mapped the pre-afforestation extent of the 48 mires managed by the Committee. The survey also mapped catchment areas. The results of this project have had a major impact on future management and are discussed in detail later.

1994 Hydrology and management options for a selection of Border Mires (Newson and White, 1993)

Recommendations for treatment in four situations - open mire, former mire now afforested, afforested soligenous margins and afforested input slopes. An important recommendation from this report was a classification of tree crops according to their position, (former mire, input zone), and their size/age which determines when they are removable/harvestable and at what cost.

Spadeadam Mires redefinition of hydrological boundaries (White, 1994a)

Similar to Lowe (1993) dealing with recently notified sites in Cumbria.

1994 Assessment of the full extent/former extent of ombrotrophic mires in the Border Mires area (White, 1994b)

A survey of all peatlands within Kielder forest. This identified a number of Cumbrian mires (not covered in previous surveys) including some open mire areas as well as some planted up areas.

1995 Ditch survey and mapping of the Northumbrian Border Mires and Cumbrian Spadeadam Mires (White, 1995)

A catalogue of all ditch blocking completed on the open mires plus a list of ditch blocking and tree pulling still required with an assessment of the priority for work on each mire.

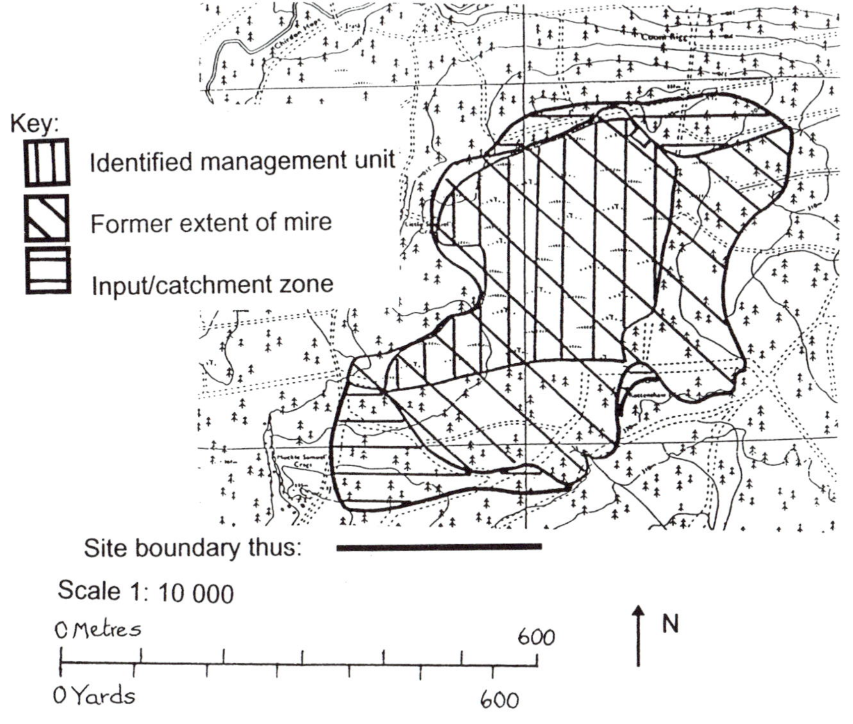

Based on the Ordnance Survey 1:10 000 map with the permission of the Controller of Her Majesty's Stationery Office. Crown Copyright reserved

Fig. 29.1. Conservation unit map for Comm Rigg Moss.

Hydrological units and conservation units

Lowe (1993) and White (1994a) defined the boundaries of the mires under Forest Enterprise management at Kielder. For each mire, a map was produced showing the current open area and the former extent of the mire (now part afforested) - see Fig. 29.1 In addition the maps show the input/catchment slopes surrounding each mire which although not directly affecting the mire, may have an impact on the mire edge or lagg area (the soligenous margin), an area which - in effect - supports the hydrological unit of the bog. Hence forestry drainage within the catchments may have an indirect effect on the mire.

The terminology adopted throughout the surveys and now used by the Border Mires Committee is:

Mire/peat lens = hydrological unit
Lagg area/soligenous margin + catchment = input area
Hydrological unit + input area = conservation unit

Allocation of tree crop types

Newson and White (1993) classified forest crops as follows:

1. Type One; where trees have been planted on the mire and are either of no value (i.e. they are of poor stunted growth) or are young enough/small enough to be removed at low cost. There is an urgent priority to remove these trees and block up any drains.
2. Type Two; where trees have been planted on the mire but are growing well and are too large (>3 m) to be removed at low cost. In many situations these crops are quite large and will have to wait until they are of harvestable size to be removed. There is an urgent need to assess the potential for mire rehabilitation on these sites.
3. Type Three; many sites have a prominent soligenous margin which is of conservation interest and is best left unplanted. When trees are removed this area should not be replanted.
4. Type Four; the precise links to the mire margin of the remainder of the input area is not clear. Forest drainage within these areas should be carefully considered (due to its likely impact on the soligenous margin) but the input can remain afforested and be re-afforested following felling.

Statistics have been calculated for the conservation units and Newson-White crop types for the Northumbrian Mires (see Tables 29.1 and 29.2).

Table 29.1. Conservation units.

	Area (ha)		
	Hydrological unit	Input area	Total conservation units
Open	637	66	
Afforested	556	673	
Totals	1192	739	1931

Table 29.2. Newson - White crop types.

	Area (ha)		
	Hydrological unit	Input area	Total conservation units
Open Mire	637	66	
Afforested			
Type 1	147		
Type 2	409		
Type 3		211	
Type 4		462	
Totals	1192	739	1931

Future management

The three main operational objectives of the recently re-written Border Mires management plan are:

1. To maintain and enhance the open mire communities by maintaining suitable conditions for active *Sphagnum* growth.
2. To re-establish the pre-afforestation hydrological units of those areas felled during the life of the plan (five years - 1996-2001), and to establish methods of mire rehabilitation.
3. To manage the input areas in a manner sympathetic to supporting the hydrological units and determine the best method of dealing with forest drainage water.

To achieve these objectives the main thrust of management during the next four/five years will cover two issues - completing the programme of ditch blocking/rehabilitation on the open mires, and, re-establishing the pre-afforestation hydrological units.

Achieving these aims will be constrained by two factors: firstly, the limitations imposed by silvicultural considerations and the timing of felling, and secondly, the cost of the operations. As there is a substantial programme of work, some of it urgent, there is the additional issue of deciding on priorities.

Silviculture - timing of felling

The extensive peaty and peaty gley soils, frequently waterlogged, mean that many forest crops are prone to windthrow. The philosophy that underpins the felling process is that relatively small blocks (30-80 ha) are felled, each designed to end at a windfirm boundary - if possible a 10-year age gap is left before felling adjacent blocks which will eventually create an uneven aged mosaic which will be a silvicultural advantage in reducing the risk of windthrow. This process is termed 'restructuring'.

Felling part block or coupes is not possible as the exposed edges will blow over. Changing the sequence of felling is not easy as this can have a knock-on effect over a wide area.

Wherever possible felling plans and timing will be adjusted to accommodate mire clearance but the process may take as long as 10-15 years.

Costs

1. White (1995) identified the need for some 5000 dams on the open mires. Average costs of dam installation are around £15 per dam.

2. Table 29.2 identifies 147 ha of Type One forestry requiring urgent clearance. Current costs for cutting and removal of such small trees are around £600-£700 ha^{-1}.

3. It may be necessary to modify standard harvesting techniques on the Type Two and Three sites, or perhaps brash removal will be necessary from these sites. In this case, costs could rise by as much as £10 m^{-3} which - with an average yield of 300 m^{-3} ha^{-1} - would be £3000 for every ha cleared.

4. The commitment not to restock the mire areas can be evaluated in revenue forgone as a cost contribution towards conservation. An approximate costing for revenue forgone is as follows:

Average Yield Class of crops currently growing on mires		10
		i.e. volume production is 10 m^3 ha^{-1} $year^{-1}$
	Current timber value	c. £25 m^{-3}
	Harvesting cost	£10 m^{-3}
	Net Revenue	£15 m^{-3}
	*Revenue ha^{-1} $year^{-1}$ (10 m^3 * £15)*	*£150*
	Restocking cost	£1000 ha^{-1}
40 year rotation:	annual restocking cost (£1,000/40)	£25
	Yearly maintenance cost	£5
	Total annual restock and maintenance ha^{-1}	*£30*
	Net revenue ha^{-1} $year^{-1}$	**£120**

i.e. each ha of cleared mire represents £120 per year in revenue foregone.

Priorities

With such a large programme of work to achieve, much of it quite urgent, it is necessary to develop a system of priorities. This has been achieved by combining three sets of information.

Firstly, the statutory designations: Special Area of Conservation (SAC) and SSSI take precedence over the non-designated sites.

Secondly, removal of small crops - Type One - is considered more urgent than ditch blocking. This assumes that the damage from growing crops is likely to cause a more rapid decline in the quality of a mire.

Thirdly, White (1995) gave a subjective assessment of the urgency for ditch blocking and removal of small trees based on the quality of the vegetation.

The combination of these facts has enabled a priority list to be drawn up. There may be other ways of looking at priorities but, with the information available, this was considered a useful guide to where the most urgent work is to be carried out.

Conclusion

During the early years of the management of the Border Mires by the Border Mires Committee, a clear picture emerged of the original/natural extent of the mires and this has helped in formulating objectives for long term management which include repairing damage to open mires and clearing and rehabilitating afforested areas.

The partnership of agencies has been a key factor in achieving a substantial work programme.

Silvicultural and cost restraints will influence how quickly the objectives can be achieved but the long term future of the Border Mires is now assured.

Chapter thirty:

A Hydrological Approach to Bog Management

Mr Christian Ginzler
Vienna University, Austria.

Introduction

The Puergschachenmmos is one of the last valley-bottom bogs in the inner Alps. It is located in the Enns valley in the county of Styria.

The central part of the bog is owned by the monastery of Admont and from 1976 it was leased by the WWF to prevent peat extraction and cultivation. In 1992 it was declared a Ramsar site. Because of continuing drainage and agricultural intensification around the marginal parts, an initial, expert, study was carried out in 1993 (Bragg *et al.*) and management plans to preserve the outstanding value of the site were suggested. This chapter reports the results of a Masters thesis which was a continuation of this expert study and which calibrates the theoretical model used therein.

Methods

The hydrological conductivity was surveyed using Kirkham's seepage tube method (Luthin and Kirkham, 1949). The measurements were taken at a depth of 120 cm. The tubes had a diameter of 50 mm and the cylinder at the bottom was 100 mm long. A certain proportion of water was added and the time to reach equilibrium was measured.

Theodolite survey was carried out using a basic grid of 50 m wide mesh and readings were intensified towards the ditches. To get a digital elevation model (DEM) the surveyed points were interpolated using kriging methods.

Dipwells (16 mm diameter) were installed at every grid point. The readings took place between April and September 1994. Peat cores were taken from 20 survey points using a Russian peat borer.

The seepage through the catotelm was estimated using the water balance, which was calculated from meteorological data. The following equation (Ingram, 1983) expresses the water balance of an ombrotrophic bog:

$$P - E - U_{cat} - U_{acr} - G - \Delta W = 0$$

where: P = precipitation;
E = evaporation;
U = lateral seepage;
G = vertical seepage;
ΔW = increased storage.

Precipitation data was available from the meteorological station in Admont. However, the potential evaporation is measured on only a few sites. In this study Thornthwaite's formula (1948) was used to calculate potential evaporation. For simplification, G and ΔW is set at zero, assuming that there is neither vertical seepage into the mineral base nor a change in the water content of the bog. Ivanov (1981) defined the border between acrotelm and catotelm as the lowest water table over a year-long period. Thus, the year with the lowest net-precipitation between 1913 and 1991 was taken as the year with the lowest water table. The amount of this net-precipitation should be the amount of seepage through the catotelm (Bragg, 1992)

Results

General features and installations on the site

The whole area is shown in Fig. 30.1. It is bounded to the south by the river Enns, to the east and west by two small brooks, and to the north by a road. The central part of the bog is surrounded by a network of ditches which discharge into the river. The drains are between 2 and 4.5 m deep, and there are some buried field drains.

Water balance

The precipitation data and the calculated values for the potential evapotranspiration revealed 1934 to be the year with the lowest net precipitation. Fig. 30.2 shows the period from 1970 to 1991 as an example of the estimated water balance.

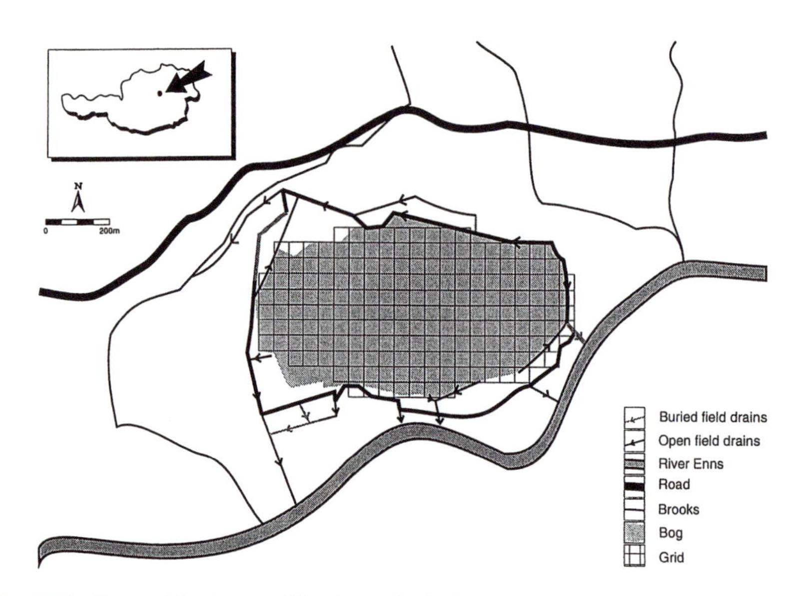

Fig. 30.1. General features of the investigated area.

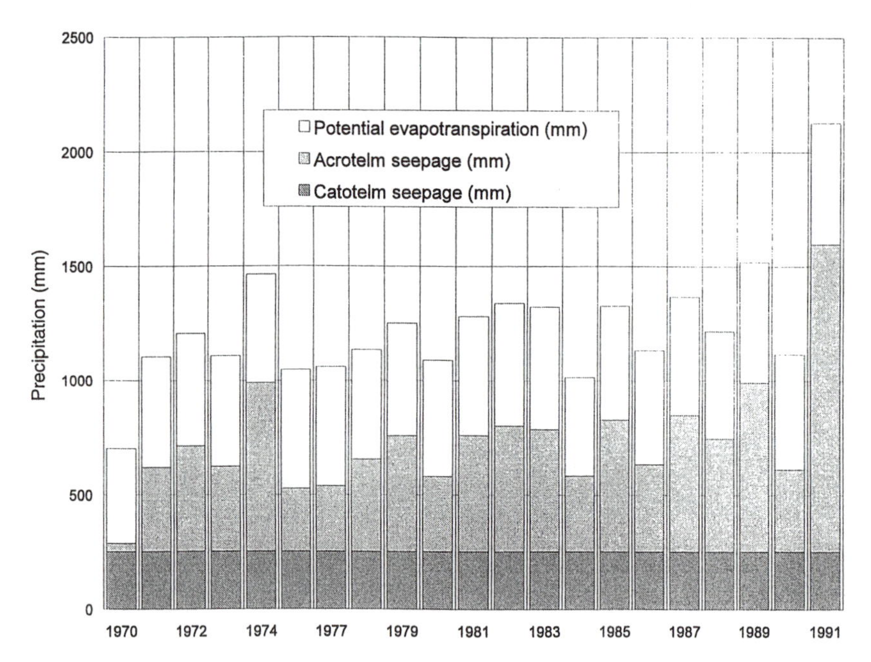

Fig. 30.2. Estimated annual water balance for the Puergschachenmoos (1970-1991).

Hydraulic conductivity

The results of the seven measurements of hydraulic conductivity are typical for bog peat (Ingram, 1983; Ivanov, 1981). The mean value of 2.86×10^{-2} cm s^{-1} was used as one of the four parameters of the groundwater mound formula.

Ground water mound concept

One way of understanding the hydro-ecology of a raised bog is the concept of the groundwater mound theory. The groundwater mound is maintained by a dynamic equilibrium between the water supply and lateral seepage through the low permeable sub-surface peat. The hydrological limit is set by the requirement that peat can only be accumulated, if it is perennially waterlogged (Bragg and Steiner, 1996).

The shape of a raised bog is a function of:

1. water supply (P)
2. permeability of the subsurface peat (K)
3. plane shape and dimensions of the bog (A, B)
4. distance of the water table at the edges of the dome to the mineral base (Z_0).

Change of any of these factors will alter the shape of the groundwater mound. This concept is illustrated by Fig. 30.3. Adapting the general solution to the groundwater mound expression (Childs, 1969) for a raised bog with an elliptic base, using the calculated seepage in the subsurface peat and the field-measured hydraulic conductivity, it was possible to establish a model representing the equilibrium of the present situation and of hypothetical conditions.

Groundwater mound modelling

Present situation

The Puergschachenmoos is a domed peat bog. Its plane shape approximates to an ellipse with a major axis (A) of 550 m and a minor axis (B) of 325 m. The highest point is 4.7 m above the base of the ditches around the margins. The drains are up to 3.2 m deep. Cores at the bottom of the drains showed an additional 3 m of peat to the mineral ground. A cross section of the bog is shown in Fig. 30.4. The Puergschachenmoos is the remaining centre of a former much larger bog. The disturbance is most visible at the north-eastern edge, where one finds the deepest drains. Due to increased decomposition at the surface and the resulting loss of height, the elevation towards these deep drains is higher than that towards the western drain. The water table remains close to the surface (mean depth in the driest month = 21 cm) but changes in vegetation indicate changes in hydrology.

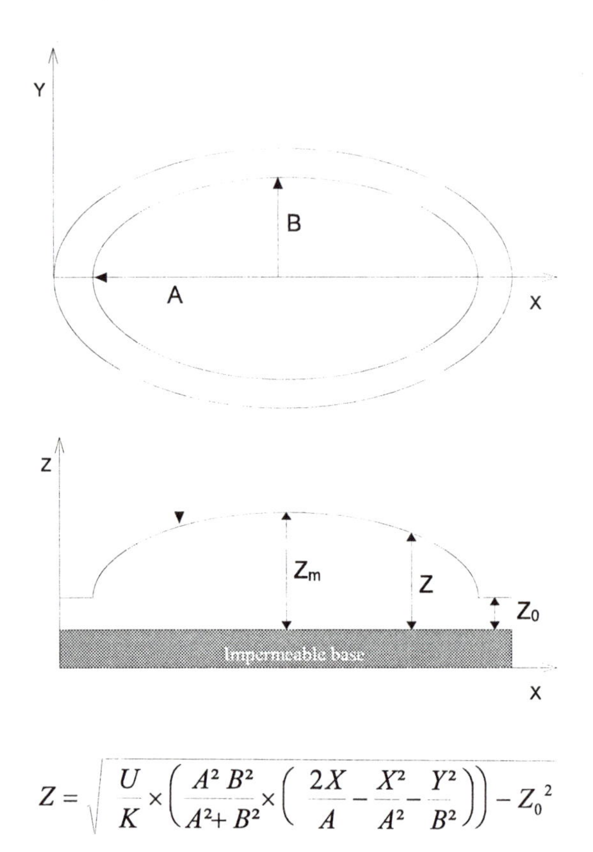

$$Z = \sqrt{\frac{U}{K} \times \left(\frac{A^2 B^2}{A^2 + B^2} \times \left(\frac{2X}{A} - \frac{X^2}{A^2} - \frac{Y^2}{B^2} \right) \right) - Z_0^2}$$

$$Z_m = \sqrt{\frac{U}{K} \times \frac{A^2 B^2}{A^2 + B^2} - Z_0^2}$$

A: Major axis of the elliptic base
B: minor axis of the elliptic base
X, *Y*: plane co-ordinates
Z: height at any point of the dome
Z_m: maximum height
Z_0: distance from the water table in the lagg to the mineral base
U: seepage through the catotelm
K: hydraulic conductivity of the catotelm
▼: water table

Fig. 30.3. Equation of the groundwater mound with elliptic plane shape.

Estimation of original bog-expanse

The thickness of the catotelm in the centre of the bog is about 7.5 m. Taking this value and the climatic conditions of this century into account, the elliptical plane shape of the original dome would have had a major axis of 900 m and a minor axis of 450 m (Fig. 30.4a). This conforms to both the area's soil map and expert estimations (Bragg *et al.*, 1993).

Scenario of future development without any management

Incorporating the hydraulic conductivity (K) of 2.86×10^{-5} m s^{-1}, a lateral seepage (U_{cat}) of 9.7×10^{-9} m s^{-1} and the dimensions of the present plane shape into the groundwater mound equation provides a maximum height for the bog (Z_m) (under present conditions) of 5.9 m. As shown in Fig. 30.4b, a new groundwater mound will establish. Due to the lowering of the water table and the decline of the dome, there will be peat wastage of more than 2 m all over the bog.

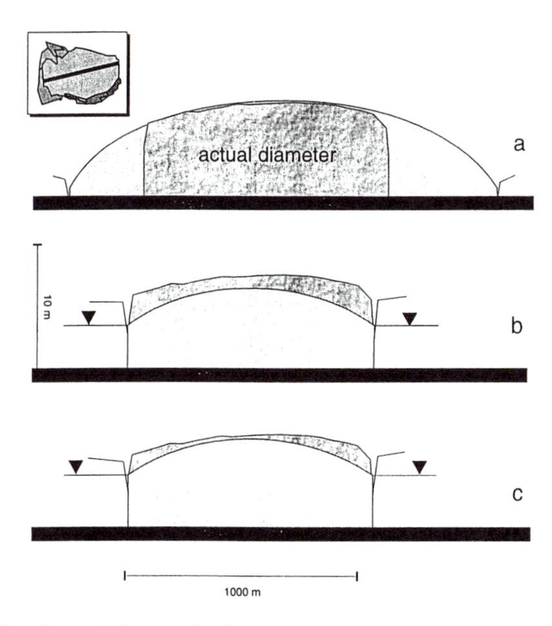

a: Estimation of the original bog expanse.
b: Scenario of future development without any management.
c: Raising the water table in the surrounding pasture by 1.5 m.
▼: water table

Fig. 30.4. Cross sections through the Puergschachenmoos (actual diameter) and possible scenarios.

Scenario of future development with management

Z_0 is the only parameter of the model which can be manipulated. To keep the water table close to the surface of the bog, the base of the groundwater mound has to be raised. This can be achieved by managing the boundary drains, but - even more importantly - the drains in the surrounding pasture. If it is possible to raise the pasture by 1.5 m (Z_0 = 4.5m) the calculated height of the groundwater mound would be 6.8 m (Fig. 30.4c).

Discussion

Although the concept of the groundwater mound theory incorporates simplifications, it remains of threefold importance (Ingram, 1992). Firstly, it provides a simple explanation for the morphology of raised bogs. Secondly, it suggests how their stability is maintained. Thirdly, it indicates how one might predict the consequences of human cultural interference.

This study shows how the prediction of future scenarios is useful for management recommendations. The results of the calibrated model are similar to those made following the initial study (Bragg *et al.*, 1993). The data required to adapt the groundwater mound model to certain sites, hydraulic conductivity of the catotelm and standard meteorological data, are quite easy to obtain.

For conservation purposes it is important to understand that conservation of ombrotrophic raised bogs implicates management and conservation of the surrounding land. The application of the groundwater mound theory gives a valuable tool to establish such management concepts.

Acknowledgements

Thanks to Dr G.M. Steiner who introduced me to the subject and gave me the possibility to present my results at the Peatlands Convention.

Chapter thirty-one:

Rehabilitation Work on Post-Harvested Bogs in South Eastern Canada

Dr Line Rochefort and Ms Suzanne Campeau
Université Laval, Canada.

Introduction

While approximately 12% of Canada's landscape is covered by peatlands (Zoltai, 1988), the impact of agriculture, drainage, urbanisation and peat harvesting on bogs and fens has mainly been concentrated in the southern portion of the country. In southern Québec and in New Brunswick, peat harvesting dates back to the beginning of the century and is an important local industry for some communities. In the sixties and seventies, vacuum-harvesting gradually replaced the labour-intensive manual block-cutting technique.

In contrast to what can be found in other regions or countries, only the moderately decomposed *Sphagnum* peat was, and still is, mined from Canadian bogs. Peat mined in Canada is used mainly for horticultural purposes (Keys, 1992) and, more recently, for other industrial uses such as the production of absorbent boards and biofilters. Once peat mining reaches layers of peat that are too decomposed for the intended use, harvesting stops and peat fields are abandoned. In the last few decades, a number of sites were also abandoned for reasons not related to the status of the peat itself (e.g. a small company closing down its operations after a few years). Therefore, most abandoned peat fields are still covered by variable thickness of residual peat. Approximately 30 abandoned sites are found in Québec and New Brunswick (Line Rochefort, unpublished data). The size of these sites varies greatly and many sites are parts of larger peatlands that are still currently being harvested.

Abandoned, vacuum-harvested sites are characterised by relatively flat surfaces bordered by drainage ditches and are generally fairly young, having been abandoned between one to 15 years. In contrast, sites that were abandoned after mining with the block-cutting method are older and have characteristic alternate baulks and trenches. Revegetation patterns of sites harvested with the two methods also differ. Abandoned sites mined with the block-cutting technique seem to revegetate with peatland species more easily than do vacuum-harvested sites, although a *Sphagnum* moss layer may not necessarily be present in the newly established vegetation (Lavoie and Rochefort, 1996). Vacuum-harvested surfaces, in contrast, can remain almost bare even ten years after peat mining stops, and thus are likely to need some form of intervention to ensure the return of a typical peatland flora.

A multidisciplinary and collaborative research project

In 1993, a collaborative research project on peatland rehabilitation was launched in Québec. The objectives of the project were:

1. To develop restoration techniques that would allow abandoned peat fields to once again become functional wetland ecosystems.
2. To further build our knowledge of peatland fauna, flora and ecology in Québec.
3. To produce a restoration field guide for peat producers.

The project is a collaborative effort between university researchers, governmental and non-governmental agencies and the peat industry. A smaller research project involving the New Brunswick government and a peat company is also taking place in north-eastern New Brunswick.

Peatlands are complex ecosystems and research aimed at restoring abandoned sites to functional wetlands necessitates a multidisciplinary approach. This is well reflected by the research team which includes not only plant ecologists, but also bird, arthropod, microbiology and hydrology specialists. A number of graduate and undergraduate students and several research assistants are also involved. The multidisciplinarity and partnership components of the projects ensure a healthy exchange of information and opinions, and allows for a productive sharing of financial and material resources as well as expertise.

The ongoing projects also encompass a broad geographical range. Restoration experiments were conducted at three different vacuum-harvested sites in Québec and at one vacuum-harvested coastal site in New Brunswick (Fig. 31.1). In addition, comparison of the chemistry, fauna and flora of natural and post-harvested sites of different ages were conducted at several more vacuum-harvested and block-cut sites. Finally, extensive vegetation surveys of abandoned sites were conducted throughout south-eastern Canada.

During these projects, efforts were concentrated on developing restoration techniques adapted to vacuum-harvested peatland, not only because they seem to revegetate less easily than block-cut sites but also because these sites represent the type of residual surfaces that will be dominant in the future. Therefore, the planned restoration field guide will deal more specifically with the restoration of vacuum-harvested peat fields.

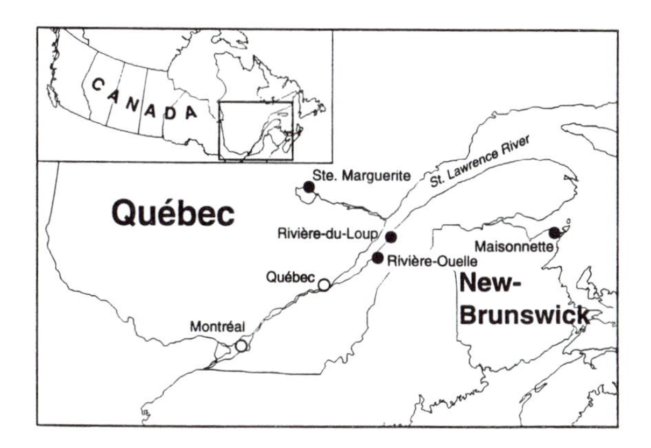

Fig. 31.1. Location of the vacuum-harvested peatlands used for restoration experiments in Québec and New Brunswick (indicated by the filled circles).

Approaches to vegetation re-establishment on cutover surfaces

A number of experiments throughout the project have specifically addressed the question of promoting moss and vascular plant recolonisation on to the remaining bare peat of cut-over surfaces. Indeed, the re-establishment of typical mire species, *Sphagnum* mosses in particular, is an essential element to the re-establishment of a functional peatland ecosystem (Wheeler and Shaw, 1995b).

A major hindrance to typical mire vegetation re-establishment on abandoned peat fields is the lowered water table, caused by drainage associated with peat mining (Wheeler and Shaw, 1995b). Another possible reason for the low rate of natural revegetation of these sites is that the remaining peat is almost devoid of plants.. Some form of plant reintroduction thus seems necessary to ensure recolonisation of the surface by vascular plants and mosses. In contrast to situations found elsewhere, where peat can be harvested almost down to the mineral layer, an important peat layer remains on the abandoned sites in Canada. Water and peat chemistry in the four experimental peat fields were shown to resemble bog or poor fen conditions, with the exception of being somewhat nutrient enriched (H. Wind-Mulder, unpublished data). It was hypothesised that direct reintroduction of bog or poor-fen vegetation should be possible without retracing all the successional vegetation stages leading to peatland formation.

Plant re-introductions

Researchers have shown that almost any portion of a living *Sphagnum* plant, possibly with the exception of leaves, can regenerate a new individual when isolated from the parent plant (Poschlod and Pfadenhauer, 1989; Cronberg,

1991; Rochefort, Gauthier and Lequéré, 1995). Can this regeneration potential of *Sphagnum* be used to re-establish a moss carpet on bare peat?

We first attempted to answer this question in a greenhouse trial where the recolonisation potential of *Sphagnum* fragments exposed to different water levels was examined (Campeau and Rochefort, 1996). The plants were collected from the top 10 cm of the peat column, cut into fragments, and hand spread on a series of peat-filled containers where the water level was maintained at either 5, 15 or 15 cm below the peat surface. The containers were watered regularly to imitate, in quantity and quality, the water input received from the rain in the field. This experiment demonstrated that it is possible, starting from fragments sparsely distributed on the peat, to re-establish a complete *Sphagnum* cover. Results also showed that recolonisation success of *Sphagnum* strongly depends on the water level in the peat substrate (Campeau and Rochefort, 1996). At high water level, for most species tested, the fragments produced a complete *Sphagnum* cover within six months. In drier conditions however, only 10-20% of the surface of the peat-filled containers was covered by *Sphagnum* after six months (Campeau and Rochefort, 1996). Similarly, other researchers have also suggested that the water table in a cut-over site needs to be maintained very near the pear surface for *Sphagnum* mosses to have a chance to establish (Money, 1995; Wheeler and Shaw, 1995b).

Rewetting the cutover surface

In an attempt to rewet the four experimental post-mined peatlands bulldozed peat was used to block drainage ditches. Generally, blocking the ditches successfully raised the water table close to the surface in the spring and fall. In the summer, however, the water level tended to drop further, and to vary more, in the experimental sites than they would in an undisturbed bog (Fig. 31.2). These observations are not unique to North American cutover peatlands. Other researchers have encountered similar difficulties when raising the water table of post-harvested sites to try and ensure favourable conditions for peatland vegetation re-establishment (Schouwenaars, 1988; Wheeler and Shaw, 1995b). Large scale increases in the water table of a post-mined area may be further complicated by the fact that, in eastern Canada, these sites often lie adjacent to areas where peat is still being extracted. In other countries, like Switzerland, raising the water table to an effective level for restoration purposes may also be very difficult in cases where the remaining bog forms an elevated island surrounded by intensive agricultural land (Grosvernier *et al.*, 1995). On surfaces that are not permanently inundated, the peat surface might not only dry up on a warm summer day but often form a crust preventing *Sphagnum* fragments from gaining access to the water in the underlying wet peat (Grosvernier *et al.*, 1995; L. Rochefort and J. Price, unpublished data). Even when high water level conditions are achieved, such as in shallow pools, research suggests that only lawn forming species of Sphagna would be able to recolonise readily (Money, 1995).

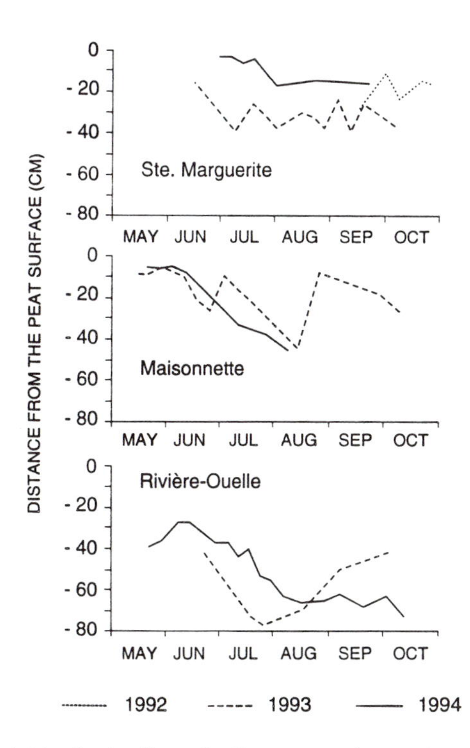

Fig. 31.2. Water table fluctuations in three experimental cut-over sites where drainage ditches were blocked with bulldozed peat.

Sheltering the reintroduced diaspores

Hence, *Sphagnum* fragments need to be kept relatively wet in order to survive and regenerate new individuals. In the field however, it is difficult to raise the water table to a level that will be sufficient to promote *Sphagnum* regeneration and growth. Our experiments thus concentrated on finding an alternative to a high water table in order to provide appropriate humidity conditions for *Sphagnum* and other bog plants to re-establish. This approach parallels that proposed by Grosvernier *et al.* (1995), who suggested that water table level, although it is a fundamental factor, may not be an exclusive necessity for the re-initiation of *Sphagnum* growth.

The project explored various methods to prevent the plant diaspores from drying up. Using an irrigation system composed of a pump and sprinklers, we attempted to wet the diaspores regularly during warm summer days (D. Bastien, MSc thesis in preparation; F. Quinty, unpublished data). The effect of enhancing the microtopography of the site (in order to create protected microhabitats that would be more suitable to diaspore regeneration) was tested, as was the use of covers, straw mulches and companion vascular plant species to shelter the diaspores against solar and wind desiccation. Finally, we synchronised plant re-introductions with the wettest period of the year (spring and fall) and thus provided a 'shelter in time' to the diaspores at the onset of their re-establishment.

Overview of results

Enhancing microtopography

The objective of an enhanced microtopography is to create a variety and abundance of protected micro-sites in which the diaspores stand a better chance of surviving and regenerating. It can be achieved by harrowing, ploughing or simply by repeatedly passing heavy machinery over the wet bare peat in the spring. Quinty and Rochefort (1997) showed that *Sphagnum* re-establishment is favoured in the depressions of an area with enhanced microtopography in comparison to flat areas seeded with *Sphagnum* diaspores at a similar density. Other field trials conducted at different sites corroborate these findings (J.-L. Bugnon, MSc in preparation; C. Ferland, MSc in preparation; F. Quinty, unpublished data).

Using a protective cover

The various field trials we have conducted with artificial covers, straw mulches and companion plant species repeatedly demonstrate the positive effect these have on typical mire species re-establishment, including *Sphagnum* mosses (Quinty and Rochefort, 1997). The positive effect of a protective cover may be related to the improved humidity and temperature conditions experienced by the diaspores at the peat-air interface (J. Price and F. Quinty, unpublished data; Grosvernier *et al.*, 1995).

At Ste. Marguerite, the wettest of our four experimental sites, a *Sphagnum* ground cover of approximately 20% was obtained after two years on plots where *Sphagnum* diaspores were reintroduced manually in the spring of 1994 at a density ratio of 1:10 (i.e. taking 1 m^2 square of material from a natural area and spreading it over 10 m^2 of cut-over peatland) and then protected by a straw mulch. Interestingly enough, both lawn and hummock species re-established equally well in these experiments, in contradiction with the low re-establishment of hummock species in comparison to lawn species observed when *Sphagnum* diaspores are reintroduced in shallow water or on floating rafts (Money, 1995).

At the driest of our three sites, Rivière-Ouelle, mechanically shredded material from the surface of a bog was reintroduced in 1993 at an initial density ratio of approximately 1:10. Although recolonisation was slower at this site,

Sphagnum cushions are now developing beneath the straw, in association with other mosses such as *Polytrichum* sp. Typical vascular bog species, which were present in the reintroduced material as seeds or as pieces of roots and rhizome are also slowly becoming established. After two years, the total vegetation cover of the plots with straw mulch was 20%, including a 5% *Sphagnum* cover (F. Quinty and L. Rochefort, unpublished data). In contrast, plots with no protective cover show barely any moss or vascular plant recolonisation from the reintroduced material.

Field experiments conducted in New Brunswick demonstrated that a well established community of vascular plants can provide cover to effectively promote *Sphagnum* re-establishment (C. Ferland, MSc in preparation). These results corroborate the findings of Grosvernier *et al.* (1995) and Salonen (1992) on the positive effect of a plant cover on *Sphagnum* and vascular species re-establishment on bare peat.

Perspectives for peatland rehabilitation

Can sheltering really compensate for a water table that may not be optimal for the re-establishment of *Sphagnum*? Can bog plant fragments provided with a protective cover really be successful at recolonising peat fields where the water table, although high in the spring and fall, can drop 20, 40 or even 60 cm below the surface during the summer? And how much surface material will be needed as a source of diaspores, so that we are not destroying natural bogs in order to restore post-mined sites? To answer these questions, large scale restoration trials are currently underway in collaboration with peat producers, using the techniques which proved to be most promising in the small-scale experiments. The purpose of larger-scale restoration experiments is primarily to test the effectiveness of the proposed approaches and the stability of the re-established vegetation. These trials will also allow us to determine how restoration work can be done by locally available machinery and at what cost. In addition, they will provide the chance to assess the impact of collecting diaspores for restoration purposes on the natural source area.

Conclusion

It is clear that returning a post-harvested site to the state of functional peatland ecosystem is not only a matter of restoring a moss layer or a certain set of typical mire species. However, as these plants are a key component of bog ecosystems, their re-establishment on a cutover surface is one of the main challenges that needs to be addressed in order to succeed in rehabilitating post-mined sites. The results obtained so far in eastern Canada suggest that it could be possible to re-establish typical bog plant species, including *Sphagnum* mosses, on the seemingly harsh and inhospitable bare peat surface of a cutover peatland by using methods that are relatively simple and cheap, such as reintroducing shredded bog vegetation, using enhanced microtopography, and providing some sort of protective cover for the reintroduced diaspores. Obviously, only large-scale experiments and long-term data will demonstrate the success or failure of the approaches we are currently testing. We, however, believe that our finds to date do open up new and promising avenues for the rehabilitation of post-mined peatlands.

Acknowledgements

The projects presented in this chapter were supported by a number of public and private organisations: The Québec Ministère de l'Environnment et de la Faune, the New Brunswick Ministry of Natural Resources and Energy, the Centre Québécois de Valorisation de la Biomasse, the Canadian Wildlife Service, the Centre de Recherche et de Développement de la Tourbe in New Brunswick, the Canadian Sphagnum Peat Moss Association, the Association Québécoise des Producteurs de Tourbe, Fafard et Frères Ltée, Lambert Peat Moss Inc., Premier Peat Moss Ltd and SunGro Horticulture Inc. The authors would also like to thank the graduate students, undergraduates students and research assistants involved in the projects for their contribution.

Chapter thirty-two:

Rehabilitation of Milled Fields

Dipl Ing Jan Sliva, Dr Dieter Maas and Professor Jörg Pfadenhauer
Technical University of Munich, Germany.

Introduction

Regeneration of milled raised bog expanses with the aim of rehabilitating a functioning acrotelm has proved to be extremely difficult until now. The complete removal of the vegetation layer by the milling procedure, the heavy drainage of the peat layer and the irreversible change in hydrochemical properties by open-casting minerotrophic peat horizons are the main reasons which prohibit a successful rehabilitation of such abandoned milled fields. The peat surface left after milling is completely devoid of any seed bank so that plant recolonisation can be compared to a primary succession (Salonen, 1987), but often under adverse microclimatic circumstances. An indispensable prerequisite for the formation of a new acrotelm (be it due to long-term natural succession or due to accelerating artificial measures) is the stabilisation of the peat water table at, or slightly below, the peat surface (Pfadenhauer and Klötzli, 1996), and an oligotrophic water quality. These conditions can not always be created by filling drains or by damming ditches. The success of rewetting is also dependent on the quality of the remaining peat layer. Thus, in the north-western German extraction areas with a remaining layer of strongly decomposed and water impermeable 'Schwarztorf' (Kuntze and Eggelsmann, 1981; Eggelsmann, 1988), rewetting is more easily accomplished than in the pre-alpine regions where these damming layers are missing. In spite of the cost, intensive re-wetting measures in southern German bog areas often result in an unbalanced water regime.

Starting from this situation, possibilities and limitations of the regeneration (i.e. the restitution of more natural conditions, following Pfadenhauer, 1989) and rehabilitation (i.e. the initiation of the regrowth of a raised bog) of raised bog are investigated in a number of observations and experiments in the Kendlmühlfilzen, an important extraction site in Bavaria. An important aspect of the investigation were measures of how to accelerate the efficient re-establishment of plants on to bare peat surfaces, be it by introduction of diaspores of selected flowering plants, by transplanting of *Sphagnum* sods or the treatment of the surface by shading or fertilisation.

Site description

The Kendlmühlfilzen lie in the main basin of the former Chiemsee glacier (scraped out during the last glacial period), almost equidistant between Munich and Salzburg, directly south of the Lake Chiemsee (Fig. 32.1). The lake is situated in the glacial basin and was formerly much more extended. The extent of the lake has been successively reduced by fen formation combined with deposition of river sediment (originating from the northern calcareous Alps). Transition and raised bogs formed in the fen area due to the local climatic (mean annual precipitation 1410 mm; mean annual temperature 7.3°C), hydrological and edaphic conditions. This is a typical pattern still frequently found in the pre-alpine region today. The Kendlmühfilzen with an area of c. 900 ha are one of several raised bog cores in the peatlands south of Lake Chiemsee. Directly south of the Kendlmühlfilzen, the northernmost mountain chain of the Alps rises to maximum elevation of 1660 m a.s.l.. Several brooks originate from this mountain chain, and east of the Kendlmühlfilzen the River Tiroler Achen crosses through the Chiemsee peatlands.

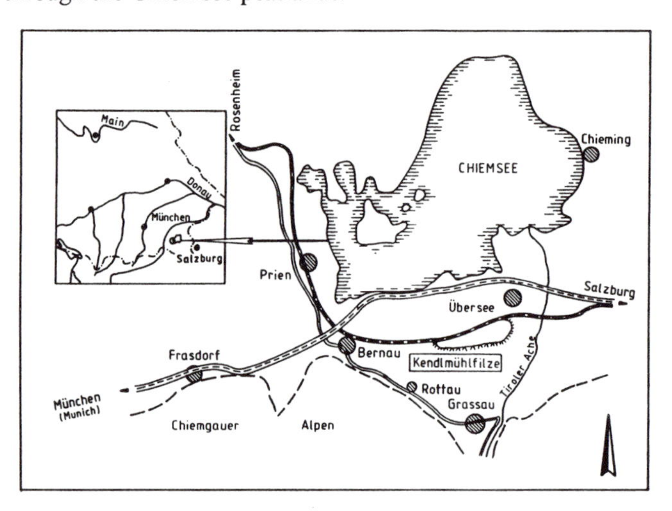

Fig. 32.1. Geographic location of the Kendlmühlfilzen.

A cross-section from south to north (Fig. 32.2) shows the peat stratigraphy and the natural slopes of the bog surface and mineral bedrock (approximately 0.5%). North of the Kendlmühlfilzen the mineral bedrock rises up to a low hill (Westerbuchberg), forming a kind of natural barrier. It can also be seen that the southern half of the Kendlmühlfilzen is nearly intact with a thickness of raised bog peat reaching 4 m. The northern half is heavily disturbed by peat extraction and immense drainage systems sometimes reaching down to the mineral bedrock. A remarkable property of the Kendlmühlfilzen is the inclusion of silt and clay layers in the fen peat. These were deposited by rivers crossing the peatland on their way from the Alps to Lake Chiemsee. A result of these mineral layers is the division of the peat water table into two partially separated layers. The deeper of these is under pressure due to the groundwater flowing in from the elevated southern neighbourhood. Since the Westerbuchberg forms a natural barrier, this groundwater table is under considerable artesian pressure. Where there are gaps in the mineral layers, and artesian pressure is high enough to overcome the downward pressure from the upper water-table and peat layers, minerotrophic water passes to the surface. This forms minerotrophic islands within the raised bog (Pfadenhauer *et al.*, 1990). A partial cross-section (Fig. 32.4; recently abandoned milling sites) from W to E shows the same phenomenon.

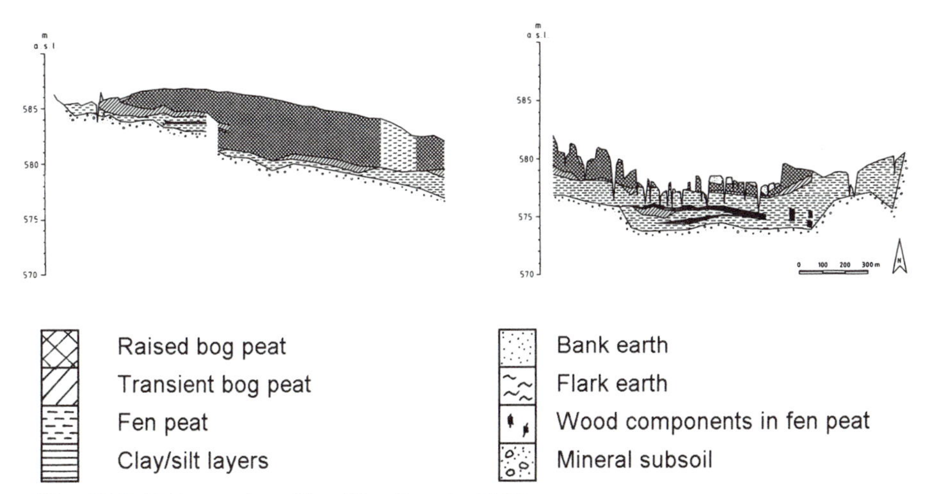

Fig. 32.2. S-N - peat profile of the Kendlmühlfilzen.

History of peat extraction

All areas within the Kendlmühlfilzen where peat extraction ever took place are shaded (Fig. 32.3). Peat extraction dates back to the middle of the last century (Gipp, 1986). Hand cutting by local people for fuel peat was common during the first half of this century and declined at the beginning of the fifties. The technique used for hand cutting is described by Pfadenhauer and Kinberger (1985). Hand cutting is restricted to private allotments near the fringes of the raised bog and in the southern half of the Kendlmühlfilzen. The northern half has been owned by the Bavarian state since 1824. Commercial peat extraction in this section began during the last decades of the last century and was continued by governmental organisations until 1971. The dotted area was later let to a private peat firm which continued extraction by milling until 1985.

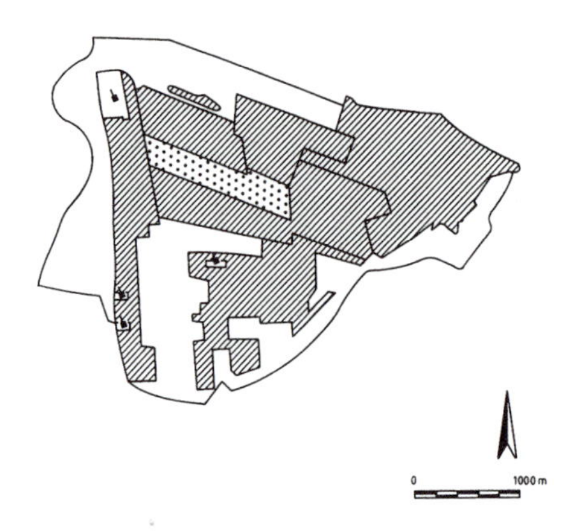

Former peat extraction   Milled area   Existing hand cuttings

Fig. 32.3. Peat extraction areas in the Kendlmühlfilzen.

The peat extraction in the northern half of the Kendlmühlfilzen was promoted by the Bavarian state with the objective of cultivating the peatland for agricultural or forestry after-use. With the experience of its economic ineffectivity, growing environmental awareness and increasing political pressure it was decided to stop peat extraction and dedicate the latest extraction fields to rehabilitation of the former raised bog ecosystem.

The milled area in the Kendlmühlfilzen is subdivided into several fields in W-E direction, separated by drainage ditches (Fig. 32.4). It has a general slope from E to W of 0.3%, and slopes towards the drainage ditches within each of the strips. Approximately 0.5 m of raised bog peat remains except at the western end of the milled area where transition bog peat forms the new surface. Such conditions are not very promising for the realisation of the pre-defined aims of rehabilitation, but may be representative for most of the remaining peat extraction areas in Bavaria where exploitation will end within the next years. The situation on milled raised bogs as compared to the undisturbed state is described by Maas and Poschlod (1991). To create conditions under which a re-establishment of vegetation, and even of raised bog species, may be at all successful, the slope left in the milled area after peat extraction was eliminated by construction of even basins surrounded by low peat walls (Fig. 32.4). The main objectives are to reduce peat erosion to a minimum, to retain water and pond it to a depth of 5 cm (which proved to be most successful for the establishment of raised bog plants) (Poschlod, 1988).

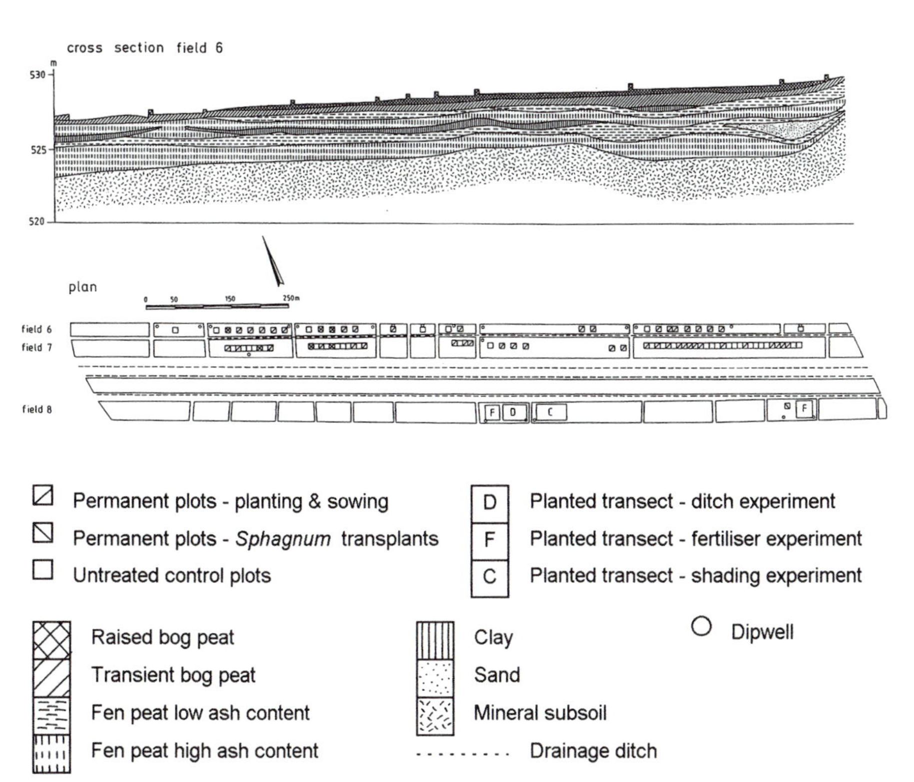

Fig. 32.4. W-E - peat profile (a) and survey (b) of the milled area in the Kendlmühlfilzen with location of permanent plots and experimental transects.

Rehabilitation experiments

Rehabilitation of c. 30 ha of milled fields began in 1986. The different experiments conducted since then, are based upon experience from a comprehensive study monitoring the biotic and abiotic conditions in numerous abandoned peat extraction sites in southern Bavaria (Pfadenhauer, 1989; Poschlod, 1990) and the results of rehabilitation experiments in north western Germany (Pabsch, 1989; Eigner and Schmatzler, 1991). Experimentation and monitoring began in 1986 on the northern half of the milled area (fields six and seven). Experiments on the southern half (field eight) started in 1990.

a) Monitoring and artificial introduction

Methods

The northern fields (six and seven) are used for monitoring the spontaneous establishment of plant species and vegetation succession as well as to test recolonisation success of introduced plants. Overall 83 experimental plots of 10 x 10 m were installed on the fields to test the efficiency of artificial introduction of typical transition and raised bog plant species to accelerate the development of the respective vegetation types (Fig. 32.4; for methods see Maas and Poschlod, 1991). The following species have been tested (nomenclature follows Rothmaler 1988 and Daniels and Eddy 1990):

- *Eriophorum vaginatum* L. - as an eurotypic tussock species in raised bogs and transient bogs.
- *Eriophorum angustifolium* Honck. and *Carex rostrata* Stokes - as successful species in transient bog and fen.
- *Andromeda polifolia* L. and *Scheuchzeria palustris* L. - as rare species of special conservation value.
- *Carex curta* Good. and *Trichophorum alpinum* (L.) Egor. - as species with expressed vegetative spread typical for areas with higher availability of mineral nutrients.
- *Eriophorum latifolium* Hoppe as fen species with vegetative spread.
- *S. magellanicum* Brid., *S. angustifolium* (C. Jens, ex Russow) C. Jens. and *S. cuspidatum* Ehrh. ex Hoffm. - used to test their survival abilities under the present conditions.

For monitoring the treated plots, as well as spontaneous vegetation development, the fields are subdivided into contiguous 10 m wide strips running perpendicular to the drain ditches. This allows the exact location of all vegetation monitoring data for a complete overview of the milled area. The cover of all plants present in each strip is estimated once a year using a cover scale with eight classes (<1%, 1-3%; 3-5%; 5-12.5%; 25-50%; 50-75%; 75-100%). The arithmetic means of these cover intervals were used. Dipwells (see Fig. 32.4 for position) are used to measure the groundwater table during three-weekly numerical analyses and for extraction of water samples for chemical analyses (pH, conductivity, cation concentration).

To define the main factors responsible for the establishment of certain species and thus to define areas within the milled field where the establishment of the most critical group of raised bog plants, (the Sphagna) might be promising, vegetation, water level and water chemistry data are subject to a detrended canonical correspondence analysis (DCCA). The increase or decrease in cover of each species in each of the monitoring strip can be followed from year to year. The number of strips were a species increases, decreases or just maintains its dominance value is calculated for the time span from 1988 to 1994.

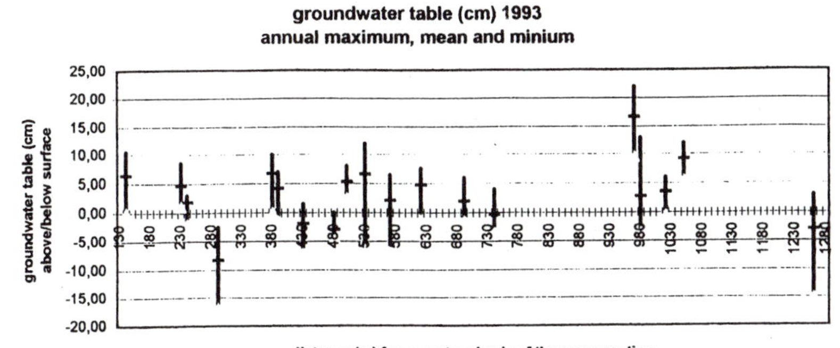

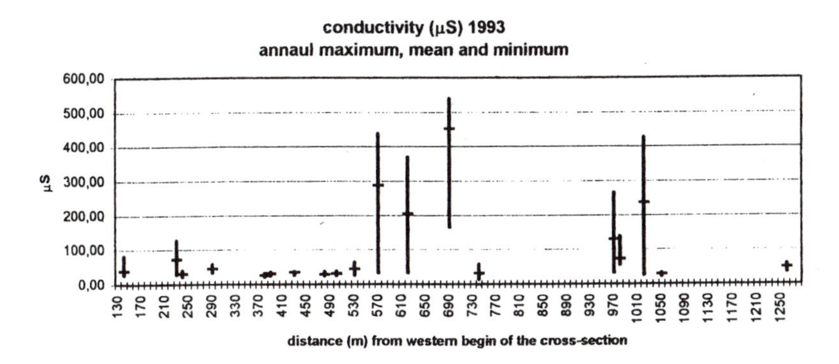

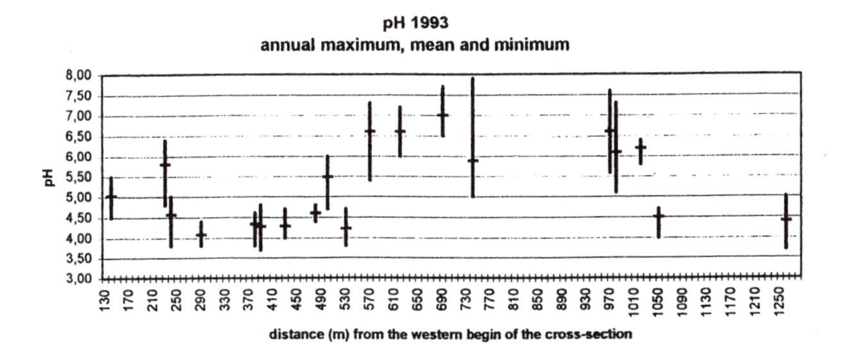

Fig. 32.5. Fluctuations of water table, conductivity and pH on field number 6 in 1993 (means, maxima and minima for the period from January to November, based on 16 recording dates). The values for the water table refer to the peat surface.

Results

The results of monitoring groundwater properties on the milled field six in 1993 are given as example in Fig. 32.5. All three factors (water table, conductivity and pH) show a considerable variation across the rehabilitation area, not only with respect to their annual means, but also with respect to their annual fluctuations. The range of the values (annual maxima and minima) must be seen as an important variable with respect to the establishment of plant species.

Two main trends can be derived from these results: firstly, there are areas where conductivity and pH are in a range which can be seen as favourable for the existence of raised bog species (conductivity ranging between 30 and 50 μS, pH ranging between 3.8 and 4.5). Such conditions are mainly found in the western part (250 and 450 m from its western end) and near the eastern end of the experimental area (from 1050 m eastward). Unfortunately, in many cases where the chemical conditions are promising, the groundwater table often does not reach the peat surface or drops below it for long periods. Only where the water table can be kept over the peat surface for most of the year (or where it only drops some centimetres below it for short time spans), can re-establishment of raised bog plants be successful in the long term.

On the other hand, the hydrological conditions are favourable over large parts of the experimental area, where permanent slight flooding was achieved by the preparation of the milled field, or where the water table fluctuates around the peat surface without dropping too deep during drought periods (from the western end of the area to 230 m, between 480 and 1050 m). Here the problem lies in the high cation content of the water, indicated by high conductivity and pH values, both with wide annual ranges. This means, that the conditions for plant growth are likely to favour those species able to cope with minerotrophic conditions and fluctuating water tables.

The plant species associated with these conditions can be derived from Fig. 32.6. The DCCA diagram for the 1993 data shows that presence and dominance of plant species across the experimental area can best be explained by the mean annual pH values and the range of groundwater fluctuation. Both environmental factors are closely correlated with the first ordination axis. The annual range of groundwater conductivity is also still slightly correlated with the first ordination axis. The annual range of pH values is correlated with the annual range of conductivity and is therefore not displayed. Neither the annual mean of conductivity nor the mean annual groundwater table are correlated with the first two ordination axes.

Three groups of plant species each associated with special site properties can be distinguished. *Lemna minor* L., *Carex pseudocyperus* L. and *Typha latifolia* L. are found in places with the highest mean annual groundwater table and pH value, a large annual fluctuation of the conductivity values but comparably low fluctuation of the groundwater table and moderate mean annual conductivity values. *Juncus articulatus* L., em Richter, *Phragmites australis* (Cav.) Trin ex

Steud. and *Carex curta* prefer more or less comparable chemical site properties but lower mean annual water tables. *Drosera* spp., *Eriophorum vaginatum, E. angustifolium, Carex rostrata, Rhynchospora alba* (L.) Vahl and *Molinia caerulea* (L.) Moench predominate over those parts of the experimental area where more or less intermediate values for all experimental factors are found. *Agrostis canina* L. prefers extremes with respect to groundwater table fluctuations as well as mean groundwater table. Sphagna are not yet included in this analysis because they have only begun to establish in very scattered spots, many of which might have been overlooked when recording the vegetation. As such, this data seems too unreliable to be included at the moment.

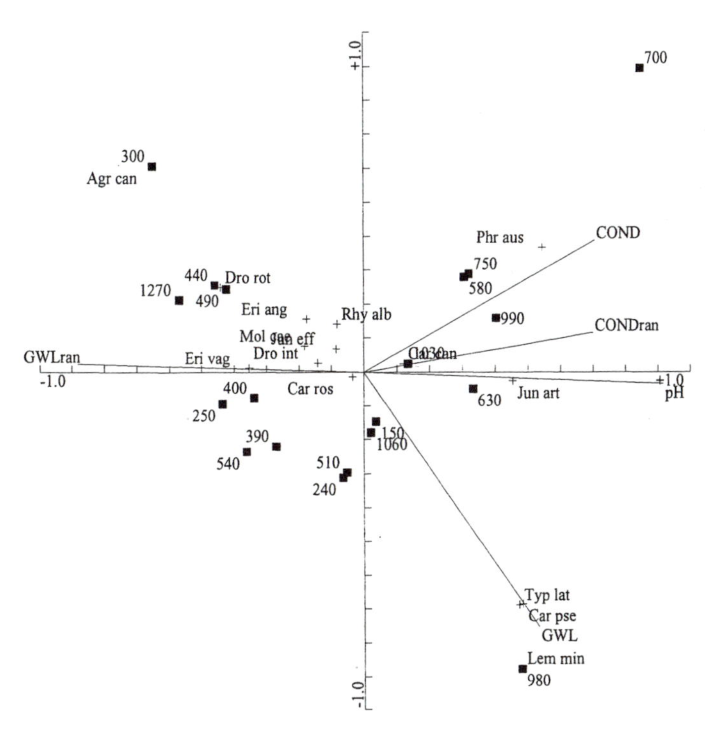

Fig. 32.6. DCCA diagram for species (+) present on sites (■) where environmental variables have been measured (arrows; COND = conductivity; CONDran = annual fluctuation of conductivity; GWL = ground water table; GWLran = annual fluctuation of groundwater table). NB Species abbreviations comprise first three characters of generic and species names.

Also of importance for the future vegetation succession and short term predictions is the colonisation success of the species present. Based on changes in cover values, the most successful species are *Carex rostrata* and *Eriophorum vaginatum* with considerable increases in cover and nearly no decrease. Most unsuccessful is *Juncus articulatus* whose dominance decreased in one half of all strips where it is present and remained unchanged in the other half (Table 32.1).

Over seven years, clearly different developments can be found concerning the sown and planted species, depending on the site conditions:

Those plots planted or sown with *Andromeda polifolia, Scheuchzeria palustris, Eriophorum latifolium* as well as with Sphagna did not develop well. The species quickly died because the unfavourable conditions of the bare peat surface prevented successful establishment. *Trichophorum alpinum* survived, although the planted individuals did not spread laterally, and, hence, plant cover always remained low.

Table 32.1. Number of increases or decreases in cover values of plant species present on the milled area in 128 vegetation relevés between 1988 and 1994. Change is calculated by subtracting number of decreases from number of increases.

Species	88=94	88<94	88>94	Mean change
Carex rostrata	52	87	0	87
Eriophorum vaginatum	52	85	2	83
Drosera intermedia	61	69	9	60
Phragmites australis	67	64	8	56
Rhynchospora alba	65	65	9	56
Molinia caerulea	45	69	25	44
Carex curta	4	81	54	27
Eriophorum angustifolium	119	19	1	18
Juncus effusus	41	56	42	14
Drosera rotundifolia	95	27	17	10
Carex pseudocyperus	72	33	34	-1
Typha latifolia	76	25	38	-13
Agrostis canina	87	18	34	-16
Lemna minor	109	6	24	-18
Juncus articulatus	61	16	62	-46

Within a few years, the planted tillers of *Eriophorum angustifolium* reached a large dispersion due to adventitious spreading and formed large monodominant stands with up to 100% cover. Similar stands developed after sowing, but over longer time scales. However, dispersal outside of the planted or sown plots was only observed occasionally (Fig. 32.7, middle).

On the contrary, after planting *Carex rostrata* reached high plant coverage on the plots and also occupied a number of large sites outside the permanent plots. The reason for this may be the high seed production of this species in the first years after planting and also the drift of the floating seeds along the central ditch - for a long time, there was no spontaneous establishment of *Carex rostrata* above the highest permanent plot on the east side of the field (Fig. 32.7, top).

Eriophorum vaginatum spontaneously occupied the milled fields outside the permanent plots by anemochorous seeds. The seed rain comes from both the planted plots and from the vicinity of the milled fields, where cotton sedge occupies the dry peat mounds (Fig. 32.7, bottom).

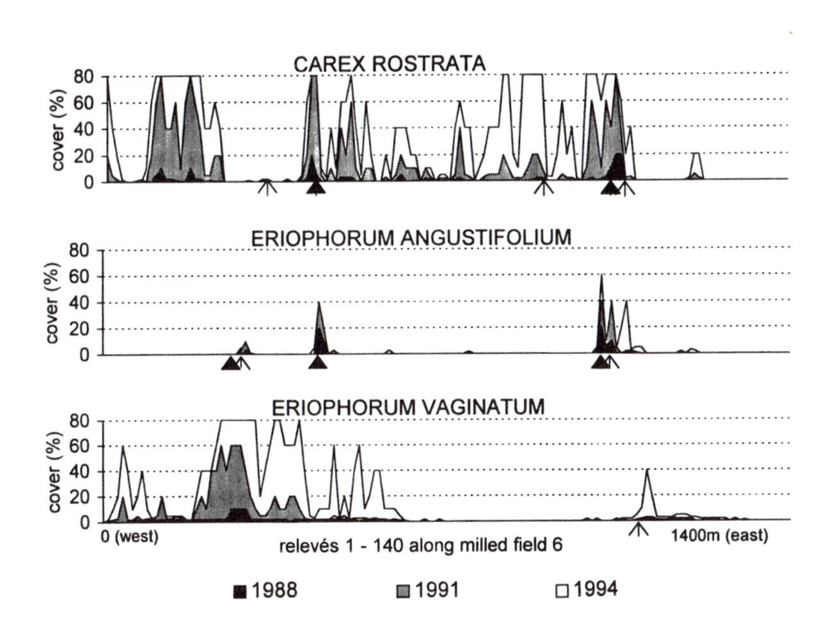

Fig. 32.7. Recolonisation of *Carex rostrata*, *Eriophorum angustifolium* and *E. vaginatum* on field six, displayed as the mean cover values in 1988, 1991 and 1994. Open arrows indicate the location of plots with initial sowing, shaded arrows indicate plots with initial planting.

b) Experiments with the artificial ditches

The abandoned milled fields frequently suffer from extremely dry conditions. In summer, the dry bare peat surface may warm up to 70°C (Schmeidl, 1965) and thus, it is considerably hostile for germination and establishment of plants. The aim of the experiments was to test the capability of three introduced vascular species with the best recovering qualities (*C. rostrata, E. angustifolium* and *E. vaginatum*) to colonising the bare peat surface under different water levels.

Methods

In 1991 extensive experimental plots were laid out on milled field 8 (see Fig. 32.3 for position). The mire water in this field is oligotrophic, but the water level fluctuates approximately -10 to -40 cm. Eight years after abandonment, the spontaneous natural cover of the field had only reached about 5%.

Under these conditions, the separated shoots of the tested species were planted in 35 m transects across milled field 8, one shoot m^{-1} with a 3 m interval between the transects. The individuals were planted on the normal basin surface as well as in shallow, 20-25 cm deep, ditches, which were dug out in order to simulate better water supply.

The number of shoots in 1 m wide strips across the transects as well as the maximum distance of the new shoots to the planted ones (maximum lateral spread) have been measured annually until 1995. The comparison for the data with non-normal distribution was accomplished using the Wilcoxon test.

Results

Figure 32.8 demonstrates the development of the transects after four years: in the ditches, the planted individuals show a significantly higher rate of vegetative reproduction (expressed by mean number of shoots per metre) and a greater rate of lateral spread.

The spatial distribution pattern of the tillers along the transects also marks the differences between the planting methods. In the ditches a well-balanced distribution of shoots predominates. Whereas on the normal surface most of the planted individuals died off and only few of them show a considerable growth rate.

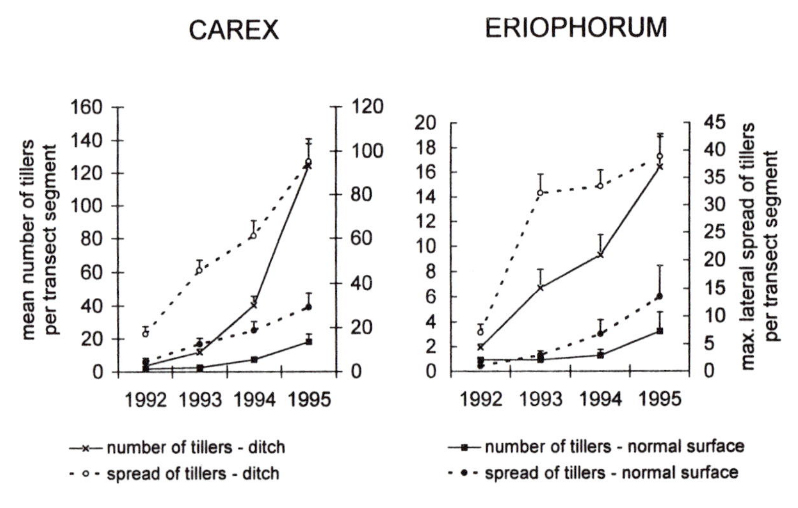

Fig. 32.8. Mean number of tillers and mean lateral expansion of tillers of *Carex rostrata* and *Eriophorum angustifolium* individuals (n=35) in 1995, planted on untreated peat surface and in artificial ditches. (Bars = standard errors; tiller number *C. rostrata* $P < 0.001$, *E. angustifolium* $P < 0.001$; lateral spread *C. rostrata* $P < 0.001$, *E. angustifolium* $P < 0.005$).

c) Experiments with the shading of seeds

Extreme microclimatic conditions on the bare peat surface do not allow the germination and the establishment of seedlings of potential colonists (Salonen, 1992). Based on these facts, some experiments covering the peat surface were carried out. They should answer the question if, and how, this would reduce the microclimatic extremes and influence the germination rate.

Methods

In autumn 1993, permanent plots on field 8 both with, and without, seeds of selected species (*C. rostrata, E. vaginatum, E. angustifolium, Calluna vulgaris* and *Molinia caerulea*) were covered with fleece (Type: Base-UV-17), wide-meshed coconut mat (Type: Terra-Safe JGO 500) and with own leaf-mulch or mulched branches in the case of *Calluna* respectively. For each species, two plots were randomly arranged on the homogenous peat surface of the basin and were subdivided into 25 sub-plots of 400 cm^2. On each plot the same number of seeds from the 1993 collection were sown. In summer 1994, the number of seedlings was counted and their vitality (height, mean number of leaves) measured. The treatments were tested for significant differences by analysis of variance and Turkey test. The normal distribution of the data was tested with a χ^2 -test.

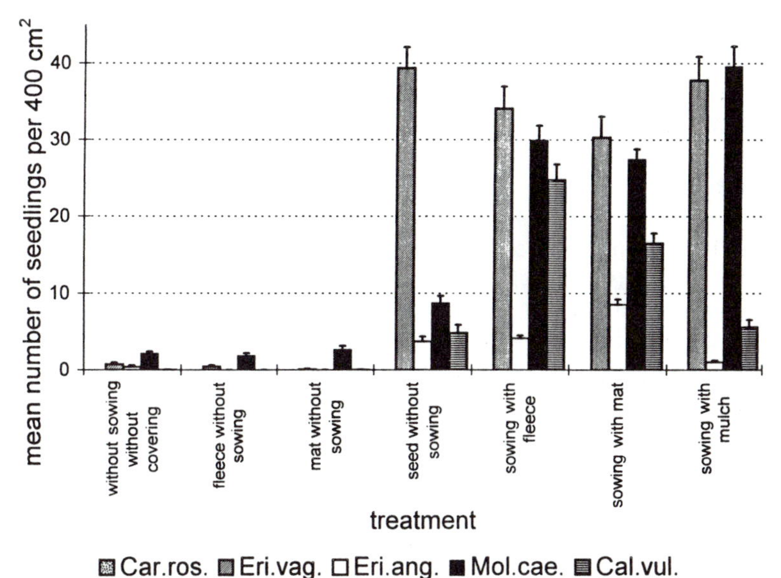

Fig. 32.9. Comparison of the number of germinated seedlings 13 months after sowing on bare peat and sown with artificial shading (Bars = standard errors).

Results

The results show that for some species the covering simulates safe sites for the seeds and thus the germination rate increases considerably (Fig. 32.9). The germination rate of *M. caerulea* under all cover materials was significantly higher than without covering ($P < 0.01$), *C. vulgaris* germinated well beneath both fleece and coconut mat. The differences between the uncovered and mulch-covered treatments were significant ($P < 0.01$). *E. vaginatum* showed no significant difference ($P < 0.01$) germinating well both with and without covering. On covered control plots without sowing, only *Molinia* and *E. vaginatum* germinated, but with significantly lower ($P < 0.001$) germination rates. *C. rostrata* has not germinated on any plot to date.

The vitality of the seedlings on these nutrient poor site is, however, very low. The mean height of the seedlings reached only 0.2 cm for *Calluna*, 3.5 cm for *Eriophorum* and *Molinia*. Thus, there is the suspicion, that the plants will not survive dry summertime and/or heaving during snowless winter periods.

d) Experiments with the application of fertiliser to milled fields

The field experiments on the dry milled field 8 as described earlier (see Experiments with artificial ditches), indicated the practical limits of the use of typical transient bog species, which suffer here from water stress as well as nutrient deficiency. Nevertheless, these species are some of the most important for the fast recovery of bare peat surfaces after milling.

To increase shoot multiplication rates, as well as the vitality of individuals, the effect of fertiliser application on the planted species was tested.

Methods

In May 1994 a long term fertiliser Osmocote 3-4M (NPK 15-11-13) was applied at 100 g m^{-2} on six selected permanent transects of 1 m width and 35 m length with planted *C. rostrata, E. angustifolium* and *E. vaginatum*. After treatment, in October 1994 and in July 1995 the number of tillers and the maximum lateral spread (see Experiments with artificial ditches) were recorded for the first two of these species. The maximum diameters of the tussock and the tussock base were measured for *E. vaginatum*. Additionally, a detailed relevé of the plant cover on every second square metre of the transects was made. The measurements have been compared with those control transects in the same basin without fertiliser, the significance was tested using a dependent *t*-test.

Results

The treatment resulted in a clear increase in the vitality and shoot multiplication rate of planted individuals of *C. rostrata* and *E. angustifolium*. In the case of *E.*

vaginatum fertiliser application appeared to have no significant influence on the development of planted individuals (see Table 32.2).

Also interesting are the side effects following fertiliser application which resulted in germination from the seed bank which had accumulated over the six years following abandonment of peat milling (see Fig. 32.10). Those species which most frequently germinated were *M. caerulea, C. vulgaris, D. intermedia, Rhynchospora alba* and *Betula carpatica* W et K.

The fertiliser encouraged development of the seedlings, so that the plants grew to a size which guaranteed survival of the individuals over winter. A lot of plants flowered in the same year. The plant cover ranged from 3 to 47% in autumn of the same year, and from 6 to 70% a year later. The highest density was reached near the surrounding peat mounds, which were already supported these plant species; a lower density was noted in the middle of the transects. These facts explain the slow rate of the re-establishment from the seed bank during the six years following abandonment of the peat harvesting.

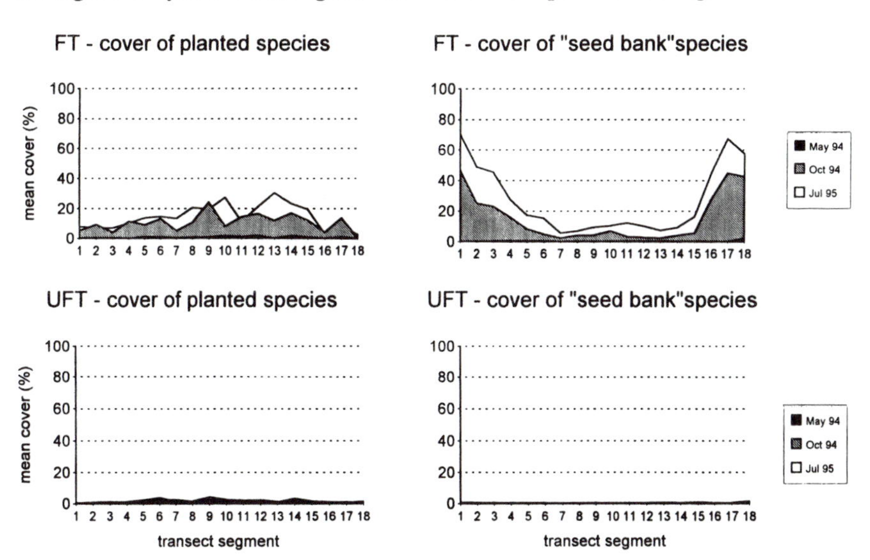

Fig. 32.10. Comparison of the growth of planted and spontaneously germinated species on transects with (FT) and without (UFT) fertiliser application (cumulative cover of all species of both groups). Segments 1 and 18 are located near peat walls, 9 is in the middle of the basin.

Table 32.2 Development of planted species on transects with (FT) and without (UFT) application of fertiliser (arithmetic means ± standard error). Thick frames indicate significance levels of $P < 0.001$, thin frames levels of $P < 0.05$.

1993	No. of tillers		Lateral spread		Tussock diameter		Tussock base	
	UFT	FT	UFT	FT	UFT	FT	UFT	FT
C. rostrata	4.8 ±0.6	4.7 ±0.4	18.2 ±2.6	24.7 ±3.2				
E. angustifolium	0.8 ±0.2	0.8 ±0.2	3.5 ±0.9	3.9 ±0.9				
E. vaginatum					37.1 ±1.4	40.6 ±1.4	12.5 ±0.4	13.4 ±0.5

1995	No. of tillers		Lateral spread		Tussock diameter		Tussock base	
	UFT	FT	UFT	FT	UFT	FT	UFT	FT
C. rostrata	47.2 ±8.1	143.2 ±19.4	64.9 ±8.5	126.1 ±11.3				
E. angustifolium	1.8 ±0.5	6.1 ±1.9	13.0 ±3.2	20.9 ±5.2				
E. vaginatum					70.7 ±3.3	69.2 ±3.9	15.0 ±0.7	17.2 ±1.0

e) Experiments with re-introduction of *Sphagnum*

All the experiments described above followed the aim of a fast, and nearly natural recovery, of the bare milled fields. But the main aim of bog rehabilitation, as well as the crucial criterion of successful bog regeneration, is the initiation of peat-forming processes. In ombrotrophic areas, this can be achieved by durable establishment of *Sphagnum* plants. In the Kendlmühlfilze experiments with *Sphagnum*-transplantation began in 1992.

These experiments investigated the extent to which selected *Sphagnum* species are able to establish and the role played by the secondary vegetation cover of the flowering plants.

Methods

Following analysis of the vegetation development and groundwater quality in the two years before transplanting (see Monitoring and artificial introduction), sites were chosen within the milled fields which appeared promising for successful establishment of *Sphagnum* spp.

Nine sites were planted, eight of which were covered with *C. rostrata, E. angustifolium* and *E. vaginatum* and the ninth of which was completely unvegetated but had comparable groundwater dynamics and quality.

Three frequently occurring test-species *S. cuspidatum* (as green hollow species), *S. magellanicum* and *S. capillifolium* (Ehrh.) Hedw. (as red hummock species) were chosen. Sods were taken with a core sampler of 12 cm diameter to a depth of c. 25 cm. Twenty sods per site and species were planted, four sods m^{-2}. From 1992 to 1994, the maximum diameter of the peat moss sods were measured and the cover of the living *Sphagnum* individuals estimated

Results

The results show that the development of the *Sphagnum* sods proceeded in clearly different ways: the best growth conditions for *Sphagnum* occurred on the sites with stable ombrotrophic conditions, a stable water table near the peat surface (mean annual values for 1993: water table -1.9 cm, conductivity 34 $\mu S\ cm^{-1}$, pH 4.3), and with a diffuse shading structure of pioneer plants (Fig. 32.11). By the beginning of 1994, the sods had already united to form large closed carpets of up to 6 * 2 m. In the ombrotrophic sites with lower groundwater level (mean water table -8.3 cm, conductivity 44 $\mu S\ cm^{-1}$, pH 4.1) *S. magellanicum* and *S. capillifolium* formed compact, typically red coloured, moss pads with a definitely slower increase in diameter; under these conditions, the growth of the third species is less satisfactory due to stress during the summer drought. Though the site with planting on the bare peat surface (Fig. 32.12) is infiltrated by ombrotrophic water and the ground water dynamics shows tolerable fluctuations (mean water table -5.7 cm, conductivity 31 $\mu S\ cm^{-1}$, pH 4.5), all transplants died due to a combination of drying out and of the total covering of the sods by mud.

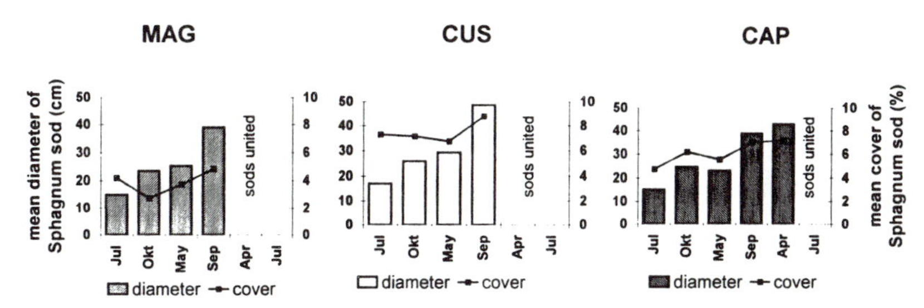

Fig. 32.11. Development of transplanted *S. magellanicum* (MAG), *S. cuspidatum* (CUS) and *S. capillifolium* (CAP) sods during July 1992 - July 1994 on a *Carex rostrata* covered plot. Bars = standard errors.

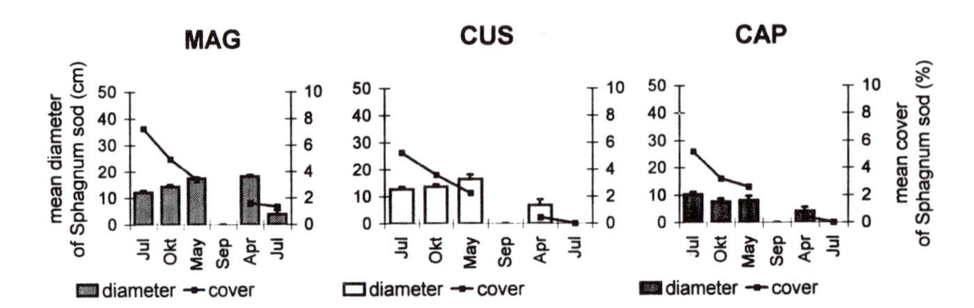

Fig. 32.12. Development of transplanted *S. magellanicum* (MAG), *S. cuspidatum* (CUS) and *S. capillifolium* (CAP) sods during July 1992 - July 1994 on a bare peat surface. Bars = standard errors.

Discussion

The field experiments indicate that long-term monitoring of an area to be rehabilitated is an indispensable prerequisite before realisable aims for rehabilitation and the techniques to be applied can be defined. Optimally, monitoring of the abiotic site factors should begin before peat extraction ends, so that the peat firm can do all the necessary site preparation (optimal extraction depth, terracing) according to the special site situation. This is also recommendable with respect to the costs, since terracing is one of the most expensive factors in rehabilitation (Maas and Poschold, 1991), and often beyond the financial means available afterwards. Additionally, it is often impossible to work with heavy machines on re-wetted sites.

Based on the hydrodynamic and hydrochemical data the extracted area can be subdivided into parts with differing properties with respect to the aims of an effective rehabilitation, and which will often require different measures to be applied.

In areas with oligotrophic conditions and a peat water level nearly constant near the surface, the conditions for the rehabilitation of raised bogs are fairly good. Here, it is most effective to accelerate the re-establishment and growth of Sphagna by spreading vegetative shoot particles onto the surface (Campeau and Rochefort, 1996).

The situation frequently found in former extraction sites, is much more complicated with the oligotrophic peat water table fluctuating a greater range, (i.e. the surface is temporarily inundated and will subsequently dry out again). Here, a matrix of taller growing, flowering plants like *Carex rostrata* or *Eriophorum angustifolium* plays an important role. The peat moss carpets can then survive the regular, partially high inundation due to the support of their vertical growth by this 'climbing frame'. A span of 20 to 25 cm between the lowest and the highest capitulae of *Sphagnum* carpets under such conditions was regularly found. On unvegetated sites, without such support, the peat moss sods were continually weakened by the permanently changing water supply conditions, by regular covering with eroded peat mud, and by exposure to high irradiation without protection from vegetation cover. The whole transplants are always affected by these adversities so that no initial cells for peat moss re-establishment can form under these conditions.

Raised bog rehabilitation on a large scale is impossible where the water table drops below a tolerable level for Sphagna during the summer. Here, the peat surface should be covered with flowering plant species of transient and raised bogs as fast as possible. Our experiments indicate that planting often results in a more rapid recovery compared with sowing. This is particularly true for plant species with strong vegetative spread. For species with less tendency towards vegetative lateral spread - like *Molinia caerulea, Calluna vulgaris* or *Rhynchospora alba* - sowing may be a more effective option. A deep water table and the extreme nutrient poverty may also hamper the establishment and dispersal of the introduced flowering species. Shading of the surface can significantly increase the germination rate (see also Salonen, 1992), but the vitality of the seedlings is reduced. Thus, the long-term establishment prospects for these individuals is low, restricting the applicability of this method. On areas with permanently low peat water levels, the homogeneity of the surface (which is a positive attribute in the case of sufficient water supply (Eggelsmann, 1987)) is a disadvantage with respect to the re-introduction of plants. Artificial hollows and ditches can create a higher diversity of micro-sites thus favouring plant establishment. The deeper parts, with their enhanced water supply, serve as initial cells for recolonisation. Deeper and larger ditches can simulate conditions comparable to hand peat cuttings, where the small-scale establishment of *Sphagnum* may be possible. Such measures can also be applied to dry extraction sites already abandoned for longer time spans.

Application of fertilisers results in more rapid seedling development and seems to break the dormancy of seeds spontaneously imported from the neighbourhood. Further investigation is needed to verify these assumptions and to clarify optimal concentrations for single nutrients (P, K, micronutrients). Nevertheless, since a seed bank on bare peat surfaces will only form over long periods, a combination of sowing and initial fertiliser application can be an effective method for revegetating abandoned milled fields.

It has to be stressed that the succession of plant stands on insufficiently rewetted sites will not develop towards the former raised bog vegetation, but can be seen as an anthropogenic stage during the development towards bog forests, e.g. with birch or pine, like in southern Germany. In these cases it should be discussed during the formulation of the aims of rehabilitation, whether it would not be more promising, to create a situation where a fen can develop. This can be accomplished by continuing the peat extraction down to deeper layers, or by introduction of (minerotrophic) water from the vicinity. Under these minerotrophic conditions, fen vegetation will establish spontaneously and it can be assumed that, under humid conditions, a raised bog can perhaps develop over several centuries (Pfadenhauer and Klötzli, 1996).

Acknowledgements

This study was financially supported by the Bavarian State Ministry for Environmental Affairs and Landscape Development and the German Science Foundation. We have to thank all our colleagues and a large number of students who assisted during the field work.

Chapter thirty-three:

Tree Removal at Langlands Moss

Mr Stuart Brooks and Dr Rob Stoneman
Scottish Wildlife Trust, UK.

Introduction

'In Reporting Scotland to-night, we will be looking at an operation, near East Kilbride, to remove over 10,000 trees from a bog.' This rather unusual item appeared on BBC Scotland in November, 1994. The reporter asked a representative from the Scottish Wildlife Trust why they were taking trees off the bog especially given the fuss SWT had made over cutting down trees in Glasgow for a new motorway. This chapter attempts to answer just that.

The Habitats and Species Directive calls for Europe's raised bogs (and other threatened habitats) to be brought into favourable conservation status. The conservation status of a habitat is considered favourable when:

1. 'its natural range and the areas it covers within that range are stable or increasing'; and
2. 'the specific structures and functions which are necessary for its long term maintenance exist and are likely to continue to exist for the foreseeable future'.

Stopping the area of raised bog from decreasing will be difficult enough (see Stoneman, Chapter 45, this volume); increasing the area is almost impossible given the many thousands of years it takes for a raised bog to form. Nevertheless, in some parts of Europe, raised bogs have become so rare, attempts are being made to rehabilitate or restore sites which have had much of their peat removed.

It is a difficult task. The conditions required for raised bog vegetation are maintained by a ground (perched) water mound which also dictates the size and shape of the peat bog (Ingram, 1982). In essence, the vegetation of a bog is primarily controlled by hydrological conditions which are inextricably linked to the peat itself. Removal of peat upsets this balance leading to the development of very different plant communities (often heath or birch woodland).

However, careful manipulation of water across cutover sites has been used to encourage colonisation by poor fen and bog vegetation. These techniques were pioneered in Holland and Germany from the 1970s and are detailed by Wheeler and Shaw (1995b). On Bargerveen (Holland), rehabilitation measures put in place over 20 years ago have been successful in revegetating bare peat surfaces with bog vegetation. At Bargerveen, mats of *S. cuspidatum* peat have formed. Vegetation on these mats is now succeeding to *S. recurvum* with an occasional birch seedling or *Molinia* tussock. As yet, there has not been establishment by hummock building Sphagna. This may be due to: atmospheric pollution, lack of an adequate source of *Sphagnum* propagules or, the 'grounding' of the thickened mats onto the base of the pool causing water levels to drop during the summer (Brooks *et al.*, 1994). This one example, out of many, demonstrates that the rehabilitation of cutover bogs remains an experimental discipline. Even if bog vegetation can be encouraged to recolonise cutover peat surfaces, it does not necessarily recreate the structure and functions of a raised bog. However, the options for raised bog conservation in England and Holland are limited due to large-scale habitat destruction. Cutover bog rehabilitation must be seen in the light of highly restricted opportunities.

In Scotland, Stoneman (Chapter 45, this volume) details a slightly less bleak, though alarming, picture of raised bog habitat decline. Nine percent (2,300 ha) of the original area of raised bog still retains 'near-natural' vegetation on un-cut surfaces. Furthermore, 23% (6,200 ha) is in a degraded state though still recognisable as bog. The priority lies with rehabilitation of these sites. Commercially worked sites only cover 1,090 ha (4%), limiting the opportunities for bog rehabilitation on post-extraction sites.

The largest class of bog (30%: 7,900 ha), outlined by Stoneman (Chapter 45, this volume), is wooded. Many of these sites were afforested in the 1960s and 1970s after technological advances allowed tree-planting on deep peats and before the nature conservation value of the mosses was fully appreciated. In the main, afforestation has been on uncut, often large, bogs such as the Lochar Mosses, Racks Moss and West Flanders Moss. It is here proposed that the prospects for rehabilitating these sites are potentially quite good.

Effects of forestry on bogs

Whilst trees can be a natural component of bog flora under more continental climates; in Britain they are rarely (if at all) a natural component of bog vegetation today. However, with adequate drainage, fertiliser application and careful management, coniferous plantations can be established on raised bogs.

The first effects of afforestation relate to drainage. The upper peat layer has to be comprehensively drained to allow trees to establish. Intensive networks of drains lower water-levels leading to a rapid loss of the wetter elements of bog vegetation communities (most *Sphagnum* species for example). Drainage also allows upper peats to oxidise whilst fertiliser applications further enhance chemical changes. As trees mature to form a closed canopy, up to 30% of rainfall is directly intercepted on leaf surfaces from where it can potentially evaporate back to the atmosphere (Gash *et al.*, 1980). Trees have relatively higher transpiration rates than typical bog vegetation and this exacerbates drying and wastage of the upper peat. In addition, shading by a closed canopy leads to a loss of any remaining ground vegetation leaving bare peat surfaces carpeted with dead needles. Any vegetation left along forest rides or open areas may also be affected (Smith and Charman, 1987). This may be the direct result of adjacent forestry, or in some areas due to cessation of traditional grazing and burning after afforestation (Charman and Smith, 1992).

These impacts all relate to the surface of the bog and in effect destroy the semi-aerated upper layer (the acrotelm). However, the bulk of the peat-bog, which is composed of permanently saturated peat (the catotelm) remains. It is the catotelm that fundamentally controls peatland hydrology (Bragg and Brown, 1996). In theory, then, following tree removal, the rehabilitation of a functioning acrotelm on to the remaining catotelmic peat should recreate the structures and functions of a bog. The simplicity of afforested uncut bog rehabilitation contrasts sharply with the complexity of rehabilitating cutover bogs. However, recolonisation of the surface by *Sphagnum* mosses is likely to be hampered by:

- the remaining forest drains;
- altered chemical conditions;
- the effects of peat loss through oxidation (whilst under forestry);
- deep cracking of dried peats forming a reticulate drainage network (Anderson *et al.*, 1995);
- damage caused to the bog surface during tree removal operations and;
- shading, interception and nutrient enrichment from brash left on site following tree-removal.

Anderson (Chapter 25, this volume) outlines the Forestry Commissions' involvement with bog rehabilitation. This has come about as a result of enlightened Forestry Commission guidelines (FC, 1990; 1993; 1995) to maximise environmental dividends within forests. Few of these schemes have been comprehensively monitored however, making it difficult to assess their success. At present, government policy presumes against converting forest to other land-uses except where there is a high public (including environmental) benefit. More research is, therefore, required to demonstrate the environmental benefits of afforested bog rehabilitation (Anderson, Chapter 25, this volume).

The Langlands Moss experiment

Introduction

Despite the lack of monitoring relating to *Sphagnum* recolonisation - and the re-creation of an acrotelm - on previously afforested bog surfaces, there is good anecdotal evidence to suggest success is likely. On the Border Mires (Northumberland) tree clearance of the former hydrological unit is being undertaken (Burlton, Chapter 29, this volume). Where trees have been cleared, *Sphagnum* is recolonising - particularly in plough furrows. A spectacular example is found on Dun Moss (Fife, Scotland). The moss is mainly uncut although one side was afforested in c. 1900 with Scots pine. Trees were removed in 1993, using a cable crane winch, to pave the way for natural recolonisation by native birch under a woodland grant scheme (WGS). Unfortunately, the very high nature conservation value (and potential for rehabilitation) was under recognised when granting the WGS. However, water levels appear to have risen back to the surface and *Sphagnum* recolonisation is rampant. All the usual ombrotrophic species are present recolonising from rides and open areas within the former woodland.

Evidence such as this, and a strong theoretical basis for success, led to a scheme to rehabilitate Langlands Moss.

Tree removal

The 20 ha moss, lying close to East Kilbride, is typical of many small Scottish central belt bogs having been variously damaged by peat cutting, drainage, burning and, in this case, afforestation. Despite these damaging activities one half remains relatively intact with *Calluna vulgaris* and *Sphagnum capillifolium* dominating the vegetation mosaic. Pools (probably unnatural) of *Sphagnum cuspidatum* indicate a generally high water table. The other side of the moss was afforested in the 1960s by the East Kilbride Development Corporation to form a shelter-belt for the southern edge of the town (see Fig. 33.1). Fire and poor drainage led to part of the forest failing although subsequent self-sown pine was dense in places.

The moss was identified by East Kilbride District Council (1992) as a potential Local Nature Reserve (LNR). Furthermore, the woodland on the moss was included as part of the Development Corporation's Woodland Regeneration Project. As a consequence, a steering group comprising of the Scottish Wildlife Trust (through the Scottish Raised Bog Conservation Project), East Kilbride District Council, East Kilbride Development Corporation and Scottish Natural Heritage was established in January 1994 to oversee the designation of the site through the preparation and implementation of a management plan. One of the objectives of any LNR management plan is the maintenance of the site's special interest. At Langlands, the priority lay with removing the trees and rehabilitating the site. In addition, it was recognised that success at this site could pave the way for a re-assessment of government policy towards whole-scale rehabilitation of afforested raised bog leading to a dramatic expansion of the resource (Stoneman, Chapter 45, this volume). Obviously, the parties involved were keen to maximise the chances of success; a comprehensive monitoring programme was instigated to assess progress on the site (Brooks, 1995).

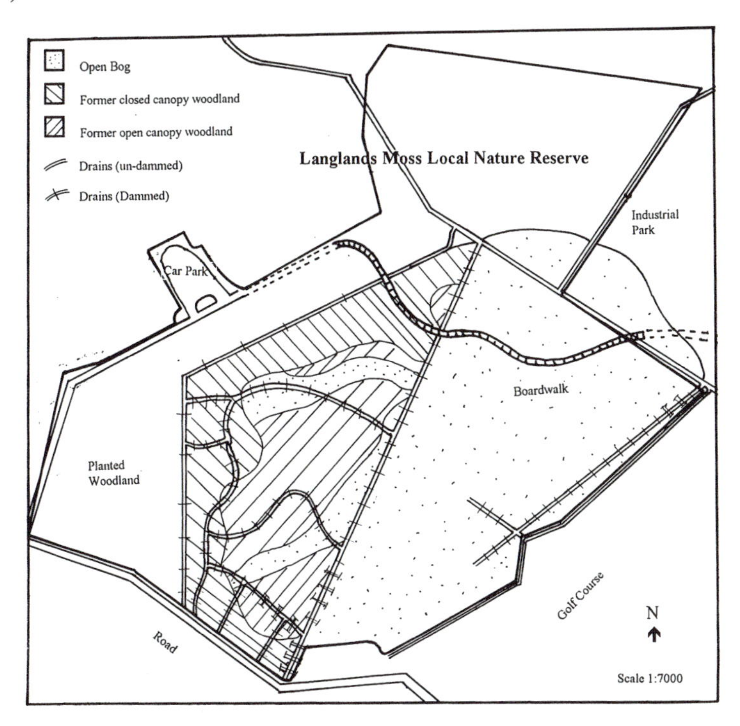

Fig. 33.1. Site map for Langlands Moss.

SWT (1994a) investigated the likely impacts of removing trees using standard harvesting techniques. The arguments for removing brash were also considered. It was suggested that whole tree removal using a cable crane would cause the least damage and allow *Sphagnum* to recolonise across a largely brash-free surface. However, the forestry contractors decided that using a helicopter would be a simpler and more effective method of tree removal to achieve the same objectives. Consequently, trees were initially felled close to the surface using chain-saws and then air-lifted to an adjacent, temporary processing yard (see Fig. 33.2). Trees were mainly processed for wood-pulp and the brash was chipped for footpath construction.

This method has the advantage of reducing damage to the peat surface and leaving it almost brash-free; this has another advantage in that the site is more aesthetically pleasing after tree removal (important for its other LNR functions). The use of a helicopter also generated valuable publicity for the project (hence the TV coverage). Its disadvantages lay with its cost, its dependency on weather conditions and its experimental nature.

Fig. 33.2 Helicopter airlifting trees felled on Langlands Moss.

Not surprisingly, there were some problems:

- A remote hydraulic grabber imported from Canada broke in the first hour of operation. Instead, ropes were fastened around the ends of 2-3 trees for airlift. Some ropes could not be retrieved increasing operational costs.
- Snedding of the lower branches - normally undertaken during felling - was not planned into the processing operation. This incurred extra delay.
- The amount and quality of the extracted material had not been assessed properly. Using standard forestry tables, 10,000 trees were estimated weighing 650 tonnes. In fact, the variability of the plantation meant that 25,000 trees were extracted which weighed only 350 tonnes. The lower quality of the wood and longer length of removal time incurred extra costs.
- Machinery used to process the material could not keep up with the amount of trees dropping from the helicopter (which was very efficient). Build-up of material incurred extra time and cost. A particular problem related to the chipper which could not handle large volumes of brash effectively.

The total cost of the operation was c. £40,000 which, when offset by £8000 of income generated from timber sales, represents a cost of £3200 per ha of the 10 ha wooded area. Clearly, this method is expensive, although this was exacerbated at this site by the poor quality of the timber crop. Costs for other schemes could be reduced in the light of this experience, and can be offset to a greater extent where a plantation has been more successful.

Damming programme

The second element to the rehabilitation of the whole bog involved damming of the drainage network. This involved installing 150 dams. Smaller dams were constructed using exterior grade plyboard sheets, whilst larger drains were blocked with interlocking PVC plastic piling. The latter is a relatively new approach (Brooks *et al.*, 1996), although it has, so far, been successful.

An area to the west of the moss is also included within the LNR. This allowed a peripheral drain to be dammed in an attempt to re-create the long-drained lagg fen. A road and a golf-course precludes this option on the other sides of the moss.

Monitoring programme

Alongside the rehabilitation work, a comprehensive monitoring programme was devised and installed (SWT, 1994b).

The programme began one year before felling took place allowing the pre-management situation to be compared with data after the felling operation. All methods used were planned in an integrated manner allowing results from hydrological monitoring to be tied to botanical data. The following approaches were taken:

- Photographic monitoring - a series of fixed point locations were selected to provide a stereo-photographic record of change on the moss.
- Permanent vegetation plots (1.8 m x 0.45 m) were set up along dipwell transects to record detailed vegetation change every six months.
- The site was split into 13 different compartments in which random quadrats were thrown for vegetation recording. This gives a better picture of the way the whole site is responding.
- Two dipwell transects running across the open bog and the forest were set up. Dipwells have been measured weekly over a 1½ year period.
- In addition, six Walrags (Bragg *et al.*, 1994) were installed to measure maximum and minimum water-levels each month.
- Hydrological data is supplemented by taking monthly rainfall readings.
- The site is also regularly checked to assess all equipment.

Postscript

At the time of writing, the trees have been off Langlands Moss for over a year. The extremely dry summer of 1995 allied to removal of tree-cover paradoxically killed off some of the *Sphagnum* which had remained under the trees. However, re-wetting over the winter has led to a notable increase in *Sphagnum* cover. *S. cuspidatum* can now be found floating in the ditches whilst *S. capillifolium* and *S. recurvum* have started to colonise ditch furrows. *Sphagnum tenellum* has started to colonise on to bare peat areas and over the remaining dead needle carpet.

Hydrological monitoring data is being collated by the Forest Authority and these results will be published elsewhere.

Other aspects of the site's management plan have also been achieved. In March 1996, the nature reserve was opened. As part of the desire to realise the site's educational and recreational potential a car-park and picnic area has been built at the old processing site. A boardwalk crosses the site and at various points interpretation boards have been erected. The site is now linked by a network of footpaths to Calder Glen Country Park and the River Calder.

This project begins to fill a research gap which could be crucial to the implementation of the Habitats Directive for raised bogs in Britain and just as important, there is now a wonderful nature reserve on the edge of East Kilbride.

Chapter thirty-four:

Monitoring Peatland Rehabilitation

Dr Brian R. Johnson
English Nature, UK.

Introduction

Conservation organisations have been monitoring the gross effects of drainage on bogs for many years by surveying the extent of vegetated areas and measuring trends in the hydrological condition of various peat bodies.

Techniques used for this task were mostly developed to be sensitive to declines in site quality. Monitoring devices such as dipwells and fixed quadrats were often placed on bogs where they would pick up changes in areas of high biodiversity or sited across boundaries created by drainage ditches. Habitat mapping using aerial photography and ground recording has also been used to monitor gross changes.

The results of this monitoring effort led to the current concern about the conservation status of lowland raised mires and are summarised well in documents such as *Out of the Mire* (RSPB and Plantlife, 1993). This information led to the NGO peat campaign and to action by government and both the statutory and voluntary conservation bodies.

In England, the statutory conservation agency, English Nature, set up the Lowland Peatland Programme in 1992. This 4-year programme set out to acquire management control of some of the largest and most important peatland sites of special scientific interest in the UK. A programme was set up to conserve existing flora and fauna on vegetated areas and to rehabilitate damaged areas, most of which had been subject to peat excavation and drainage. Acquisition and management control of the vegetated areas of the largest bogs was achieved in 1992 and rehabilitation started immediately.

The objectives of this rehabilitation programme were to put in place the hydrological and vegetational foundations for active peat-forming ecosystems to recolonise. More specifically we set out to achieve the following conditions on each site:

1. Raise captured rainwater to within 10 cm of surface and minimise seasonal fluctuation;
2. Develop simple cost-effective techniques for damming;
3. Control invasive vegetation on each SSSI to allow peat-forming systems to spread.

In order to achieve these objectives sites were chosen where there was existing peat-forming vegetation *in situ*, and there were real opportunities to carry out rehabilitation work. A survey of English bogs was commissioned by English Nature and carried out by Sheffield University, and a database (BOGBASE) constructed which allowed decisions on rehabilitation priorities to be made.

Rehabilitation plans were developed for each site and work on damming and vegetation control is now largely completed on all six sites in Phase 1 of the programme and is well under way on seven sites in Phase 2. By the end of 1996, we expect to have 18 of the largest and most important English bogs under rehabilitation management.

This is a major area of investment for English Nature, using considerable funds and staff time. Clearly we believe that the effort will be successful, and already the restored wetland character of these sites is obvious to even the most casual observer. But casual observation is not enough; we need to have supportable and more objective information on these sites for the following reasons:

For national needs:

1. Site evaluation to discharge national statutory duties SSSI, National Nature Reserves (NNR), favourable conservation status for Special Areas of Conservation (SAC) , Species etc.;
2. Add to national overview of resource - define and monitor critical natural capital;
3. Add to knowledge (for example of successional stages);
4. Evaluate management efficacy - value for money;
5. Set objectives for other sites;
6. Stimulate research into causes of trends;

For local needs:

1. Assessment of wildlife value of site compartments for intra-state and local reasons;
2. Evaluation and feedback into management activities - early warning system to improve techniques;
3. Provision of information to neighbours and visitors.

To some extent, our current monitoring effort achieves these objectives, although the site programmes were not entirely designed for these purposes.

Fortunately some baselines were set up prior to rehabilitation work on Fenns and Whixall Mosses, the Solway Mosses (Wedholme Flow and Glasson Moss), Shapwick Heath and Westhay Moor, and Thorne Moors. Most of these sites have had dipwell transects installed, and dipwell-linked permanent quadrats set up since the 1980s. Although dipwells were largely standardised, the sizes of permanent quadrats varied between 0.5 m^2 and 2 m^2. Some fixed-point photography was available and on some sites long-term data were available on bird and invertebrate populations.

Site managers have continued the pre-rehabilitation monitoring throughout the rehabilitation period to date. So what have we found?

So far as vegetation is concerned, we expected plant communities to respond rather slowly to remedial measures on damaged lowland bogs for at least two reasons:

1. Peat-forming vegetation, such as M18, may now be confined to small refugia on large sites such as Thorne Moors. We know little about recolonisation rates on acid, vegetated peats, but there is some evidence from continental Europe that this is a slow process.
2. It is thought that cut and bare peat surfaces are hostile to recolonisation (even by Sphagna) because they are potentially more acidic than uncut peats.

It is also very early in the rehabilitation programme to expect a detectable response by vegetation. Nevertheless, our site managers report rapid establishment of *Eriophorum* spp., possible precursors to the bog mosses, and of *Sphagnum cuspidatum* in the rewetted areas. In the palaeoecological record, *S. cuspidatum* is often an initial component of the re-establishment of wetter plant communities after natural rewetting of a dry bog surface (Barber *et al.*, 1994c) and its increase now is an encouraging sign.

Since our first objective was the hydrological rehabilitation of sites in the programme, we have some long data-sets from dipwells installed early in the programme or, in some cases, before rehabilitation began. This is particularly so in the case of NNRs where site managers installed hydrological monitoring in the early 1980s. We expected sites to respond quickly to damming and bunding operations and we have not been disappointed.

Data analysis at Wedholme Flow and Glasson Moss (Figs. 34.1 and 34.2) shows a rapid response to damming and bunding, with a significant rise in water tables and attenuation of seasonal fluctuations. These rises in water table cannot be explained by variation in annual rainfall (see for example Fig. 34.3). The rewetting effect is more marked on those parts of the sites which had not been cut over in the past (the 'primary' surfaces) but which had been drained, usually in preparation for peat extraction. When dammed, the rise in water tables on these areas was more obvious and showed less fluctuation than on rehabilitated cutover areas (Fig. 34.4).

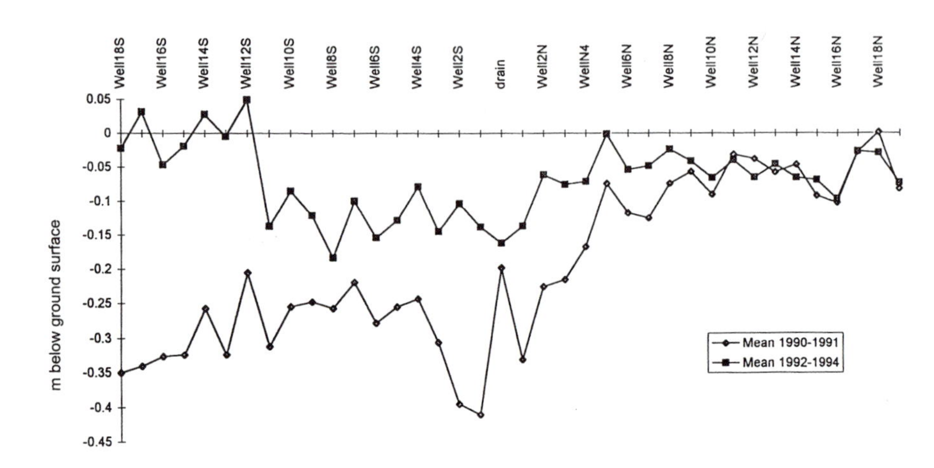

Fig. 34.1. A typical dipwell transect across Wedholme Flow, Cumbria showing mean water table depth before (1990-1991) and after (1992-1993) damming (from Wheeler and Shaw, 1995a). The transect runs from south (wells with S suffix) to North (wells with N suffix). The southern area was drained and cut before rehabilitation, whereas the northern area is primary bog.

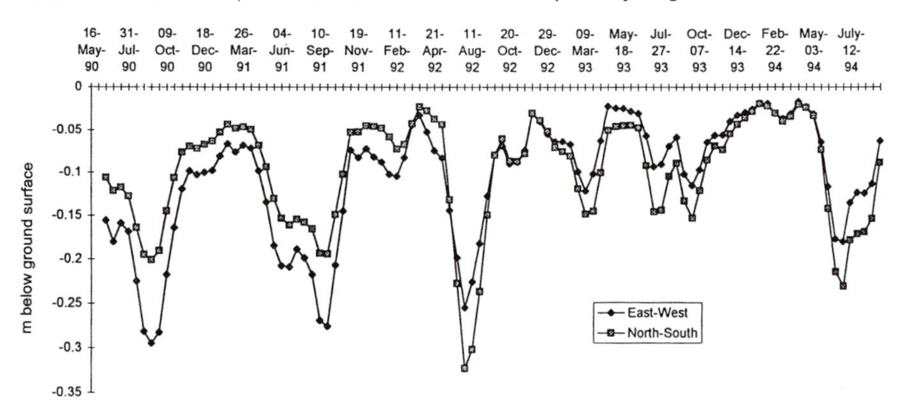

Fig. 34.2. Mean water table depths from two dipwell transects on Glasson Moss, Cumbria (from Wheeler and Shaw, 1995a). Damming operations were completed in 1992, since when there has been a marked attenuation in seasonal water table fluctuations.

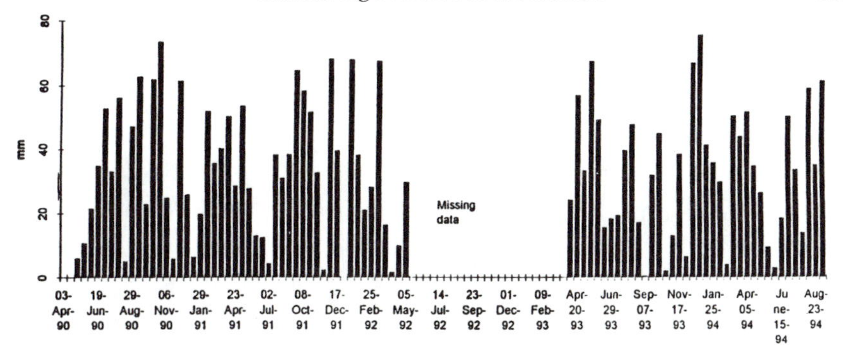

Fig. 34.3. Fortnightly rainfall data from Wedholme Flow, Cumbria. This site is immediately adjacent to Glasson Moss. Annual rainfall has remained similar throughout the rehabilitation period of 1991 to 1994.

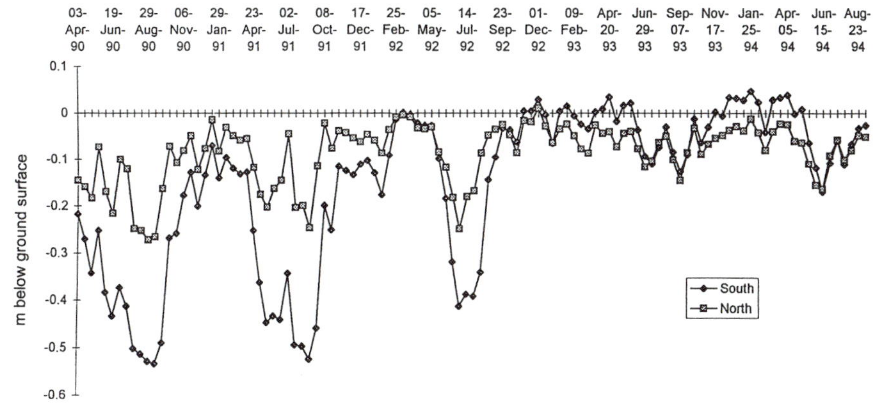

Fig. 34.4. Mean water table depths from rehabilitated uncut areas (North) and cutover areas (South) of Wedholme Flow, Cumbria (from Wheeler and Shaw, 1995a). Damming took place in 1992. Although final water tables were higher, the previously cutover areas still show more seasonal fluctuations after rehabilitation than the dammed primary uncut bog.

Dipwells on Thorne Moors situated in dammed and bunded areas are showing similar responses (Fig. 34.5) (Johnson and Rowarth, 1994).

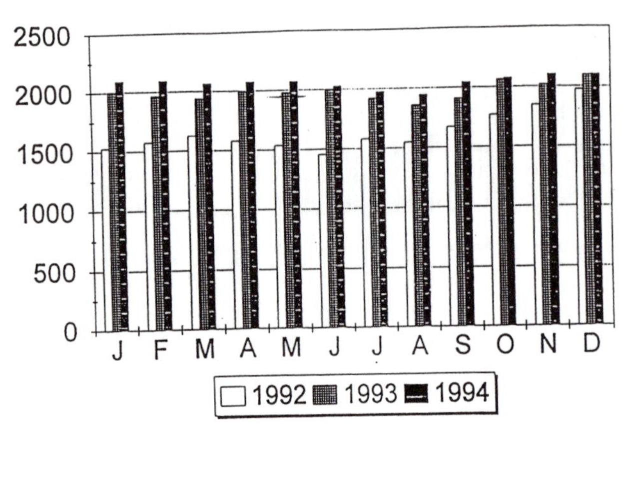

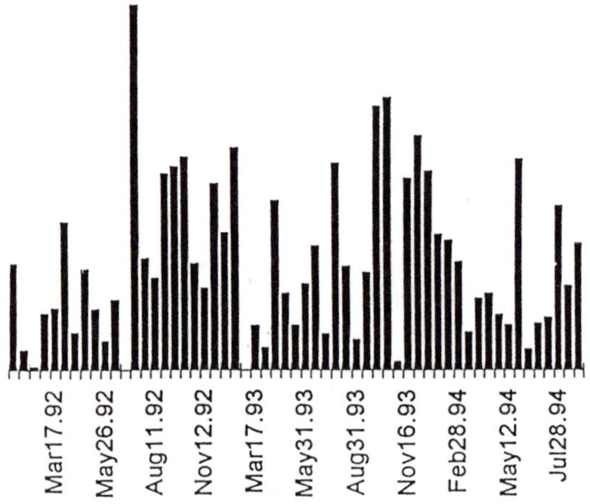

Fig. 34.5. Water levels in mm above Ordnance datum (AOD), measured from 1992 to 1994 in a dipwell on a rewetted part of Thorne Moors, South Yorkshire (from Johnson and Rowarth, 1994). Rewetting took place in 1992, since when levels have been consistently 0.5 m higher and close to the peat surface, which in this case is at 1583 AOD. Rainfall records (below) show a slight decline in annual precipitation during this period, with a prolonged summer drought in 1994.

On several sites rehabilitation has resulted in shallow (less than 30 cm) water conditions in winter. These have in some cases (notably on Wedholme Flow, Glasson Moss and Thorne Moors) attracted large numbers of wetland birds, including gulls. At Thorne a small colony of breeding black-headed gulls has established, causing localised eutrophication of the peat surface, with increases of *Juncus* spp.

Monitoring of the severely cutover excavations in the Shapwick Heath area in Somerset has revealed that the 2.5 m deep excavations (down to the clay substrate) filled with water within one year of switching off the pumps. With rainfall over the period not exceeding 1.4 m, it is clear that ground-water ingress must be partly responsible for the rapid rise in water levels. This has resulted in the formation of several large (>40 ha) bodies of open water up to 2 m deep containing up to 50% reed-swamp. There are indications here that the adjacent, 'perched areas' of acidic peats, which had previously suffered from desiccation, are beginning to revert to poor-fen vegetation. Invasive trees on these areas are showing signs of distress and in some cases appear to have succumbed to high water table, anoxic conditions within 2 years.

If we combine our knowledge about the way water tables respond to damming and vegetation clearance with similar work carried out elsewhere in Europe (Fojt, 1992), it is clear that site managers and ecologists have a reasonable understanding of the hydrological behaviour of rehabilitated mires. Vegetational responses are less clear, although data emerging from monitoring and research over the next few years is likely to give an insight into the successional sequences occurring on rehabilitating bogs.

Setting up monitoring arrays and collecting data is both expensive and time-consuming. At a time when resources for monitoring are limited we must try to pool data, developing standardised formats for data collection and analysis. Monitoring on each site would then be contributing to a much wider effort to further understand the effects of rehabilitation work.

However, given our present state of knowledge, it can be argued that we can, to some extent, predict the effects of rehabilitation work. I suggest that for sites subjected to a rehabilitation programme, we adopt a minimalist approach to monitoring, which could be termed 'confirmational' monitoring.

This would involve a minimum sampling regime using simple techniques, data collection and storage. Data analysis needs to be designed from the start, using standard spreadsheet management software such as Paradox or Excel. A good example of this approach can be found in the Scottish Wildlife Trust guide to integrated monitoring of raised bogs (Wilson, 1995).

English Nature has recently published a review of monitoring on raised bogs throughout the UK (Wheeler and Shaw, 1995a) and held a joint workshop for the statutory agencies and NGOs to explore issues arising from the review (English Nature and Scottish Wildlife Trust, 1995).

As a result of these activities, we can propose a future monitoring programme which conforms to the minimalist approach. The most important monitoring activities would be for national purposes, enabling statutory agencies to assess the conservation status of sites of international value. The proposed programme would also assess rehabilitation work in terms of value for money.

For national purposes:

1. No more than 15 rehabilitation sites should be monitored, chosen for their size and nature conservation value. All Special Area of Conservation (SAC) sites should be included.
2. Sites should represent a wide geographical range.
3. Five integrating dipwells (such as Walrags (Wilson, 1995)) with peat anchors should be installed on each site. These should be sited in rehabilitation compartments as far from drainage channels as possible. Readings should be taken monthly. This should be regarded as the highest priority monitoring on the site.
4. Management activities, costs and events should be recorded using the project-based Countryside Management System (Hirons *et al.*, 1995)
5. Aerial photographs showing standing water should be taken every two winters.
6. Climatic data should be obtained from nearest existing stations wherever possible.
7. Permanent quadrats (1 m^2) should be set up using the SWT method (Wilson, 1995) in rehabilitation compartments containing Walrags to assess the relative cover of *Sphagnum* and other peat-forming vegetation once every five years. The extent of peat-forming vegetation should be mapped at the same time, using the standard technique developed at Thorne Moors (Dargie and Tantrum, 1995). Mapping should be performed using GIS techniques to produce digital output.
8. If resources allow, an assessment of the extent of indicator invertebrate species could be undertaken. These could include the large heath butterfly (*Coenonympha tullia*) and the dolicophodid flies.
9. The only bird monitoring should be mapping of gull colonies. Further bird monitoring might be undertaken for Special Protection Area (SPA) purposes.

This proposed programme would not preclude setting up the same regime on further sites if resources were to become available. Clearly the greater the number of data sets available, the more valuable the whole exercise becomes. Most of these activities are already taking place on NNRs and SSSIs throughout the UK, although the Sheffield review (Wheeler and Shaw, 1995a) has shown that methods are not standardised and data collection and analysis are inconsistent. Re-direction of existing resources and adoption of standardised techniques would greatly enhance the value of existing monitoring.

Above and beyond the national monitoring programme there would continue to be a need for collection of data for intra-site purposes. These might include feedback and 'fine-tuning' into site management, together with the collection of information needed for publicity purposes.

For local management needs:

1. Hydrological monitoring to assess the efficacy of different designs of dam and bund;
2. Fixed point photography for assessment of vegetation structure and surface water;
3. Birds (breeding waders and wintering wildfowl);
4. Rare species;
5. Other monitoring (including archaeological and industrial heritage) to increase the value of site interpretation. This is particularly important on sites with large numbers of visitors and a high public profile.

Collecting data for both national and local purposes is nothing new, although it has largely been done on an *ad hoc* basis for many years. For data to be useful at a national level there is a clear need for collection, analysis and promulgation of the results to be performed by a central body; a lead agency. This can only be done effectively if a standardised, IT-compatible format is used for data capture. An example of such a format is given by Wilson (1995), although this may need to be expanded to fulfil national needs.

Apart from data analysis, perhaps the most neglected side of monitoring in the past has been presentation of results and their promulgation to potential users such as the statutory agencies, government departments and the NGOs. In recent years, there has been a move within the statutory conservation bodies towards producing regular 'state of the environment' reports, particularly for ecosystems listed in Annex 1 of the Habitats and Species Directive. Producing simple, readable outputs from a monitoring programme such as that put forward in this chapter would be a significant contribution to reporting the success or otherwise of the efforts currently being put into bog management and rehabilitation.

Part Ten

Legislative Protection

An Introduction

Dr Rob Stoneman
Scottish Wildlife Trust, UK.

Much is made of the voluntary principle in the UK and the Peat Producers Association's (1990) Code of Practice is a good example of how industry may regulate itself. However, it is particularly galling to see these principles breakdown as demonstrated by the present tussle to save Rora Moss Site of Special Scientific Interest (SSSI) from peat extraction by a PPA member. The need for legislation is clear.

From the selection of legislative approaches shown here, the Swiss and Northern Irish approaches can be particularly applauded. In Switzerland, Mr Nowak shows how the almost uniquely Swiss system of direct, decentralised democracy has been used to enact strict protective measure in the face of government pressure to destroy valuable peatland habitat - a remarkable 'phoenix from the ashes' story. In Northern Ireland, Messrs Seymour and Corbett relate a genuine appreciation by the Department of the Environment for Northern Ireland for peatland's natural and cultural heritage. Their wide ranging strategy sets out a strong basis for the conservation of the resource although, in some respects, measures in Northern Ireland mirror some of the problems in mainland Britain (see Bain's contribution, Chapter 36, this volume).

Mr Bain reflects on the inadequacies of some forms of legislative protection in the rest of Britain. Indeed, it is interesting to note the stark contrast between the view of the RSPB (Bain) and current policy thinking within the UK Department of the Environment (Ward *et al.*, Chapter 35, this volume). On the one hand, Bain argues that 'legislation has failed to prevent losses to one of Britain's rarest habitats - raised bog' whilst Ward *et al.* argue that 'a good and balanced way forward for peatlands…' has been achieved.

Other chapters in this compendium point to particular failures in legislative protection. Firstly, there is scant protective regard for the potential archaeological and actual palaeoenvironmental resource of peatlands (Charman, Chapter 6, this volume). Secondly, the potential wildlife value of badly damaged sites (Steel, Chapter 40; Stoneman, Chapter 45, this volume) is often not protected and, thirdly, the holes in the SSSI system for bogs, identified by the UK government (DoE, 1994c) in the Biodiversity Action Plan, have not yet been filled - particularly in Scotland.

Many of these gaps in legislative protection - common across many countries - could be filled with an imaginative and flexible implementation of the Habitats Directive (detailed by Raeymaekers, Chapter 38, this volume). Despite the problems of the CORINE classification and the paucity of funds allocated to its implementation, the Directive has the potential to shift the focus from largely site-based protection (cf the British SSSI system) to protection of the whole resource. Conservationists look forward to the definition and implementation of the 'wider measures', invoked in Article 10 of the Directive, to bring about whole-scale habitat protection across the Union.

Chapter thirty-five:

Legislation and Policies for Peatlands in Britain with Particular Reference to England

Miss Anne E. Ward, Mr Graham Donald and Dr Tom Simpson
Department of the Environment, London, UK.

Introduction

By the early 1990s, there were considerable concerns about damage which commercial peat extraction was causing to the nature conservation interest of some peat areas - particularly raised bogs. In response to these concerns, the Department of the Environment (DoE) convened a Working Group in June 1992 to consider the key issues in the balance between nature conservation and mineral extraction interests affecting peatlands. The Group's membership was widely drawn from those organisations who had a statutory involvement in the key issues; other organisations were invited to submit views and factual information to the Department, for consideration by the Group.

In its published report (DoE, 1994b), the Working Group made recommendations for a future policy approach towards peatlands in Great Britain. This approach should ensure protection and conservation of representative areas of peatland biotopes (of a quantity and quality which represent the country's 'critical natural capital') while also meeting the needs of the horticultural industry, and other peat users, using supplies of domestic or imported peat. This forecast incorporated a contribution to overall market requirements predicted to be made by peat alternatives.

This chapter summarises the Government's policies for the identification and protection of habitats (including peatlands) for nature conservation. It describes the policies and legislation which have been developed on the basis of the Working Group's recommendations. These make provision for development whilst also ensuring effective conservation of wildlife and natural features. The context of policies on both habitat conservation and development is increasingly an international one, founded on European law and the obligations which this country has entered into under nature conservation Conventions.

Identification and protection of the habitat for nature conservation

Since 1947, conservation legislation in England, and Britain as a whole, has developed alongside a comprehensive system of town and country planning legislation. Its essential features are the wise management of the nation's land (and freshwater) resources and the safeguarding of key areas of national importance for nature conservation. This is achieved primarily through the designation of areas as National Nature Reserves (NNRs) and Sites of Special Scientific Interest (SSSIs).

The SSSI system is the foundation stone for nature conservation. The system has been closely linked with town and country planning since the 1940s, when local planning authorities were formally notified of SSSIs in their area and required to consult with, and take into account the views of, the then Nature Conservancy, when considering planning applications likely to affect a protected, designated site.

A requirement for local planning authorities to take account of nature conservation interests whenever they considered development proposals, proved to be an effective means of safeguarding important areas. It nevertheless had a basic weakness because in Britain activities such as agriculture and forestry are not classed as 'development' of land and hence are not subject to planning control. Many SSSIs were damaged or destroyed by these activities. The Wildlife and Countryside Act 1981 introduced a system designed to give greater protection to SSSIs. The national conservation body, which was then called the Nature Conservancy Council (NCC), was for the first time required to notify the owners and occupiers, as well the local planning authority and the Secretary of State, of any land which in their opinion was of special interest (i.e. a SSSI) because of its flora, fauna, geological or physiographical features.

In turn, owners and occupiers were required to consult with the NCC, before carrying out any of the 'potentially damaging operations' (PDOs) which were identified within the notification. Landowners could, however, enter into a management plan which the NCC backed, where appropriate, with a management agreement providing financial assistance for any work necessary to enhance the interest.

Thus, notification of SSSIs now identifies important wildlife habitats and protects them from inappropriate management and development.

Since the early 1990s, the NCC has been subdivided into three Country agencies for England, Scotland and Wales, but they continue their responsibilities under the 1981 wildlife legislation.

In England, at 31 March 1995, there were 3825 SSSIs covering 893,335 ha. The total area with some form of damage declined from 16,470 ha in the previous year to 2,266 ha. In percentage terms damage affected 2.9% of sites by number or 0.3% by area. Almost all such land can regain its special interest.

Figures published by the Joint Nature Conservation Committee (JNCC) show a continuing decline in habitat loss and damage throughout Great Britain.

Nature conservation is increasingly being seen to be important at both the national and international level, and some sites in Britain are recognised to be of wider than national importance. Our approach has been, and continues to be, to use the national SSSI series as the starting point from which to assess and select sites of international importance (as shown in Table 35.1).

Table 35.1. International and national site designations for nature conservation.

Importance	Site Designation	UK Statutory Designation
Sites of International Importance	RAMSAR - wetlands	SSSI
	SPECIAL AREAS OF CONSERVATION (SACs) - Habitats Directive	SSSI:SAC
	SPECIAL PROTECTION AREAS (SPAs) - Birds Directive	SSSI:SPA
Sites of National Importance	National Nature Reserves (NNRs)	SSSI
	Sites of Special Scientific Interest (SSSIs)	SSSI
Sites of Regional/Local Importance	Various - e.g. Local Nature Reserves (LNRs)	LNR

The Habitats Directive, and National Policy Guidance

The Directive* provides for the creation of a network of protected areas across the European Union to be known as 'Natura 2000'. Natura 2000 will consist of Special Areas of Conservation (SACs) designated under the Habitats Directive and Special Protection Area (SPAs) under the Birds Directive. The Habitats Directive also requires all Member States to set up an effective system to prevent the killing, injuring or damaging disturbance of endangered species.

The main aim of the Directive is to promote the maintenance of biodiversity, taking account of economic, social, cultural and regional requirements. It makes a contribution to the general objective of sustainable development.

Member States were required to submit a list of candidate SACs to the Commission by June 1995. The Government consulted on a list of 280 possible sites recommended by the conservation agencies in April 1995. A first tranche of 136 land based sites was sent to the Commission in June. These sites include peatland habitat sites from all parts of the UK. A first tranche of marine sites and further land based sites will follow soon.

* Directive on the Conservation of Natural Habitats of Wild Fauna and Flora (92/43/EEC)

As a result of the UK signing the Convention on Biological Diversity in Rio de Janeiro, the Government published Biodiversity: the UK Action Plan in January 1994. This contains specific commitments to the conservation of species and habitats. Part of this is the planned publication, in December 1995, of a set of costed action plans for key species and habitats together with directional statements of what is needed to conserve all habitat types found in the UK.

In England we have also prepared a series of guidance notes on Government policies and legislation, to help local planning authorities (and other interested parties and organisations) to carry out their functions in respect of town and country planning. A recent Planning Policy Guidance Note (PPG), PPG9 (DoE, 1994e), is specifically concerned with Nature Conservation.

PPG 9 explains the key concepts as 'to ensure that the Government's policies contribute to the conservation of the abundance and diversity of British wildlife and its habitats or to minimise the adverse effects on wildlife where conflict of interest is unavoidable'.

The PPG goes on to describe the main statutory nature conservation obligations, under both domestic and international law; how nature conservation objectives should be reflected in development plans; and describes the controls that help to protect SSSIs, including those of international importance. In particular it takes account of the implementation of the EC Habitats Directive.

Since peatlands also contain important archaeological and palaeo-ecological records, it is relevant to note that the Department has also published guidance on the identification and protection of archaeological sites, in PPG 16 'Archaeology and Planning' (DoE, 1990).

Policies and legislation to balance the need for development and the conservation of the natural environment

In 'the real world' Governments, and indeed many people, have to deal with the question of how to provide for necessary development and economic growth without harming nature conservation interests. In England, and in the UK as a whole, such decisions are chiefly made through the town and country planning system.

Both of these aspects were key parts of work conducted over the last 2-3 years, to develop new policy guidance for peatlands in England - and indeed in other parts of Britain. This work, and the publications which resulted from it, form the second section of this chapter.

The Peat Working Group

The Working Group considered the key issues set out in Fig. 35.1.

Fig. 35.1. Peat Working group: Key Issues.

- The extent and condition of raised bogs
- Requirements of the Habitats Directive & other nature conservation policies
- Peat production and uses
- Types and qualities of alternatives
- Future trends in the user markets
- The extent of present peat workings, and planning controls
- The impact of peat working on sites of nature conservation, archaeological value and on the landscape
- Rehabilitation of existing workings.

The Group was able to use a draft of the emerging 'Raised Bogs' part of the National Peatland Resource Inventory (being carried out by Scottish Natural Heritage) to consider *the extent and condition of the habitat,* and also the extent to which important conservation sites were affected by commercial peat extraction. Whilst there were some important sites where there was a conflict of interests, there were many other sites where this was not, and is not, the case.

The Group also looked carefully at *the usage of peat in the range of horticultural sectors; and at the range, properties and usage of alternatives.* Specific inquiries were carried out by the Ministry of Agriculture, Fisheries and Food (MAFF) within the professional user sectors; and DoE commissioned consultants to investigate the usage of materials by amateur gardeners, the landscape industry and local authorities (Aspinwall and Company, 1994). Both studies confirmed the importance of making a distinction between 'soil improvers' and 'growing media' - since the technical requirements for the latter are much more demanding.

The Working Group's report was published in autumn 1994, and should be consulted to see how these key issues were assessed in detail, and what conclusions and recommendations were reached.

New planning guidance for peatlands in England

At the same time as publishing the Working Group's Report, the Department issued the draft of a new 'Minerals Planning Guidance Note' (MPG) on peat in England, for public consultation. This draft was informed by the outcome from the Working Group report. In total DoE received comments on the draft guidance from about 70 respondents, including local authorities, nature conservation and amenity groups, the minerals industry, and others such as professional institutions. We completed our assessments of the responses to this draft MPG in the spring of 1995, and had a series of meetings with some of the key respondents - including the statutory and voluntary nature conservation organisations, archaeological interests, and the peat industry. We took account of these meetings and responses in revising the MPG for publication.

In July 1995, the new guidance was issued as MPG13 - 'Guidelines for Peat Provision in England, including the Place of Alternative Materials' (DoE,1995d). The new guidance contains four main aspects of Government policy for peatlands in England - see Fig. 35.2.

Fig. 35.2. Government policy for peatlands in England.

The Government's policy for peatlands in England is to:

- conserve a sufficient range, distribution and number of all peatland habitats, representing part of the critical natural capital of the country; & promote the wise use of the wetland resource within the nation's peatland heritage;
- avoid wherever practicable the destruction of important archaeological remains in peatland;
- enable the horticultural industry to continue to be supplied with peat; and also to encourage the development and use of suitable alternatives so that market needs can be met in different ways;
- provide a suitable framework for updating old permissions for peat extraction, especially in respect of rehabilitation of sites.

The first section of this chapter provides a summary of how the selection and protection of sites for nature conservation takes place. There are also a range of measures, legislation and procedures in respect of archaeological sites, which are referred to in the MPG.

In respect of usage of peat and other materials, the Government considers that, whilst increasingly greater proportions of the full range of horticultural markets should be supplied by alternatives to peat, there will still be a need for peat-based products. Some of that peat may need to come from new working sites in England and elsewhere in Britain. However, the new policy guidelines set out strict criteria for selection of any new sites for peat extraction, in accordance with the principles of 'sustainable development' to ensure that future workings do not damage sites of nature conservation or archaeological importance. Such new sites should be sought from '*Areas significantly damaged by recent human activity and of limited or no current nature conservation or archaeological value*'.

The guidance sets a new Government target for peat alternatives to meet 40% of the demands of the total horticultural and landscaping markets in the next 10 years.

Framework to update old permissions, including the rehabilitation of sites

A very important aspect of the guidance is that it sets out agreed principles on how to deal with updating the operational controls over older permissions where some peat will continue to be extracted - especially about restoration or rehabilitation of the sites. The basis for these principles in respect of peat sites was first put forward by the Working Group, but at that stage it would have been implemented voluntarily by the industry and planning authorities.

The Environment Act 1995, which received Royal Assent on 19 July 1995, includes a new statutory framework for updating all of what are known as 'old mineral permissions' in Britain. These are sites which can continue to be worked, but where the planning permissions may date from between July 1948 and February 1982. The Act provides a new means to update the operational and restoration controls over many current workings. The key elements of this updating are:

1. New statutory requirement to update all permissions granted between July 1948 *and* February 1982.
2. Oldest sites, and workings in SSSIs, National Parks and Areas of Outstanding Natural Beauty (AONBs) to be tackled first.
3. Operators to submit schemes of updated conditions for approval of mineral planning authorities (MPAs).

The new MPG 13 guidelines spell out some agreed special emphasis on how this updating will be carried out for peat sites, such that it will be on schemes which provide for rehabilitation to benefit nature conservation.

To assist in providing guidance on how such rehabilitation may be undertaken, the Department commissioned research in 1992 at the University of Sheffield. The report was published in April 1995 under the title 'Restoration of Damaged Peatlands' (Wheeler and Shaw, 1995b). The following broad principles were identified (and are set out in more detail in Annex D of the MPG):

1. maintenance of critical refugia areas,
2. buffer zones of undisturbed peat,
3. phasing of operations/rehabilitation,
4. the retention of *in situ* peat for rehabilitation,
5. hydrological control.

English Nature are developing practical guidance for rehabilitation at the peat sites which they manage and their experience contributed to, and is also informed by, the Sheffield report.

In October 1995, post-dating the Peatlands Convention, the Department and the Welsh Office issued new joint guidance about the general implementation of these new mineral working provisions in the Environment Act, as MPG14 'Environment Act 1995: Review of Mineral Planning Permissions'.

Conclusion

It is the authors' view that, with the publication of the new MPG13 and the existence of PPG9, we have achieved a good and balanced way forward for peatlands in England. We should be able to fulfil both important objectives which are set out at the start of the chapter:

1. identification and protection of a representative range and distribution of the habitat for nature conservation; and
2. make provision for development and economic growth whilst ensuring effective conservation.

****
© Crown Copyright.
Whilst any views expressed in this chapter are not necessarily those of the Department of the Environment, they do reflect current policy thinking and published advice.
****

Chapter thirty-six:

Legislative Protection for Bogs: Does the British Approach Work?

Mr Clifton Bain
The Royal Society for the Protection of Birds, Edinburgh, UK.

Introduction

The Royal Society for the Protection of Birds is Europe's largest voluntary wildlife conservation body, with a £30 million annual income and over 925,000 members. We own and manage over 100 nature reserves throughout the UK including 7000 ha of blanket bog in the Flow Country. At this Forsinard reserve, the RSPB employs local staff to help encourage positive conservation over a much wider area thanks to a £1.4 million grant from the EU LIFE fund (supported by Scottish Natural Heritage and Caithness and Sutherland Enterprise) and generous donations from our members (who raised over £1.6 million for this project). Evidently, the public values peat bogs. Much of our work is involved in influencing legislation and policy decisions at a national and international level to ensure important wild birds and their habitats are conserved.

The RSPB's conservation campaigns have included a focus on peat bogs as many of these sites are of international importance for birds. Together, with our partners in the Peatlands Consortium, over the last five years we have sought to highlight the problems facing raised bogs in particular and the urgent need for legislative change to tackle the threats facing this habitat. The issues presented in this chapter draw largely from information contained in a document *Out of the Mire* (RSPB and Plantlife, 1993) which was produced on behalf of the Consortium by the RSPB jointly with Plantlife and from a review of statutory site protection published by Wildlife Link *SSSIs: A Health Check* (Rowell, 1991).

Conservation legislation in Britain

In Britain, our habitat conservation legislation centres on statutory designation of the most important sites. Sites of Special Scientific Interest (SSSIs) are notified under the Wildlife and Countryside Act 1981 (as amended). Their stated objective is to form a national network of areas representing Britain's natural heritage where it occurs in the greatest concentration or highest quality to encompass the 'necessary minimum' which must be safeguarded if we are going to pass that heritage on to future generations. It is certainly questionable whether maintaining only the vestigial remnants of something we have now is an adequate objective for the future. With over 90% of our raised bogs damaged or destroyed (Plantlife, 1992), we must surely look to repairing some of the past damage.

Site notification

It is widely accepted that the remaining area of near natural raised bog is below the critical levels necessary for safeguarding this habitat's future survival (DoE, 1994b). The statutory conservation agencies have sought to address this issue and, in their guidelines for selecting raised bog SSSIs (NCC, 1989; JNCC, 1994), they state that all remaining natural areas should be notified and also make provision for including damaged bogs capable of rehabilitation.

The statutory agencies have yet to notify many of the raised bogs as SSSIs. In Scotland, in particular, there is a huge shortfall with only 46 out of 293 remaining primary raised bog sites having been designated (DoE, 1994b). Furthermore, the boundaries of existing SSSIs often do not extend to include the full functioning hydrological unit of the bog. Of particular note is Flanders Moss, one of the best remaining primary bogs in Britain, where part of the site proposed for peat extraction was not included in the original SSSI notification. This has recently been rectified with a revised notification of the whole site.

The statutory agencies have pointed to limited resources as the reason why designations have had to wait whilst other work priorities proceed. With raised bogs clearly identified under the EU Habitats and Species Directive as priority habitats requiring urgent conservation action, Government and the statutory agencies have a responsibility to ensure that sufficient resources are available to ensure designation does proceed.

Another possible reason for the delay in notification is the apparent public opposition to SSSIs being 'forced' on to their land resulting in objections and lengthy negotiations. In Scotland an appeals procedure was established to enable owners and occupiers to have their objections to notification heard by an advisory committee. The objections should only be considered on scientific grounds and not on socio-economic issues such as possible limitations on development opportunities. None-the-less, non-scientific objections are made and further delay results as discussions take place.

Notification is simply a statement of scientific value which does not carry prohibitive powers. Under the Wildlife and Countryside Act protection of SSSIs depends on owners or occupiers voluntarily consulting with the statutory agency over operations likely to damage the site. The problems lie, not in the notification itself, but in the measures applied to conserve the sites. The methods of reaching conservation agreements and the limited incentives available to resolve disputes are where the real debates exist.

Protecting SSSIs

The Wildlife and Countryside Act requires the statutory agencies to identify potentially damaging operations when notifying owners of SSSIs and the owner must then consult with the agency before undertaking any such damaging activity. The statutory agencies can then enter into a management agreement to prevent damage. This process has been known to drag on for some considerable time. At Flanders Moss for example, discussions had been underway between SNH and the developer for over two years before a solution was found. Once agreement is reached the compensation payments can be very large. The peat extraction rights at Flanders Moss were bought out by SNH as part of a £1.8 million management agreement.

A major criticism of the legislation is that it offers little incentive to positively manage neglected or previously damaged sites. It largely focuses on preventing new damage with landowners being paid for not doing something. The majority of raised bogs have been damaged in some way and are in urgent need of conservation action to stem further loss and begin the process of rehabilitation. Drainage of bogs arising from water loss through old ditches and scrub encroachment, for example, need to be tackled through positive management.

The current powers of the statutory conservation agencies do not encourage positive management on SSSIs where agreement cannot be reached with a landowner. There is no statutory requirement for ditches to be blocked up and the landowners are unlikely to undertake the work at great expense to themselves. Under the Wildlife and Countryside Act, the only final course of action is the use of compulsory purchase powers where agreement cannot be reached. Such drastic measures are seldom employed by the statutory nature conservation agencies.

SSSI notification has an important role in safeguarding sites against potentially damaging new development proposals. The Government has put great emphasis on the need to protect SSSIs and has advised local authorities to have regard to considerations of nature conservation as necessary in determining planning applications (DoE, 1994e; Scottish Office, 1991). Local authorities are also required to consult with the statutory conservation agencies before determining planning applications in SSSIs. Unfortunately the backlog of sites yet to be notified will have a lesser degree of protection. An example of one site not yet notified and under threat is Airds Moss (Strathclyde) which is an intermediate bog of international importance for birds and which has been granted a permission for open-cast coal mining with further developments proposed.

For existing developments, planning permission under the Town and Country Planning Acts overrides the Wildlife and Countryside Act provisions. SSSIs can therefore legitimately be destroyed by activities with planning permission. In some cases the planning permissions were granted in the 1950s and 1960s when the conservation importance of peat bogs was not fully appreciated and prior to the introduction of heavily mechanised, industrial scale peat extraction. Vast areas of Thorne and Hatfield Moors were mechanically stripped of bog vegetation whilst the site was an SSSI resulting in the well documented loss of several important plant, bird and invertebrate species (Eversham, 1991).

The Government has sought to protect peat bogs from such damaging development by issuing new policy guidance to planning authorities. For peat extraction, the Mineral Planning Policy Guidance for England, July 1995, gives planning authorities a clear message that Government policy is to restrict peat extraction to areas which are of limited or no conservation value. In Scotland the National Planning Policy Guideline, Land for Mineral Working, April 1994 only goes so far as to require planning authorities to consult with the statutory agencies before granting planning permission for peat extraction. It gives no indication as to the level of priority to be given to protecting peat bog sites.

Mineral planning policy guidance also aims to minimise the impacts of existing peat extraction on a site of conservation importance. However, in those situations where even the most tightly constrained peat extraction operation would still be damaging to the conservation interest of the site, the guidance offers no solution. Local authorities do have powers to revoke planning permissions, but the legislation requires compensation for profits foregone. The high potential costs of compensation to be met by limited local authority budgets discourages the use of such powers. The Government has accepted this point to some extent in implementing the Habitats and Species Directive and has stated in the Habitats Regulations, October 1994, that where a local authority is minded to revoke a planning permission which clearly threatens the conservation interest on a designated Special Area of Conservation (SAC), any compensation paid to the developer will be reimbursed by national funds. Unfortunately, this will not apply to many peat bogs since the current list of proposed SACs excludes several important raised bogs where planning permission exists (RSPB, 1995).

Another failing of our conservation legislation is that it has limited ability to address dispersed threats such as atmospheric or agricultural pollution, overgrazing by deer, acid rain or water abstraction. Wider countryside measures are available to help address these issues but arguably these should be encompassed in legislation to act as a backstop if the more voluntary approach should fail.

Habitats and Species Directive

The EU Directive on the Conservation of Natural Habitats and of Wild Flora and Fauna, 92/43/EEC, the Habitats and Species Directive provides the opportunity for tackling some of the shortcomings identified above. It requires Member States to avoid deterioration of natural habitats and identifies raised and blanket bogs as priorities for conservation.

The British Government has chosen to implement the Directive largely through the existing conservation legislation described above. SACs will therefore be selected from sites already designated as SSSIs and will be protected through the SSSI mechanism. The Government has recognised that the proposed list of SACs falls short of what is required to bring about favourable conservation status of peat bog habitats. This is largely because the SSSIs have not yet been designated.

The Directive encourages management plans to be produced for all designated sites. Once achieved this will be a great improvement on the existing situation in which few peat bogs notified as SSSIs have management plans. There will also have to be greater emphasis on positive management within SACs if the terms of the Directive are to be met. As explained previously there are difficulties with existing British legislation on this matter. A solution might be to encourage greater links between the wide range of countryside management funds available through, for example, Agri-environment Regulations, which have far greater budgets attached to them than is available within the statutory agencies' own budgets.

Financial support from Europe is available for the conservation of priority habitats. Designation of peat bog sites should be seen as a positive attribute which can attract funding for good management practice.

The three year Peatland Management Scheme being operated by SNH and RSPB in the Flow Country and which is part-funded by the EU LIFE scheme is an example of an initiative which offers payments for positively managing bogs in environmentally sensitive ways. It also offers capital payments for reversing damaging activities, i.e. blocking ditches, tree removal and reducing grazing pressure.

Conclusion

This chapter has only briefly summarised key elements of the legislative approach to peat bog conservation in Britain. There are quite clearly a number of major improvements required if we are to meet our international responsibilities to safeguard this habitat. If the Government takes an imaginative look at how our legislation could be used, not only to protect peat bogs but also to repair some of the past damage, perhaps we can bring about the 'favourable conservation status' which the Habitats and Species Directive seeks.

Chapter thirty-seven:

The Conservation of Peatland in Northern Ireland

Mr Paul McM. Corbett and Mr Graham R. Seymour
Department of the Environment for Northern Ireland, UK.

Introduction

Peatland is of enormous importance to the stability and general well-being of the environment in Northern Ireland. It influences both upland and lowland landscapes, plays a major role in conserving biodiversity and affects river catchment hydrology. The importance that Government attaches to peatland conservation was reinforced by the recommendations made by its official advisory body, the Council for Nature Conservation and the Countryside (CNCC, 1992).

In 1993, in response to the CNCC's Peatland Strategy, Government published a Policy Statement on Peatland Conservation. This sets out Government policy on a range of issues affecting the conservation of peatland habitats and the use and exploitation of peat. In recent years Government has made great strides towards effective conservation of peatland, principally through the identification, declaration and protection of the most valued sites by its Environment Service.

The peatland resource

Northern Ireland is well endowed with peatland, because of its climate, soils and altitude. Even the drier areas in the east have an annual rainfall of 800 mm, over an average of 195 rainy days. In the west this rises to over 1000 mm over 250 rainy days (Betts, 1982). Soils are generally poorly draining glacial tills, which favour bog formation. Much of Northern Ireland is at an altitude of over 200 m, with major upland blocks in Antrim, Londonderry, Down, Tyrone and Fermanagh.

Although peatland covers a high proportion of the land area, much of the resource has been destroyed or damaged by a variety of means. The survey and protection of peatland has therefore been given a high priority by Environment Service.

The standard classification of bogs into lowland raised and upland blanket is generally recognised as an oversimplification. These habitats represent the opposite ends of a continuous range of variation. At the two extremes, the differences are clear; within the transitional zone classification is much more difficult, as the sites may have characteristics of both lowland and upland bogs. The most recent attempt at classification (Lindsay, 1995) has recognised the two major types and has described a transitional 'intermediate' type, which is akin to the 'ridge-raised' bog of previous literature (Moore and Bellamy, 1974; Bellamy, 1986). In Northern Ireland the situation is rather complex (see Fig. 37.1), for the reasons discussed below.

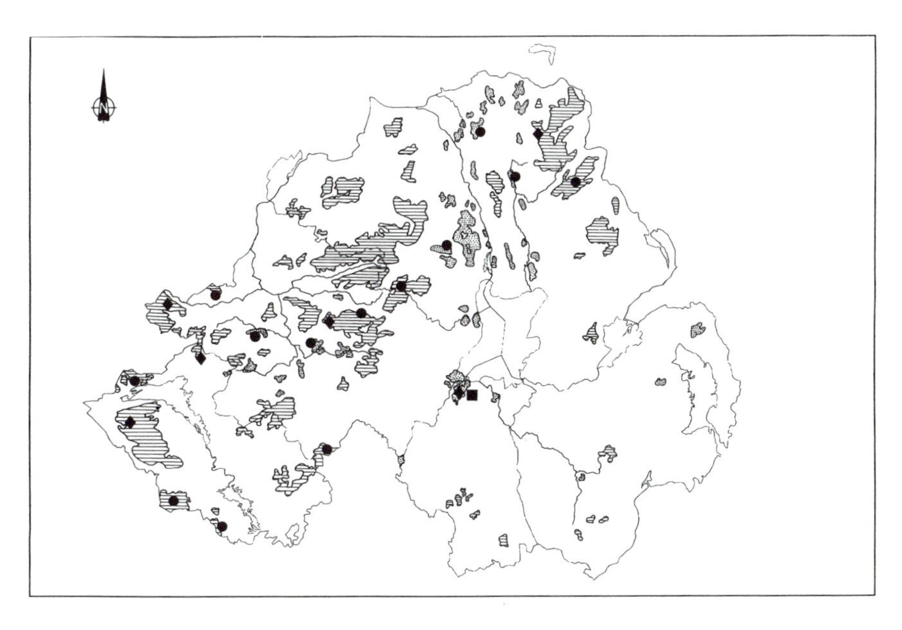

Fig. 37.1. Peatland areas and protected sites in Northern Ireland.

Lowland raised bogs

These occur all over Northern Ireland, but are generally very scarce in Counties Down and Armagh, where the majority have been totally destroyed by a combination of turf-cutting and agricultural reclamation (e.g. Brown, 1968). They tend to be of two types, with the 'typical' raised bog (Lindsay,1995) generally confined to flood-plains in the major river valleys in Antrim and Londonderry, and along the southern shores of Lough Neagh. In fact, the topography and climate of Northern Ireland have made this a rather uncommon type. The majority of lowland bogs share some, but not all, of the characteristics of intermediate mire (Lindsay, 1995), where peat formation has occurred in depressions and spread out from these to coalesce in the form of large and irregularly shaped bogs. This is particularly apparent in the drumlin belt of Tyrone and to a lesser extent, Fermanagh, but is also a feature of many sites in Antrim and Londonderry. It is considered that the majority of such sites in Northern Ireland have more affinities to lowland bog than to blanket bog, and they are treated as such in this chapter.

Blanket bog

Again, this is a broad term which covers a number of different peatland types. Although extensive areas of blanket bog are generally found at altitudes in excess of 200 m, a number of intermediate bogs are also included in this category, because they have more affinities to blanket bog than to lowland raised bog. These generally occur in the altitude range 150 to 200 m, but may be found at lower elevations in the extreme west, where extensive blanket bog development occurs even as low as 90 m a.s.l. Areas of raised bog within blanket bog - 'Atlantic raised bog' (Moen, 1985) or 'unconfined raised bog' (Lindsay, 1995) - are also included. The most extensive areas of blanket bog are to be found on the Antrim Plateau, the Sperrin Mountains and in County Fermanagh.

Resource survey

Work on the distribution and condition of Northern Ireland's bogs has been carried out through a combination of commissioned research and by 'in-house' survey by Environment Service staff. A major resource inventory was completed by Queen's University of Belfast in 1988 (The Northern Ireland Peatland Survey, Cruickshank and Tomlinson, 1988). Using aerial photographs, the distribution of different types of peatland (lowland raised bog and blanket bog) was recorded and mapped and an indication of the condition of each site was made.

The Northern Ireland Peatland Survey (NIPS) defined peatland as wetland still covered with bog vegetation or with clear evidence of peat accumulation. It excluded land reclaimed for agriculture or under forest cover. Hence the overall figures for each category are to some extent an underestimate of the total resource. Although the most recent aerial photographs were used, for some parts of Northern Ireland coverage was only available from the 1960s, making accurate assessment very difficult in those areas. However, the survey is the only one that gives a complete picture of the total peatland resource in Northern Ireland.

NIPS found that some 167,580 ha (12%) of Northern Ireland is covered by peatland, with lowland raised bog accounting for 25,196 ha and blanket bog making up 142,384 ha. It was estimated that 2270 ha of lowland raised bog (about 9% of all lowland raised bog) was still intact, with most of the remainder being cut, principally by hand. For blanket peatland, the intact figure is 22,175 ha (15% of all blanket bog), with 56% having been cut or drained. Again, the majority of the latter has been cut by hand, but there is an increasing amount of machine cutting.

A site survey of lowland raised bogs with an intact surface of 10 ha and over was undertaken by Environment Service staff in the late 1980s (Leach and Corbett, 1987). This was a preliminary survey, aimed at identifying the most important sites for nature conservation. Detailed work on a number of sites has been carried out since then. To date, 83 lowland raised bogs have been surveyed.

Environment Service surveyed the majority of Northern Ireland's blanket bogs between 1988 and 1993. Detailed reports have been compiled for the four major areas of blanket bog with an intact area of 1000 ha and over - Garron Plateau, Cuilcagh Mountain, Slieve Beagh and Pettigoe Plateau. More recently some smaller and more dispersed areas of blanket bog (with an intact surface larger than 50 ha) have been examined.

In-house survey has given a detailed impression of the condition of individual sites, thus enabling Environment Service to prioritise its designation programme. In addition, the work has contributed to our understanding of the regional variability of the peatland resource. Lowland raised bogs in the west display subtle differences from those in the east as a result of topography and climate. As discussed earlier, many of Northern Ireland's lowland raised bogs do not conform to the 'typical' flood-plain situation with their origins in a single peat body, but rather are the result of two or more peat bodies coalescing. The west, with its higher rainfall, provides even more favourable conditions for peat growth than the remainder of the country. In addition, the topography of the region is rather more varied. These factors accentuate the tendency for bogs to merge together and become more complex in shape. This is accompanied by floristic changes, as a number of plant species occur more frequently on bog surfaces in the west, notably carnation sedge *Carex panicea* L., white beak-sedge *Rhynchospora alba* (L.) Vahl, the moss *Racomitrium lanuginosum* (Hedw.) Brid. and the liverwort *Pleurozia purpurea* Lindb.

A similar transition from east to west in response to climatic gradient has been identified for blanket bogs, with the Pettigoe Plateau showing distinct affinities to low-level Atlantic blanket bog (Moore, 1968; Doyle, 1982). The peat here has formed at a much lower altitude than in the remainder of Northern Ireland's blanket bogs. Once again, there are corresponding floristic differences, with such species as purple moor-grass *Molinia caerulea* (L.) Moench and black bog-rush *Schoenus nigricans* (L.) becoming more abundant on the bog surface.

Site protection and management

The protection of peatland habitats, particularly in the form of raised and blanket bogs, is regarded as a very high priority by Environment Service. The greatest importance is attached to the conservation of sites which have remained largely free from disturbance.

Before 1985, the conservation of peatland by the Department of the Environment for Northern Ireland was accomplished under the Amenity Lands Act 1965, either through outright purchase or by management agreement as National Nature Reserve (NNR). Several peatland NNRs have been established. In addition, Forest Service (part of the Department of Agriculture for Northern Ireland) has designated several peatland areas as Forest Nature Reserves (FNR) (see Tables 37.1 and 37.2).

Table 37.1. Protected Lowland Raised Bogs in Northern Ireland.

Site	Designation	Total area (ha)	Intact area (ha)
Garry	ASSI/FNR/pSAC	121	55
Dunloy	ASSI	108	60
Moninea	ASSI	45	30
Fairy Water	ASSI/pSAC	230	118
Deroran	ASSI	75	50
Ballynahone	ASSI/pSAC	244	95
ASSI TOTAL		**823**	**408**
Aghagrefin	FNR	40	0
The Argory	UWT	17	0
Inishargy	UWT	8	0
Mullenakill	NNR	20	16
Annagarriff	NNR	35	16
Peatlands Park		95	0
TOTAL		**215**	**32**
GRAND TOTAL		**1038**	**440**

Note: The Argory and Inishargy are Ulster Wildlife Trust reserves.

The Nature Conservation and Amenity Lands (NI) Order 1985 (and subsequent amendment in 1989) introduced more effective means of site safeguard, principally the Area of Special Scientific Interest (ASSI). Environment Service immediately initiated a process of site survey and declaration, with the aim of establishing a comprehensive ASSI network by the year 2001 in which peatland was identified as a priority habitat.

The cornerstone of Government policy towards protecting peatland sites is thus the declaration of ASSIs. The ASSI network for peatland, as for any other habitat, must encompass a sufficient area and a sufficient number of sites to:

1. give protection to a sufficient amount of the total resource concomitant with its rarity and fragility; and,
2. ensure that the full range of variation is represented.

The current criteria for selecting peatland Sites of Special Scientific Interest in Great Britain have recently been published (JNCC, 1994). Environment Service has adopted these criteria for peatland Areas of Special Scientific Interest in Northern Ireland. However, these are considered to be *minimum* criteria only, and Environment Service will select the best examples from those which qualify to be ASSI.

Although several sites have already been declared, a review of conservation needs is currently being carried out, and further peatland designations will be made in the future.

Lowland raised bogs

A number of ASSIs from both eastern and western types have been declared. Together with a number of other protected sites (Table 37.1), these account for a total of 1038 ha, with a combined intact area of 440 ha. This represents 19.4% of the total intact area remaining in Northern Ireland.

Blanket bogs

Several ASSIs have been declared, including the four largest areas of intact blanket bog (which have been designated in the last two years). The total area of blanket bog protected (Table 37.2) is over 11,000 ha, taking in an intact surface of 6473 ha (29% of the intact area).

Habitats Directive

Active bog is one of the priority habitats in the Habitats Directive, and the importance of lowland and upland bog is reflected in Northern Ireland's proposed network of Special Areas of Conservation (SACs) . Over 11,000 ha of blanket bog and 595 ha of lowland raised bog have been put forward to the Commission for consideration as SAC. Conservation plans are being prepared for all peatland sites which have been proposed as SAC.

The Department of Environment has recently revoked an outstanding planning permission for peat extraction at Ballynahone Bog, which is widely regarded as one of the top lowland raised bogs in the Province. The decision was taken once the Department was satisfied that Ballynahone Bog would qualify as a candidate SAC.

Table 37.2. Protected blanket bogs in Northern Ireland.

Site	Designation	Total area (ha)	Intact area (ha)
Moneygal	ASSI	122	42
Teal Lough	ASSI/ FNR/pSAC	196	196
Black	ASSI/ FNR/pSAC	194	140
Pettigoe Plateau	ASSI/pSAC	1,270	950
Garron Plateau	ASSI/pSAC	4,650	1,416
Cuilcagh Mountain	ASSI/FNR/pSAC	2,750	2,327
Slieve Beagh	ASSI/FNR/pSAC	1,900	1,116
Teal Lough Part II	ASSI/pSAC	44	44
ASSI TOTAL		**11,126**	**6,231**
Murrins (part only)	NNR	20	20
Killeter Forest	NNR	22	22
Lough Naman	NNR	41	30
Meenadoan	NNR	20	20
Slieveanorra Forest	NNR	49	49
Slieveanorra	FNR	176	65
Bolusty Beg	FNR	7	7
Killeter	FNR	16	16
Mullyfamore	FNR	13	13
TOTAL		**364**	**242**
GRAND TOTAL		**11,490**	**6,473**

Peat extraction

Historically the most damaging activity to have affected bogs in Northern Ireland is turbary. It has been practised for at least 1000 years. Maximum production of peat for domestic fuel was probably reached in the early nineteenth century prior to the arrival of the railways and the spread of coal. The twentieth century has seen a steady decline, only interrupted by wars and oil crises. It is estimated that as much as 80% of lowland peatland has been cut for turbary with major and serious implications for the viability of the surviving fragments of bogs.

The winning of turf for domestic purposes takes place by virtue of traditional turbary rights or 'banks' let to individuals by bog owners. Planning permission is not required for peat cutting which is solely for domestic purposes, whether the cutting is done by hand or machine. Many owners, however, do take steps to ensure that peat cutting is not done in a haphazard way and it is not uncommon for machine-cutting to be forbidden. In some Environmentally Sensitive Areas farmers entering the scheme are forbidden by the Department of Agriculture from extracting peat from previously uncut land.

Planning permission is required if peat cutting is to be undertaken for commercial purposes. Since 1979, 12 applicants have been given planning permission covering an area of 1034 ha. Another four applications are currently under consideration.

The Department of Environment has set out its policy on peat extraction in the Policy Statement and in its Rural Planning Strategy. The Department will oppose peat extraction on sites which are designated as ASSI or which are likely to be so designated in the future. Indeed planning permission will only be granted where there is little nature conservation value and where other environmental criteria can be satisfied, for example, the effect on watercourses, landscape impact etc. Where peat extraction is likely to have a significant effect on the environment the Department requests an Environmental Statement. In practice, few applications are now proceeding without such a statement, which among other things, should consider mitigating measures and rehabilitation to bog or other semi-natural habitat.

Peatland and forestry

Peatland areas have traditionally been seen as suitable for forestry in Northern Ireland. Up to 40% of the state-owned forest estate has been planted on peat deeper than 0.5 m. This is a result of a combination of factors, including land values and availability, and a belief that deficiencies in deep peat as a growing medium could be overcome.

Government policy on afforestation has changed in recognition of the value that society now places upon peatland as a habitat and resource in its own right. It is current policy to encourage afforestation on land where forestry is judged to be the most appropriate long-term land use. In effect, this means bringing forestry 'down the hill' on to marginal agricultural land on mineral soils and away from deep peat. This policy extends to private forestry seeking grant aid as well as acquisition of new land for afforestation by the state sector.

The interpretation of this policy means that for undisturbed lowland raised bogs, or those which are cutover but judged capable of regeneration, there is a presumption against planting. The major forms of blanket peat are classified in forestry terms as mesotrophic, oligotrophic and dystrophic in decreasing order of fertility. Dystrophic peat should never be planted on silvicultural grounds and there is a general presumption against afforestation on oligotrophic blanket bog. It is likely that any major proposal for planting on blanket bog would need to be accompanied by an Environmental Statement.

Peat and horticulture

Despite its attractions to various sectors of the horticultural industry, the continued harvesting of peat from Northern Irelands' bogs is not sustainable. It was thus considered important that Government in Northern Ireland should set an appropriate example in its use of peat. The local horticultural industry is not extensive and the bulk of the horticultural grade peat harvested in the Province is exported. Nonetheless, Government has taken the view that to continue using peat as a soil conditioner or mulch would be wasteful and unnecessary. Government Departments and Agencies thus no longer make use of peat in open ground situations and will encourage the appropriate use of alternative materials.

Government is firmly of the opinion that raising public awareness of the importance of the peatland habitat and the consequences of its continued loss and damage is a vital element in its campaign. It is important that the message comes across clearly that peat is a finite resource and one which supports a unique wildlife for which Northern Ireland has a particular responsibility.

Environment Service has directed much of its effort in this respect towards the establishment of Peatlands Park near Dungannon (see Stanfield, Chapter 42, this volume). This is a country park situated amidst 250 ha of bog and woodland, parts of which are NNRs. A recent addition to the facilities at the Park is a fully-equipped classroom for use by visiting school groups. A video has also been made on the use and exploitation of peatlands. Environment Service is also supporting other organisations through grant aid to develop educational material.

Conclusion

Like many other organisations involved in peatland conservation the Environment Service is committed to ensuring that the habitat is adequately and effectively conserved across its natural range. It has used its position within Government to exert influence over policy as well as ensuring an adequate and representative series of sites with statutory protection.

Bogs are an integral part of the culture and heritage of Northern Ireland and it is important that that they remain a major feature in the local environment. Government has set out to achieve a balanced approach between the need for conservation and the legitimate use of a natural resource.

Acknowledgements

The authors would like to thank Mr J.S. Furphy, Chief Conservation Officer, Countryside and Wildlife, Environment Service, for his valuable comments on the text and Miss Bobbie Hamill, for producing the map.

Chapter thirty-eight:

The Habitats Directive: Centrepiece of the European Union's Nature Conservation Policy. Its Implications for Peatland Conservation

Dr Geert L.M. Raeymaekers
Ecosystems Ltd., Belgium.

Introduction

The Habitats Directive is a legal instrument of the European Union for the conservation of natural habitats and of wild fauna and flora. In order to better understand the objectives, possibilities and limitations of this Directive, the Directive is presented in the context of the Union's organisations (Commission, Parliament, Council) and its environmental policy.

The European Union

The origin of the European Union can be traced back to a number of international treaties which started in 1951 with the creation of the first European institution, the 'European Community for Coal and Steel' by the founding countries: Belgium, The Netherlands, Luxembourg, Federal Republic of Germany, France, and Italy. The same countries set up the European Economic Community (EEC) in 1957 by signing of the Treaty of Rome. Other countries later joined the European Economic Community: Denmark, Ireland and the United Kingdom in 1973, Greece in 1981, and Spain and Portugal in 1986.

The EEC was transformed into the European Union as a result of the Treaty of Maastricht, which came into effect in November 1993. After this Treaty, Finland, Sweden and Austria joined the European Union (1995). Under the Maastricht Treaty, the European Union has a much wider range of competencies, including economic and monetary policy, transport, foreign policy, environment, and social policy. Also the influence of the European Parliament is increased, now giving it the ability to veto legislation.

The main institutions of the European Union constitute:

1. The directly elected *European Parliament* which approves the Union's budget and has, since the Maastricht Treaty, more influence in the decision-making process;
2. The *Council of Ministers* which formally adopts new legislation and is composed of one representative (minister) of each of the Governments of the Member States;
3. The *European Commission* which has the sole power to propose legislation. It also enables the implementation of the legislation and enforces it;
4. The *Court of Justice* which ensures that the Union's law and Treaties are respected.

Each of the above-mentioned institutions of the European Union have played a crucial role in nature conservation. For example:

1. The *European Parliament* decided in its meeting of 24 November 1993 to raise the budget of LIFE, the financial instrument for the environment, to 95,500,000 ECU and to earmark 50% of this budget for the protection of natural habitats.
2. The *Council of Ministers* approved in its meeting of 21 May 1992 the Council Directive on the Conservation of Natural Habitats and of Wild Fauna and Flora, the so-called 'Habitats Directive', thereby providing us with a very important piece of legislation for nature conservation.
3. The *European Commission* is the initiator of most legislation and financial support for nature conservation and is responsible for the daily management of these instruments. It organises the meetings of the Habitats Committee and of the scientific working groups concerning its conservation policy.
4. The *Court of Justice* decided on the 2nd of August 1993 to condemn Spain for not having classified the region 'Marismas de Santoña' as a Special Protection Area (SPA) under the Birds Directive, thereby implying that Member States should classify all Important Bird Areas (IBA) (Grimmet and Jones, 1989) as SPAs. The IBA is a list of important bird areas drawn up on request of the EC by the International Council for Bird Preservation (ICBP) and the International Wetland and Waterfowl Research Bureau (IWRB). In addition to the protection of bird species, this was also a potentially important decision for the conservation of several important bryophyte sites covered by the IBA list.

The European Commission has organised itself into 23 Directorate-Generals, the legal service and the Secretariat-General. The activities of some of these Directorate-Generals (or 'DGs' as they are commonly called) directly affect the natural environment, rural areas or regional policies. For instance:

DG XI: 'Environment, nuclear safety and civil protection'

Since 1973, over 200 legal texts covering the environment have been adopted. They concern air, water, soil conservation, nature protection etc. Additionally, DG XI manages research contracts for the support of its environmental policy, and it follows up the implementation of the Union's environmental legislation. Within DG XI, the unit for nature conservation, coastal protection and tourism is mostly involved with the implementation of the Habitats Directive, the Birds Directive, the EU-CITES regulation and with relations with the Council of Europe's Bern Convention.

DG VI: 'Agriculture'

As the largest DG, it implements, monitors and enforces the Common Agricultural Policy. For nature conservation, the activities of this DG are often far reaching, both positively and, more often, negatively. For example, the nitrate pollution from intensive farming activities or global change directly affects natural habitats both inside and outside of protected areas; control, or lowering, of the water table has desiccated several wetlands, including bogs. On the other hand, the EU Agri-environmental Regulation has provided financial support to the farming community for extensification programmes, some of which have been important in the preservation of hay-fields on peaty soil.

DG XII: 'Science and Technology'

This DG develops important research programmes in its 4th Framework Programme, setting aside a substantial budget for biodiversity-related research.

DG XVI: 'Regional Policy'

DG XVI takes daily care of the Union's regional policy, in part by means of the Structural Funds which are an assemblage of financial support instruments to implement the Union's policy in a structural manner. One of the objectives of these Funds ('the 5th objective') is crucial for nature conservation since it concerns (a) the acceleration to adapt agricultural structures and areas, and (b) the development of rural areas. A number of actions financed by the Structural Funds (for example land re-allotment, irrigation, or drainage schemes) have too often been implemented without environmental consideration. As a result, several non-governmental organisations (e.g. WWF) have criticised these funds for neglecting the environmental consequences. Under the new Structural Fund Regulation (CEC, 1993), Member States or regional authorities have to inform the Commission of the potential negative environmental effect of activities financed through the Structural Funds.

DG XVII: 'Energy'

This is an important Directorate because of the Unions' dependence upon external energy resources. Given the use of peat as both as a culture medium in horticulture and as an energy source, a policy to control the extraction of peat should fall within the remits of this DG, DG VI (Agriculture) and DG XI (Environment). This DG, together with the International Peat Society, organised a International Peat Conference in Brussels in March 1994.

The EU Environmental Policy

Since the adoption of the First Environmental Action Programme (1973), the Commission has formally included environmental protection in its Common policy. Since then, more than 200 legal instruments in the field of the environment have been approved by the Council of Ministers.

Legislation has been developed and proposed within the framework of 'Environmental Action Programmes' which set general principles for the Union's environmental policy. They are reviewed and adopted for periods of four years. The ongoing Environmental Action Programme (1992-1996) 'Towards Sustainability' is the fifth in the row and focuses on sustainable development and the integration of the environmental policies in five economic target sectors: agriculture, industry, transport, energy and tourism.

The European Union can adopt several legal instruments:

1. non-binding **recommendations** and **resolutions**;
2. **regulations** that are binding and directly applicable in all member states;
3. **decisions** that are directly binding on the person to whom they are addressed, including Member States, individuals and legal persons;
4. **directives** which must be implemented by the laws or regulations of the Member States within a designated time (usually 18 months or 2 years).

Of the many legislative acts concerning the environment, two specifically relate to nature conservation and which have received widespread attention are the Birds Directive (79/409/EEC) and the Directive on the Conservation of Natural Habitats and of Wild Fauna and Flora or Habitats Directive (92/403/EEC). Both Directives are the major EU legal instruments for the conservation of mires or peatlands.

The Birds Directive

The Birds Directive (79/409/EEC), was adopted by the Council of Ministers on 2 April 1979 and came into force on 6 April 1981. As a pioneering Directive this has given protection to bird species but also to other animals, flora and threatened habitats types.

Under the Directive, Member States must, apart from certain exceptions, protect wild birds from killing or capture, deliberate damage to - or removal of - nests, eggs or young. Hunting is allowed under the Directive, but is subject to controls.

Member States have a parallel obligation to 'preserve, maintain or re-establish a sufficient diversity and area of habitats' for all wild birds. In addition to this general provision, Member States must protect sufficient habitat to ensure the survival of endangered and threatened species (which are listed in Annex 1) and of all migratory species. Member States have to classify the most suitable territories for these species (Annex 1 species and migratory species) as Special Protection Areas (SPAs) and notify the Commission of what measures they have taken. (Annex 1 at present contains 178 taxa). Because of the designation of Special Protection Areas, important mire areas have become protected areas in the EU (in particular in Ireland, where blanket bogs have been protected since they are important habitats for the Annex I bird species).

There are over 1000 SPAs declared under the Birds Directive so far, covering more than c. 7 million ha (the area of the Netherlands and Belgium combined). Conservation groups estimate that there are about 1500 sites in the Community that qualify as SPAs*, hence although the number and total area of SPAs declared is encouraging, the task is by no means complete, and apart from Belgium and Denmark, the other Member States should designate further areas as SPA.

The Habitats Directive

Origin and aim of the Habitats Directive

The origin of the Habitats Directive (92/43/EEC) lies in the Community's 4th (1987-1992) Environmental Action Programme. This programme, outlined the need for a new legal instrument for protecting wildlife in the following words: 'What essentially is needed is a Community instrument aimed at protecting not just birds but all species of fauna and flora; and not just the habitats of birds, but the habitat of wildlife - animals and plants - more generally. Such a comprehensive framework should ensure that, throughout the Community, positive measures are taken to protect all forms of wildlife and their habitat; such measures should be aimed at the three main objectives of the World Conservation Strategy: the maintenance of essential ecological processes and life support systems, the preservation of genetic diversity, and the sustainable utilisation of species and ecosystems.'

The aim was to provide a legal instrument to implement the Bern Convention across the Union, to give the opportunity for Union funds to be made available for this purpose and to provide enforcement possibilities through the

---

* Special Protection Areas. March 1993. Commission of the European Communities.

European Court of Justice. In particular, a new instrument would tighten up some of the provisions of the Bern Convention, make improvements in the light of the experience with the Bern Convention, and strengthen the capacity for its enforcement.

The Directive was adopted by the Council of Ministers in May 1992 and published in the Official Journal of the European Communities (OJ L 206 pp. 7-49) of 22 July 1992. After the adhesion of the three new Member States, Austria, Finland and Sweden a number of habitat types and species have been added to the annexes of the Directive.

Description

Together with the Birds Directive, the Habitats Directive plays an important role in the maintenance of biodiversity in the European Union. Additionally, if used effectively, it can help to promote sustainable and integrated land-use. As such, it attempts to move away from the concept of creating small islands of strictly protected areas for high-profile species. The Habitats Directive strives towards a wide-spread system of protection where human and environmental interactions are in balance. This objective must be achieved by the creation of a coherent European ecological network of Special Areas of Conservation - called NATURA 2000 - to maintain or restore species and habitats of Community interest to a favourable conservation status.

The Directive consists of four parts:

1. In the first part (Art.1 - Art.2), definitions are given of the most important concepts which are used in the Directive (e.g. natural habitat, species, priority species, favourable conservation status, etc.). In this regard, the definition of a *favourable conservation status* for a species is defined as:

(a) population dynamics data on the species concerned indicate that is maintaining itself on a long-term basis as a viable component of its natural habitats, and;

(b) the natural range of the species is neither being reduced nor is it likely to be reduced in the foreseeable future, and;

(c) there is, and will probably continue to be, sufficient habitat to maintain its populations on a long-term basis.

2. The second part (Art.3 - Art.10), defines the procedures, criteria and time frame for the implementation of the Directive and in particular for the creation of the Natura 2000 Network.

Art.4 defines the procedures for selecting SACs for habitats listed in Annex I and species in Annex II (see Table 38.1). This is a two-stage process with an assessment at the national level followed by an assessment at the European level.

Art.6 requests the designation of SACs. It obliges Member States to establish all necessary conservation measures and to take steps to avoid deterioration of the habitats and disturbance of the species in so far as such disturbance could be significant in relation to the objectives of the Directive.

Table 38.1. Annexes I and II of the Habitats Directive.

ANNEX I Habitat types	ANNEX II Species		
A total of c. 200 habitat types listed in this annex. About 50 priority habitat types (indicated with a '*'). The following are important peatland habitat types:	193 animal species and 432 plant species[†]. By far the largest part of the plant species are point endemics (or species with a very restricted distribution area). Fungi, algae or lichens are not yet included in the annex.		
* Aapa mires	**Group**	**Priority**	**Other**
* Palsa mires			
* Active raised mires	Mammals	8	27
Degraded bogs	Reptiles	3	16
Blanket bogs (* active only)	Amphibians	3	16
Transition mires & quaking bogs	Fish	56	5
Depression on peat substrates (*Rhynchosporion*)	Arthropods	4	32
	Molluscs	0	22
* Calcareous fens with *Cladium marisci* & *Carex davalliana*	Angiosperms	237	157
* Petrifying springs with tufa formations of *Carion bicoloris-atrofuscae*	Gymnosperms	0	1
	Pteridophytes	13	3
	Bryophytes	17	4
* Bog woodland			

Sites with priority species or priority habitats[‡], should receive stricter and earlier protection. Member States must establish the necessary conservation measures to maintain, re-establish or restore the areas conservation values (e.g. management plans). In addition, Member States are encouraged to structure their land-use policies in such a way as to create links between the Special Areas of Conservation. This could be done, for example, through such linear landscape features as hedges, river banks, woods or small ponds which can serve as migratory stages or places of refuge for wild species.

[†] Including three new priority species since the inclusion of Sweden, Finland and Austria: *Artemisia lacinata* Willd., *A. pancicii* (Janka) Ronn., and *Stipa styriaca* Martinovsky.

[‡] Priority species or priority habitats are those species or habitats for which the European Union has particular responsibility in view of their natural range which falls within the territory of the European Union. However, as we will indicate below, financial support is restricted to sites harbouring priority species. Therefore, the definition has somewhat been 'stretched' to allow more financial support for species conservation.

Additionally, appropriate assessment should be made of the implications of any plan or project not directly connected with, or necessary for, the management of the site, but likely to have a significant effect thereon, either individually or in combination with other plans. In the case of negative implications for the site, and in the absence of alternative solutions, the Member State shall take all compensatory measures to ensure that the overall coherence of Natura 2000 is protected.

3. The third part (Art.12-Art.16) takes up a number of strict protection requirements for the species listed in Annex IV similar to the requirements under the Bern Convention. Annex V concerns the conservation of exploited species. *Sphagnum* is listed in Annex V, probably indicating only the living (green) material. However, the wording of Art.14 places few legal restrictions on the Member States. It mentions: 'If, in the light of surveillance, provided for in Art.11 (obligation to monitor habitat types and species), the Member States deem it necessary, Member States shall take measures to ensure that the taking, in the wild, of specimens of wild fauna and flora listed in Annex V (as well as their exploitation) is compatible with their being maintained at a favourable conservation status'.

4. The last part of the Directive (Art.17-24) indicates complementary clauses for the implementation of the Directive such as information, monitoring activities, research and education.

The timetable for the implementation of the Habitats Directive and for the establishment of the NATURA 2000 network is presented in Fig. 38.1. In addition to the Special Protection Areas which are already - and will continue to be - classified under the Birds Directive, the NATURA 2000 network will be composed of sites harbouring habitat types listed in Annex I (about 200 in total) and species listed in Annex II of the Habitats Directive. Several of the habitat types mentioned in Annex I are important plant habitats (e.g. blanket and raised bog, heathland, scree vegetation, and some forest types) and can as such be used to protect several endangered plant species which are not covered by Annex II.

As a result of the adoption of the Habitats Directive, the ACNAT, and later the LIFE-Regulation was adopted in order to assist the Member States financially with the implementation of the Directive.

Implementation

As mentioned above, the Directive requests Member States to follow a strict agenda for its implementation. A *Habitats Committee*, composed of representatives of the Member States assists the Commission with the implementation and follow up of the Directive. Revision of the Annexes and Articles of the Directive must receive approval from the Council of Ministers. A *Scientific Working Group* is set up to discuss scientific issues related to the implementation of the Directive. For instance, for those cases where the

CORINE description was not precise enough, or was incomplete, the Working Group prepared a manual for the interpretation of habitat types, a map of biogeographical areas in Europe and 'NATURA 2000 Standard Data Forms' in order to inform the Commission of all possible SACs.

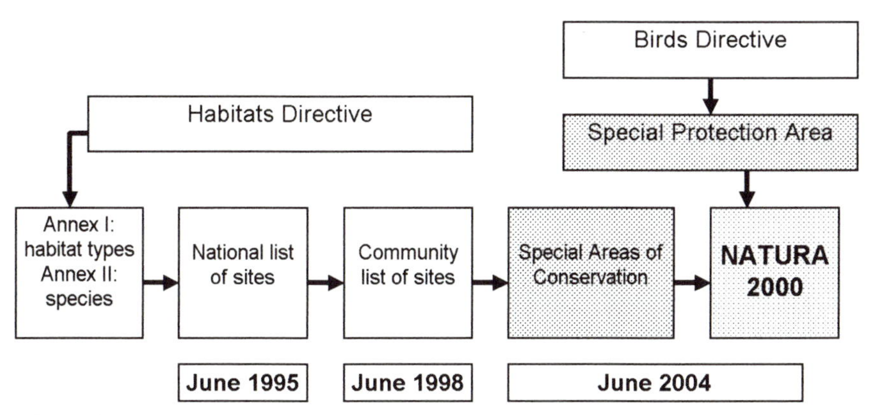

Fig. 38.1. Timetable for the implementation of the Habitats Directive and the creation of the Natura 2000 Network.

Importance of the Habitats Directive for conservation of peatland or mires

The mire types listed in Annex I are based upon an early version of the CORINE classification, which was later partially updated in the CORINE Biotopes Manual (CEC, 1991b), which until now is the only overview of European habitat types. However, it remains incomplete and the classification of mires is wrong (e.g. the subtitle of Annex I refers to 'Raised bogs and mires and fens' instead of referring to 'Mires'; this heading is then further divided in two parts '*Sphagnum* acid bogs' and 'Calcareous fens'). Problems with classification caused confusion during interpretation of the different habitat types. However, this problem was later solved by the adoption of an interpretation manual of the Annex I habitat types. The list of mire types covered by Annex I is now wide enough for the Member States to protect those in need of protection.

At this moment (June 1995) Member States are providing the Commission with the list of sites that harbour Annex I habitat types and Annex II species (list of possible SACs). It is thus premature to make an assessment of the mires in the Natura 2000 network. The UK has already presented a complete list to the Commission which includes over 10% of mires (Hopkins, 1995).

As soon as these lists have been submitted, there will be a second evaluation by the Commission (in agreement with the Member States) on the basis of the criteria set out in Annex III (stage 2) and in the framework of the biogeographical areas and the EU territory as a whole. At that moment, all priority habitat types will be considered sites of Community Importance, to be designated by the Member States as SACs. Using Art. 5, the Commission can, in exceptional cases, initiate a bilateral consultation to increase the number of sites harbouring priority habitat types (and species).

In a recent paper (Ratcliffe, 1995) already points out some of the problems concerning listing of mires in the UK: '*English Nature is believed to be under pressure to de-notify Thorne Moor; peat extraction would then continue to destroy the area with minimum awkwardness. The Flow Country (only partially included) is regarded as a sensitive area where Scottish Natural Heritage has to tread with extreme caution. Sometimes, splendid opportunities for adding priority habitat types to adjoining sites chosen for other interest have been ignored, e.g. the calcareous fen of Loch Tallant to Duich Moss.*'

Natura 2000 network combines SPAs (Birds Directive) and SACs (Habitats Directive), and both kinds of protected area are included under Art. 6 of the Habitats Directive. However, some Member States argue that in SPAs, only the birds and the habitats important for their survival deserve EU protection, and not the other habitat types, or species, listed in the Habitats Directive which happen to occur on these sites. This interpretation would mean that appropriate protection of mire types which occur in these SPAs but which are not important for the favourable state of conservation of the birds, would not be covered by Art. 6 of the Directive.

The listing of mires types, obligations for maintenance of a favourable state of conservation of future SACs (Art. 6), and for monitoring and reporting are all well covered by the Directive. It is now up to the Court of Justice to test the Directive.

LIFE Regulation

Apart from the Habitats Directive and the Birds Directive, the European Union has adopted other directives and regulations for nature conservation. Some of these have been financial instruments to support nature conservation (ACE, ACNAT, LIFE), the Agri-environmental regulation to provide financial support to farmers for extensification programmes, and the EU-CITES Regulation to solve problems related to the utilisation and trade of wild fauna and flora.

There are significant costs attached to the implementation of the Birds and Habitats Directives. This financial imperative has clearly been recognised, not only in the text of the Habitats Directive itself, but also through the creation of the LIFE-Regulation (CEC, 1992). This regulation brought together a number of budget lines from the Commission (ACE, ACNAT MEDSPA and NORSPA).

Nature conservation was given an indicative percentage of 45% of the total budget in the Regulation (c.400 MECU) which was destined not only for the maintenance and re-establishment of biotopes for priority birds under the Birds Directive, but also for other priority species and habitats of major importance under the Habitats Directive. The priorities and resource allocation for each year are approved by the LIFE-Committee, which is made up of Member State representatives. During the first three years, the LIFE-Management Committee has allocated only 30% of the budget to actions under these two Directives. In 1994, an exceptional 45% of the 98 million ECU budget was reserved for projects regarding nature conservation. So far, the LIFE-budget has been too small to co-finance all proposed projects. Consequently, there is considerable competition for EU support. The whole LIFE instrument is very ambitious and is endowed with too little resources.

LIFE support for nature conservation activities has been very important for the conservation of peatlands or mires. For example the following projects have been co-financed under LIFE: conservation of lowland raised bogs in Scotland (UK); restoration of Redgrave and South Lopham fen (UK); habitat protection in the Rhön (D). Other projects concern large-scale inventory projects which incorporate peatland sites.

Need for an integrated approach

Nature conservation has benefited from the above mentioned instruments, which are the spearheads of the Union's nature conservation policy. However, these instruments cannot operate in isolation and do not work in a number of the cases essential for habitat and species conservation. For example, the LIFE-Regulation, the Birds and the Habitats Directives primarily address protected, or to be protected, areas. Thus support for mires outside protected areas is not a priority and, as a result, these biotopes receive too little attention. Secondly, most of the existing EU-legislation on water, soil and, in particular, regarding air pollution, are not far reaching enough to reduce the pollution levels which presently affect mires, in particular calcareous mires in intensive agricultural areas. Thirdly, the existing instruments are generally of little use to countries outside the Union, which is a major disadvantage for eastern and central European countries (only 5% of LIFE is for non-EU countries).

The first programme of the European Union concerning the environment of central and eastern European countries was PHARE (CEC, 1989), which initially concerned only Poland and Hungary, but which was later extended by two regulations to include the other central and eastern European countries. One of the support programmes of PHARE (OUVERTURE) and a new programme, TACIS, for the Commonwealth of Independent States, provide in the first place technical assistance for institutional building of administrations responsible for environment and, further down the line, assistance for the improvement of the environment (CEC, 1989; 1991a). As a result, the emphasis lies on improving the quality of air, water and soil. So far, the competent authorities of these countries have not given appropriate attention to nature conservation aspects of the environment, it is only recently that 'sustainable development' has been taken into consideration. At best, air quality will improve, but not to a level to sustain sensitive peatland areas in the near future.

Looking back at the financial instrument for environment (LIFE) and the number of proposals from Member States and non-governmental organisations for financial support, one can only conclude that the budget earmarked for nature conservation is far too low. Considering the limited budget, we are forced to redefine the nature conservation strategy. In addition to working within the priority actions of the LIFE-Regulation and with projects connected to the NATURA 2000 network, more attention should be devoted to the use of other financial means for nature conservation; in particular, the set aside of the Regional Policy and Common Agricultural Policy which affect land-use in the Union. The importance of using budgets other than the environment one has been recognised by the 5th European Environmental Action Programme (1992-1996). This programme stresses the need to integrate the EU Environmental Policy into the other policy sectors (agriculture, industry, tourism, transport and energy), and also to allow the possibility of these to contribute to initiatives that will actively benefit the environment. Therefore, new regulations and procedures are being developed in order to take the environment into consideration when developing or implementing other sectorial policies. For example, this may lead to better forestry management (agricultural policy) or the conservation of ravines which are threatened by dam building for hydro-electric projects (energy policy).

The information from, and collaboration with, experts can be instrumental in incorporating EU environmental legislation into plans and programmes developed for other sectorial policies. For instance, it may be that the Commission finances (by means of its Structural Funds) the planning - or construction - of a motorway affecting an important habitat which should be, or is, protected under the Habitats or Birds Directive (the motorway would not have cut through the site but might affect the site from a distance). In such a case, any person can put forward a complaint to DG XI, indicating that the Member State is not fulfilling its obligations under the Habitats Directive, or Directive on Environmental Impact Assessments, and can ask for compensatory measures. This may lead to a long and complicated procedure, but such actions from the public have been successful.

Chapter thirty-nine:

The Rothenthurm Initiative: the Swiss Way to Mire Protection

Mr Hans Peter Nowak
Rothenthurm Initiative Committee, Switzerland.

Introduction

I have been living c. 2 km away from the Rothenthurm peatland since 1973. This is one of the largest peatlands in Switzerland - yet at 700 ha (or 7 km^2) it remains ridiculously small when compared with Scottish peatlands! But Switzerland has lost nearly all of its peatlands - they have been drained for agriculture, for building or simply to allay fears of illness (e.g. malaria). Today less than 2% of the surface of Switzerland is composed of peatlands - most of them very small.

The geography

Rothenthurm is situated between the Lakes of Zurich and Lucerne at an altitude of c. 900 m. In winter, the surface may be blanketed by up to 1 m of snow, whilst spring and autumn rains often flood the brook which passes the peatland (Fig. 39.1).

The projects in the mire

This almost pristine soil has attracted attention from many potential developers - people cannot simply leave such a place to be enjoyed as it is! In 1942, the Swiss army announced plans to build an artillery-range on the peatland, however these plans were later shelved in 1945. Later, the national aviation company, Swissair, planned to build a freight airport - but it was too foggy, too boggy and too costly so again these plans were shelved.

Fig. 39.1. The Biber brook.

In 1973, the Swiss Department of Defence considered this region for the site of an army garrison. The government of the Canton Schwyz (the canton of the village of Rothenthurm) backed these development plans. The key building contractor was owned by the brother of the privy councillor; some £80 million were to be invested in the project!

Mr Gottesmann, who lived in the same region and was at the time a government delegate for environmental issues within the Canton, gave his personal (and negative) opinion of the project to the local section of the Swiss League of Nature Conservation. The cantonal government reacted by first suspending, and soon after that dismissing, him. It seems that even in Switzerland freedom has limits!

Growing opposition

Fig. 39.2. An early campaign poster.

A local group of opponents to the project started a somewhat confused campaign whose focus was primarily *against* the Army, rather than *in favour* of peatland protection (Fig. 39.2). As I was working within Swiss television at the time, the campaigners asked me to prepare an exhibition using modern media.

In 1975, a referendum amongst the villagers of Rothenthurm came out against the Army project (87% of the population participated and 84% of them voted against the development). In 1977, the Municipal Council of Rothenthurm ceased discussions with the Department of Defence.

Special real estate situations

The real estate in this region is somewhat special: although part of the ground belongs to local farmers, a large proportion of it is owned by two corporations: The 'Oberallmeindkorporation' and the Corporation of Oberägeri, Rothenthurms' neighbouring village. The Oberallmeindkorporation is probably the oldest corporation in the world with historical sources going back to the ninth century; they participated in the foundation of primitive Switzerland and still own the foundation document (sealed in 1291). The farmers in the region are members of these corporations simply by virtue of their names - everybody named 'Beeler, Föhn, Betschart, Schuler, Inderbitzin....' is automatically a member (although this is only true for males in the population). In 1978 this corporation voted to sell an important part of its peatland ground to the Department of Defence (1150 votes in favour and 1115 against the motion).

To us this was a major failure and one which was hard to overcome.

Defining a strategy

As I was now Vice-President of the campaign committee, I realised that the onus of the campaign needed to shift away from an anti-military stance and towards the need for nature protection. Slowly discussions moved away from the peatlands' poor suitability for an army development, and towards the need to protect peatlands from any disturbance. Until that point, the campaign had not fully exploited the power of the word; we then began to tap the potential of the news media, publishing carefully timed press releases which attracted radio, television and newspaper coverage of the issue.

This approach began to have an impact - the Army's Secret Police considered me to be the campaigns' intellectual leader and began to collate a dossier on me. Such police investigations were common at that time, however subsequent developments have lead to such archives being opened to the public and it is this new freedom which enabled me to present this chapter.

Returning to the real estate situation: the Department of Defence asked the Corporation of Oberägeri to sell them their peatland ground. We began a campaign of intense opposition. As the council of the Corporation declined to provide us with the addresses of their members, we had to extract all the possible names from the local telephone directory. We sent some 7000 letters to these individuals encouraging them to vote in the forthcoming debate about this land sale. Some 1700 members attended the vote (compared with a normal representation of c. 500) and voted against selling the peatland.

Continuous efforts

This success was followed by others:

In 1979 the Committee bought a small area at the centre of the peatland.

The Department of Defence threatened to expropriate land from all owners who would not sell voluntarily. The Committee made it clear to all, that they would not sell their land-holding to the Army, and this defiance encouraged solidarity amongst other landowners who wished to resist to the offers of the Defence Department.

A form of trench or guerilla warfare evolved between the people of Rothenthurm, the Committee, the Department of Defence and the government of the Canton Schwyz. The Initiative Committee have been to the Supreme Court a total of 13 times with several successful outcomes. We even went to the European Court of Human Rights in Strasbourg on two occasions. Throughout the Swiss Department of Defence were our unwitting allies due to their clumsy and contradictory approach to the matter.

Fighting on all levels

Let us follow one of these stories:

- On April 11th 1983, the expropriation process began. Land-owners were given 30 days to present their objections to, and demands regarding, the land-sale.
- The Committee lodged a complaint to the Supreme Court.

- On May 5th the consultation phase was extended to 60 days due to a failure on the part of the Department of Defence to display the proposed garrison perimeter.
- On June 3rd the period was again extended (by decision of the Supreme Court).
- By July 26th, some 100 objections had been lodged (some of them by the WWF of Switzerland).
- The Committee lodged another complaint against the expropriation process.
- On December 9th, the Supreme Court decided against this new complaint but stated that there were legal questions to be 'checked extensively and carefully by the legal proceedings'.
- The Army started some road constructions over the peatland.
- The Committee lodged a complaint with the local court.
- On February 10th 1984, the local court forced the Department of Defence to stop the road construction immediately.
- From February 14th to February 28th, conciliation negotiations between the legal representative of the Department of Defence and the land owners were held.
- We lodged a complaint to the European Court of Human Rights in Strasbourg.
- On March 23rd, the local court ordained that the Department of Defence had to restore the primitive condition of the started road constructions.
- On April 24th, the Department of Defence asked to be put in premature possession of the ground to be expropriated - enabling initiation of the works before the expropriation process had even been concluded.
- The Committee lodged a complaint to the Supreme Court.
- On February 28th 1985, the Supreme Court refused to ordain this premature possession. The Department of Defence could not start construction of the barracks.

By this point, it was clear that the expropriation process was going to last some five to eight years.

Things get serious

In 1982, we charged a parliamentarian to officially ask the Minister of Defence what the future of the peatland of Rothenthurm would be and what the plans of the Department of Defence were. It was clear that the answer of the Minister would determine both the future of the project and our chances of success.

By this time we had discussed formation of an Initiative with the council of Rothenthurm, WWF (Switzerland) and others. This begs an explanation:

Direct democracy - our chance

In Switzerland, everybody has a right to formulate an Initiative, which is an addition to, or change in, the national Constitution. He or she has only to send the newly proposed constitutional text to the Chancellery of the Swiss Government. There, it is checked for conformity with the rest of the Constitution. If the text is accepted, 100,000 people have to sign up to the Initiative within 18 months. If these signatures are lodged at the Chancellery within this period, the Government must bring the new article to a referendum. This is one of the principles of direct democracy in Switzerland.

On a Friday morning in autumn 1982, the Minister of Defence gave the official Government response to the parliamentarian, saying that the garrison would be constructed on the peatland and the land expropriated if necessary.

On this same Friday, I phoned the national press agency to state that we were going to formulate an Initiative against the projects of the Department of Defence. This news hit the radio and television, headlines that day, and was covered by all of the Saturday newspapers.

Preparing the Initiative text

We discussed the text of the Initiative with two representatives of WWF Switzerland, one of whom - Dr Claude Martin - is presently General Director of WWF International. Our goal was not only to avoid any further constructions on Swiss peatlands, but also to protect them from any other threats. Over and over again, you find that articles in the Swiss Constitution make exception for, what they call, 'higher interests'.

What is of higher interest for the politicians - nature protection versus national defence - can easily be imagined.

The text of our Initiative which was finally accepted by the Chancellery read as follows: *'The Swiss Constitution will be supplemented as follows: Art. 24 series paragraph 5: Mires, mire landscapes of particular beauty and of national importance are considered objects of protection. In these no installations are to be constructed nor any changes in the soil may be performed. The only equipment accepted will be that necessary for the protection aims or to continue the former agricultural utilisation.'*

A transitional article was later added which stated that any construction or soil changes, instigated after June 1st 1983, would be rendered null and void. This was inserted after a Department of Defence representative stated (in official discussions) that the planned barracks would probably be built at this time, whilst the Initiative was still being voted on by the Swiss people. The main goal of the Department of Defence was to begin construction of the barracks. The rationale being that the Swiss people would never vote in an Initiative which forced the army to destroy a building constructed using public expenditure.

Fig. 39.3. People had to sign the Initiative inside this paper.

The success of the Initiative

We lodged the text of the new article on March 8th 1983. A strategy on how to collect the necessary 100,000 signatures was prepared. However, from that point, we never had time to read this paper again - we were overwhelmed by the number of people who contacted us in order to offer to collect signatures (see Fig. 39.3). After less than 3 months, we had 100,000 signatures and, after another 2 months, with nearly 170,000 signatures the petitions were closed.

Why this success? The Swiss people were becoming increasingly sensitive to questions of nature, environment and protection of living space. It was hard to understand that a mire could be suitable for a garrison. And - of course - we tried to prepare our campaign well. I had two visions in my head: One - coming from observations of the confrontation between the Department of Defence (the big power) and the resistance of the little village (Rothenthurm) was the parallel struggle of the little village of Asterix and Obelix fighting against the Roman army. The other was the confrontation of David against Goliath. Both visions bring the observer to empathise with the little village, with David.

Governmental politics

The political climate was obviously in favour of Rothenthurm; the enormous success of the Initiative forced the Federal Council and Parliament to decree important changes in the Federal Act on Wildlife, Countryside and National Heritage Protection. Although these amendments failed to satisfy our requirements, we welcomed this proposed revision to the Federal Act with the motto: 'Both are even better.'

In 1984, the Swiss Inventory of Raised and Transitional Bogs was completed. We felt that this inventory would clearly identify those sites which required protection should the Initiative be accepted. By 1986, three inventories had been set up based on the new articles within the Federal Act: one on raised and transitional bogs, another on fenlands and the third on floodplain forests. These were completed in 1989.

Preparing the national voting

Whilst the inventory progressed, the date of the national voting on the Initiative was fixed for December 6th 1987. Preparations for the referendum were time-consuming and challenging given our lack of experience in such matters. Despite limited funds, a Public Relations professional was employed; this proved to be a valuable investment by the Committee.

Despite a lot of media attention (television, radio, newspapers) and a high level of commitment, none of the Committees' four or five key organisers were confident of success. We really felt as David, challenging a Goliath - so big, so dominant and so powerful.

The success

The Swiss people accepted our Initiative by a 58% majority. It was only the third Initiative this century to be accepted by the Swiss nation*.

We assumed that the mires and mire landscapes would now be protected, however a hard lesson was still to learnt: a successful referendum is one thing, the political reality another. The Federal Government, as well as all 25 cantonal governments and almost all political parties remained opposed to our Initiative, and, after the referendum, we found in whose hands the power really lay.

The problems

As the Initiative is only one article within the Constitution, a whole body of law must accompany it. These laws have to be prepared and discussed by Parliament whose majority lay with those political parties who strongly opposed the Initiative. You may imagine how enthusiastically they approached this job....

The consultation process (which only ended in 1992) highlighted three main problems:

1. It is difficult for the public to understand the implications of strict protection of landscapes rather than just habitats. The term 'mires of particular beauty' was difficult to interpret - only a few parliamentarians realised that this term provided the flexibility to expand from a strictly quantitative view and incorporate qualitative interpretations.

---

* It is interesting to note, that another Initiative has since been accepted: the Swiss Alp Initiative which limits the amount of street traffic through the Swiss Alps.

2. The interim measurements are too rigid for many people. It is difficult to accept that legally erected installations should later be demolished again at the cost of those responsible for their erection.
3. Preventing construction of many buildings or installations is necessary for the protection of the landscape but appears to veto any activity in these regions.

People wanted a more flexible approach, allowing them to balance their interests against the objectives of mire conservation.

In Parliament, opponents of the Rothenthurm article are trying to alter its wording and, at the very least, establish a legal basis to increase the number of exemptions. Fortunately, there are still a number of parliamentarians who remember that the electorate clearly accepted the *unconditional* mire protection incorporated within the Rothenthurm Initiative and was not satisfied solely by the Federal Act proposed by the Government at the same time.

Nevertheless, industrial exploitation of peat continues on a site in Canton Neuenburg. Constructions have been, or are currently being, erected at many sites which are clearly protected within the remit of the Constitutional article.

The Initiative also had implications for tourism, however the Swiss National Tourist Office recently published a paper showing that the protection of mire landscapes and the promotion of tourism in these areas are not mutually exclusive.

Nature conservation and the farmers

One major problem which we tried to address, was the attitude of farmers against the protection of peatlands. In Rothenthurm we succeeded, during the long political fight to collaborate with local farmers. Farmers use the mire landscapes in an extensive manner, mainly to gather litter. Continuing discussions between nature conservationists and farmers aim to raise awareness of the aims of nature conservation. But, in many cases, farmers continue to perceive nature protection as a limitation of their rights.

Dramatic changes within Swiss agricultural politics suggest that these conciliation efforts might not have been in vain. The Government is starting to subsidise more extensive and organic farming techniques (rather than simply rewarding output). Such policy shifts should slowly change farmers function in society and encourage them to act in a more sustainable fashion.

If the Rothenthurm Initiative has had a role in these developments, we feel we have succeeded - even if every mire in Switzerland will not 'survive'.

Part Eleven

Raising Awareness

An Introduction

Dr Rob Stoneman
Scottish Wildlife Trust, UK.

It was particularly gratifying to show delegates from the Peatlands Convention around the Drumshangie Mosses as these wonderful bogs had been threatened with outright destruction. How could such important sites be threatened given the raft of nature protection legislation available in Britain (see Ward, Donald and Simpson, Chapter 35, this volume)? In part, it was due to yet another example of gaps in the British Sites of Special Scientific Interest (SSSI) network. Grünig's quote of Konrad Lorenz (p.221, this volume), which highlights the gap between the setting of legislation and its sustained implementation of legislation, is relevant here. Happily, at Drumshangie a greater awareness of the importance of the Moss, in part engendered by Scotland's nature conservation non-governmental organisations (NGOs), has led to the conservation of most of this site.

The tragedy of losing Drumshangie Mosses was averted. For many, many sites, the sum of many individuals buying 'Gro-bags' for their tomatoes has sealed their tragic fate and wonderfully wild bogland has been transformed to black lifeless wet deserts. For most of the individuals who caused this destruction, their part in the degradation of our countryside was largely unknown. It was interesting to challenge these very people at the recent 'Plants without Peat Easter Weekend' and explain their role. On realisation, the majority quickly turned to less damaging horticultural alternatives; after all, as many related, 'it's just a bag of dirt!'

Professor Maltby outlined the science case for conservation at the Convention. The case is strong and clear, yet peatland destruction continues apace. Clearly the science case is not enough; it is getting the message across that sets the seal on conservation action, as opposed to conservation desire.

Three approaches are outlined here:

- Campaigning has had an enormous impact in Britain and Ireland as explained by Ms Steel and Dr Foss respectively. Governments, peat suppliers and consumers are gradually becoming aware of this issue and the results have been dramatic. From selling virtually no alternatives to peat in the 1980s, 24% of growing media sold by B&Q (a large chain of hardware and garden centres) is now peat-free.
- Showing people bogs has its problems - bogs are sensitive to trampling - yet careful management has gradually allowed people to appreciate these habitats at first-hand. The Peatlands Park in Northern Ireland (see Stanfield, Chapter 42, this volume) leads the way, attracting over 100,000 visitors per year.
- Equally significant are the education initiatives at Peatlands Park and those co-ordinated by the Irish Peatlands Conservation Council (see O'Connell, 1991, 1992, 1993, 1994a).

These initiatives have wider implications. The Rio 'Earth' summit hammered home the message of a sustainable development ('*meeting the needs of the present without compromising the ability of future generations to meet their needs*'). Arguably, and certainly in the case of peatlands, this generation has largely failed to achieve sustainable development. Education unlocks the potential for sustainable living, allowing ourselves to make decisions based upon not just self-interest but planetary interest. Big themes, big tasks!

Chapter forty:

The UK Peatlands Campaign

Ms Caroline Steel
The Wildlife Trusts, UK.

Background to the UK Peatlands Campaign

Ten years ago, concern for peatland conservation in the UK was centred on the Flow Country and the damage being caused by afforestation. Conservation of lowland raised bogs was not considered to be a national issue: local problems were of local concern.

Wildlife trusts and other organisations fought against planning applications for development on bogs, and acquired some as nature reserves. The Scottish Wildlife Trust (SWT) fought for Flanders Moss as long ago as 1971. However, individual cases were not looked at in their national or international context. The conservation movement was ignorant of the scale of past habitat loss and there was little public support for bogs: lowland bogs are generally inaccessible and are unfamiliar to the majority of the population.

By the mid 1980s there was growing unease within the conservation movement. Old-style block cutting extraction techniques - which allowed some regeneration of vegetation - were being replaced by modern milling methods. For the first time vast expanses of bog surface were being stripped of vegetation, exposing the bare peat over hundreds, or even thousands, of hectares. The Wildlife Trusts (WT) and Friends of the Earth (FoE) began discussing the national implications.

The Wildlife Trusts commissioned independent research - sponsored by The World Wide Fund for Nature (WWF) - from the Environmental Consultancy University of Sheffield (ECUS) to determine whether the problem was as great as we thought. The outcome was the decision to launch the peatlands campaign.

Preparation and launch

With other organisations showing interest in peatland protection, the ECUS research had been guided by a steering group. This formed the nucleus of the Peatlands Consortium. Initially we thought primarily in terms of nature conservation and failed to identify all potential allies. The lesson was learnt, and archaeological and geological organisations were recruited. The Peatlands Consortium includes representatives from: British Association of Nature Conservationists (BANC); British Dragonfly Society; Butterfly Conservation; Council for British Archaeology; FoE; Geologists' Association; Geological Society; Greenland White Fronted Goose Study; Irish Peatland Conservation Council (IPCC); Plantlife; Royal Society for the Protection of Birds (RSPB); The Wildife Trusts; Wildfowl and Wetland Trust; WWF (UK)

Information gained from the ECUS research and from the British statutory conservation agency, the Nature Conservancy Council, was used to produce 'The Peat Report' (RSNC, 1990). This was published just before Easter 1990 (to coincide with the start of the annual gardening bonanza) and the campaign was launched by Professor David Bellamy and media gardener, Geoff Hamilton.

Soon after the launch the campaign was boosted by a statement from HRH, The Prince of Wales, Patron of The Wildlife Trusts. He announced his decision to stop using peat, and carried on to say: *'if we would like other countries to stop regarding their rainforests as useless jungle, we would do well to set an example by not treating our peatland habitat as useless bog, to be drained, dug up and scattered about in our gardens'.*

We threw a pebble in a pool back in 1990 and the ripples are still spreading.

Lessons have been learnt over the last five years. Neither the Peatlands Consortium nor the peat industry expected the campaign to run for so long. The Consortium naïvely thought that public attitude to peat-use would change as rapidly as that to CFCs in aerosols, and that the industry would respond accordingly. The Peat Producers Association (PPA) underestimated the commitment of the Peatlands Consortium: it expected a campaign lasting weeks or months, not years.

Campaign aims

The campaign aims were set out in The Peat Report (1990):

1. the development and implementation of a UK strategy for the conservation of peatlands;
2. the development of alternative materials and practices to replace peat in the horticulture, gardening and landscape industries;
3. the protection of UK peatlands of nature conservation importance;
4. the review of all planning consents for UK peat extraction;
5. the rehabilitation of damaged UK peatlands.

Before the campaign launch, the Consortium agreed to focus on lowland raised bogs because so little near-natural habitat remained and urgent action was needed as a result of an accelerating rate of peat extraction.

As discussed above, the emphasis at the start of the campaign was on nature conservation, and the original aims reflect this. Insufficient attention was given to archaeology and the functional value of bogs as carbon sinks. There are, however, important reasons for conserving bogs even if they have no current - and little potential - value for wildlife.

Over the last five years the environmental debate has moved on. The Earth Summit in Rio de Janeiro in 1992 highlighted the importance of:

1. biodiversity - lowland bogs are recognised as scarce habitats;
2. sustainable development - commercial extraction of peat is not a sustainable activity as peat cannot form on a drained and stripped peat surface;
3. climate change - bogs are recognised as important carbon sinks.

The European Habitats and Species Directive has highlighted the importance of bogs in a European context.

The Peatlands Consortium considers that, as so little lowland raised bog habitat remains, it is important not only to conserve active bog but also those sites with potential for rehabilitation, and those of archaeological and geological value.

However, the PPA, and some within the Department of the Environment and other arms of Government, have difficulty in accepting a wide definition of conservation importance. Initially they chose to consider only 'pristine' (near-natural) bogs as being of conservation value. Now there is more widespread acceptance within these sectors of the importance of degraded bogs supporting peat-forming vegetation.

Campaign targets

The campaign has proved to be considerably more complex than that to remove aerosols from supermarket shelves. Numerous agencies are involved in the 'cradle to grave' supply chains. The UK horticulture industry has become reliant on peat. The major targets for influence by the Consortium are listed in Table 40.1.

Table 40.1. Targets to influence.

Government & Government Agencies	- Department of the Environment - Department of Trade and Industry - Ministry of Agriculture Fisheries and Food - Department of Energy - Scottish, Welsh and Northern Ireland Offices - MPs and Peers - European Commission - Ecolabelling Board - Local Authorities - Statutory Conservation Agencies
Suppliers	- The Peat Producers' Association - Producers of Peat-Free Products (including PPA) - Retailers
Consumers	- The Gardening Public - Horticulture Industry - Landscape Industry - Local Authorities - Retailers

The 'consortium approach' to campaigning has many advantages. Each organisation has its own 'niche' and is able to play to its strengths, as shown in Table 40.2.

Responses to the UK Peatlands Campaign

Different target sectors have responded in different ways to the campaign.

Central Government response to the Peat Campaign

- Review of peat extraction
- Mineral Planning Guidance specifically for peat - with a presumption against extraction on bogs of conservation value in England
- Government departments reduced/eliminated peat use
- Plans for peat as energy source were scrapped
- Finance of research into
  - peat alternatives
  - peatland rehabilitation
  - composting
- Ecolabels will not be awarded to peat based soil improvers
- Post Rio recognition of value of peatlands
  - biodiversity convention
  - carbon sinks

- Recognition in the national waste strategy of the value of composted organic waste as replacement for peat

But:

No plans to revoke/buy out extraction rights, even on designated sites

Table 40.2. Outline of NGO action.

Date	Action
1980s	Growing concern
1989	Formation of Peatlands Consortium Decision to launch campaign BANC Conference
1990	Campaign launch The Peat Report (The Wildlife Trusts) Statement from The Wildlife Trusts patron HRH The Prince of Wales Debate in House of Lords leading to review of peat industry The Wildlife Trusts Peatland Protection Charter
1991	FoE guides to alternatives - amateur and professional FoE garden centre partnership Plantlife Commission of Inquiry set up
1992	The Wildlife Trusts Community Composting Guide FoE reports on peat and greenhouse effect Plantlife Commission of Inquiry report First National Bog Day (The Wildlife Trusts)
1993	RSPB/Plantlife - Out of the Mire SWT Scottish Raised Bog Conservation Project Biodiversity Challenge (Butterfly Conservation, FoE, Plantlife, RSPB, The Wildlife Trusts, WWF)
1994	The Wildlife Trusts Growing Wiser Conference and report
1995	Biodiversity Challenge II Peat Campaign Review (The Wildlife Trusts) SWT The Peatlands Convention

Local Government response to the Peat Campaign

- 59 signatories to the Peatland Protection Charter
- Massive reduction in use of peat
- Composting of organic waste
- Representation to central government

But:

- Some local authorities not vetoing development on peatlands of conservation value

No revocation of planning consents

Statutory conservation agencies response to the Peat Campaign

- Policies developed
- National Peatland Resource Inventory
- Major rehabilitation projects
- Negotiations and agreements with peat producers

But:

- Failure to defend designated bogs from continued peat extraction

Horticulture industry response to the Peat Campaign

- Research on alternatives
- Increase in plants grown in peat-free products on the market

But:

- Industry used 992,000 m^3 of peat in 1990. Estimated use in 1995: >1 million m^3

Retailers response to the Peat Campaign

- All good garden centres sell peat-free products
- Major garden centre chains not selling peat from SSSIs
- B&Q price parity
- Peat-free plants being stocked
- Promotion of peat-free products

But:

- Peat sales continues
- Peat prices too cheap

PPA responses to the Peat Campaign

- Code of practice produced
- Research and development of alternatives
- Croxden's sold Fenns and Whixall to NCC
- Bord na Móna agreed to sell all ASIs to Irish Wildlife Service
- Fisons (now Levingtons) agreed to give all of its SSSI peatlands to EN (with lease-back on production areas)
- International PPA code of practice
- Sinclairs halted plans to work Flanders Moss

- PPA agreed to review of old planning consents
- Bulrush agreed not to work Ballynahone Bog

But:

- Extraction continues on bogs of value for conservation including c. 2,500 ha SSSIs in England
- PR campaign continues

Peatland conservation is firmly on the agenda of all the above.

Campaign successes

The campaign has achieved a great deal in the last five years. Its successes are outlined in the five year review published this year (Wildlife Trusts, 1995). None of the campaign aims have been achieved in full, but significant progress has been made as a result of action taken. This can be seen by revisiting the original aims.

The protection of all peatlands of conservation importance

Some lowland raised bogs have been acquired as nature reserves and the threat of future development has effectively been removed from the most important bogs in England. Significant areas of land owned by members of the PPA have been given (e.g. Levingtons) or sold (e.g. Bord na Móna) to conservation bodies. Planning permission to develop Ballynahone Bog in Northern Ireland has been revoked and negotiations are taking place over other bogs.

The development and implementation of a UK strategy for the conservation of peatlands

Neither the UK Government nor the statutory conservation agencies have attempted to produce a UK strategy. However, the Biodiversity Challenge initiative outlines the views of leading voluntary conservation bodies. It is to be hoped that the Government's biodiversity planning process will incorporate the Biodiversity Challenge targets.

The development of alternative materials and practices to replace peat in the horticulture, gardening and landscape industries

Although over 70% of the total materials used as soil improvers are peat-free, only 9% of growing media is peat-free (DoE, 1994b; DoE, 1995d). Government is committed to an increase in composting of organic waste.

The review of all planning consents for UK peat extraction

The Environment Act 1995 requires a review of consents in England. The PPA has agreed to the updating to modern standards of permissions held by its members. However, without a robust peatlands conservation strategy it is unlikely that any review will result in significant changes in working practices to protect bogs.

The rehabilitation of damaged UK peatlands

Significant progress has been made by voluntary conservation bodies (particularly the SWT through its Scottish Raised Bog Conservation Project), statutory conservation agencies and members of the PPA.

Campaign failures

Significant progress has been made towards achieving our aims, but:

- **Peat extraction is continuing - even on designated sites.** The Peatlands Consortium has not succeeded in convincing the Government that, because there is so little active bog remaining, it is vital that badly damaged bogs are protected and rehabilitated as a matter of urgency.
- **Peat use is expected to continue to rise.** The Peatlands Consortium has not succeeded in convincing Government, the peat industry, the horticulture industry, or gardeners of the opportunities and advantages of switching from peat to alternatives. Government and industry continue to insist that there is a need for peat *per se*, rather than a need for effective growing media.

The PPA has waged an effective counter-campaign. It has succeeded in influencing sectors of Government that the Peatlands Consortium has failed to impress.

Lessons learnt

A conservation campaign has strong parallels with a military campaign. Battles and skirmishes have been won, but not the war.

Important lessons learnt from the process to date are:

- the value of the consortium approach
  - the broader the range of allied interests the better
- not to underestimate....
  - the time and resources needed to win the war
  - those with vested interests
  - in-built inertia and resistance to change
  - the attractiveness to the consumer of a very cheap product

- not to overestimate....
  - the attention span of the media.

The future

Failures may cause depression and successes may not seem great enough, but the campaign has come a long way. It is important to remember that if there had been no campaign there would have been little chance of a future for the UK's lowland peat bogs.

The Peatland Consortium will continue to campaign to achieve its aims. Initial targets for site protection and rehabilitation as set by NGOs in Butterfly Conservation *et al.* (1995) are:

1. all areas of near-natural bog must be protected;
2. all active, damaged bog must be protected and rehabilitated;
3. at least 7000 ha of other bog damaged by drainage, afforestation and peat extraction must be rehabilitated;
4. protection and rehabilitation must encompass entire hydrological units.

However, unless the demand for peat dwindles to nothing, the pressure for more extraction in the UK will continue and there will be no rapid rehabilitation of UK bogs.

In the future, imports from Ireland may be supplemented by peat from eastern Europe or elsewhere. The Peatlands Consortium does not want to export conservation problems. Reduction of the market for peat is vital, otherwise bogs will remain under threat.

Over the next year the campaign will focus on consumers - not only the gardening public but also professional growers and retailers. It is hoped that discussions with the peat industry will result in a constructive approach to marketing of peat-free products.

All the evidence shows that peat-free products can be as effective as peat. Peat is not needed by the amateur or professional. The UK has developed peat alternatives to a greater extent than any other European country. It could take the lead in marketing sustainable horticulture.

There is a future for lowland bogs in the UK. Campaigning by conservation organisations in other countries could ensure a future for peatlands throughout the world.

Acknowledgements

The campaign to protect the UK's peatlands has involved countless people over the last five years. The successes are due to their determination.

For assistance with the production of this chapter I would like particularly to thank Linda Hackney, Allison Crofts and Rob Benington of The Wildlife Trusts' national office.

Chapter forty-one:

Ten Years of the Save the Bogs Campaign

Dr Peter Foss
Irish Peatland Conservation Council, Ireland.

Introduction

Peatlands cover some 5% to 8% of the world's land surface, an estimated 409 million ha spread across five continents. Much of the world resource lies in the northern temperate zone. This concentration of peat in some of the most industrialised countries of the world has meant that vast tracts of peat landscape have vanished through development of the resource. Development has been at its most intense in Europe, where the environmental cost means that some western nations can point to a date in the near future when - without direct conservation effort - the very last natural peatlands will have vanished forever.

For others it is already too late; all natural peatlands (or bogs) in the Netherlands and Poland have been lost and Switzerland and Germany each have only 500 ha remaining. In the United Kingdom, 90% of blanket bogs have been lost leaving only 125,000 ha, whilst a 98% loss of raised bogs has left only 1170 ha.

This loss of peatland habitats has also been reflected in Ireland, contradicting the common perception of abundant peat resources only recently threatened by development pressure. In fact, there has been a 94% loss of raised bogs and an 86% loss of blanket bog in Ireland. At current rates of exploitation it has been estimated that all unprotected raised bogs will be extinct by 1997, and all unprotected blanket bogs will be extinct early in the next century (Ryan and Cross, 1984; IPCC, 1992).

Bogs once covered one-sixth of Ireland. Today - due to drainage, peat extraction and commercial developments - less than one-fifth of the original area remains. Only 4% of the original area of bogland in Ireland is earmarked for conservation, and just over one-third of the target is presently protected (see Table 41.1). Moreover, many of our protected peatlands are in urgent need of repair if they are to survive this decade. The pressure on this natural resource continues unabated and with only a slowly growing public awareness of the threat to these habitats. These facts, coupled with the threat of extinction of all unprotected raised bogs predicted for 1997, means that the 1990s represent a critical decade for Irish and European bog conservation.

Table 41.1. Ireland's protected peatland area (1995).

	Raised Bog	Blanket Bog	Fen	Totals
Original peatland area (ha)	311,300 *(100%)*	774,990 *(100%)*	92,508 *(100%)*	1,178,798
Government target area for conservation (% of original area)	10,000 *(3.2%)*	40,000 *(5.2%)*	N/A	50,000 *(4%)*
Area conserved to date (% of original area)	2,853 *(0.9%)*	23,598 *(3.04%)*	353 *(0.4%)*	26,804 *(2.3%)*
% of government target conserved	28%	59%	N/A	53%

The table shows the original area of peatlands in three main categories based on figures published by Hammond (1979), the areas earmarked for conservation by the Government, and the % of the target area conserved in 1995.
(N/A = data not applicable).

The disappearance of the Irish bogs would have serious international consequences. For various plant and animal species their last European refuge would be destroyed. Several species of birds would lose important wintering grounds. A type of landscape which once was characteristic for large parts of Western Europe and which is unique for its quiet spaciousness and its cultural heritage would be annihilated. Pollen and archaeological remains preserved in peat represent a most important archive for the history of man and the landscape since the Ice Age. The education potential of peatlands is only just being realised. An ecosystem which has only been studied incompletely to date - and which has been providing important ecological insights into hydrology, carbon fixation and environmental change - would be lost for further research.

Ireland is in a unique position in Europe in that it possesses almost 200,000 ha of actively growing raised and blanket bogs and fens which are of increasing European conservation importance. Therefore the European Union has a key role to play in conserving this resource. The importance of conserving Irish peatlands has been recognised by the international community in the Council of Europe's recommendations to member states on Peatlands in Europe (1981), the European Parliament's Resolution on the Protection of Irish Bogs (1983), and the

International Mire Conservation Group Resolutions (1986, 1988, 1990, 1992, 1994) (see Foss, 1991).

Save the Bogs Campaign

The recognition of the rapid decline in the area of Irish bogs worthy of conservation (due to exploitation) led to the formation of the Irish Peatland Conservation Council (IPCC) in 1982 (formerly the National Peatland Conservation Committee) (IPCC, 1986; Doyle, 1990a). The IPCC was established as a limited company with charitable status supported almost entirely by voluntary contributions. Those state grants that are received are usually tied to specific projects and are not given as core grant aid. The principal aim of the IPCC is to ensure the conservation of a representative example of Ireland's peatland heritage.

IPCC has been spearheading the 'Save the Bogs Campaign' since 1982, making it one of the longest running campaigns of any non-governmental organisation in Ireland. To date, we have achieved considerable success in the area of peatland protection, education and public awareness. From a starting point of just a few 100 ha of bogland protected in the early 1980's, we have now achieved a protected bog area of 26,000 ha.

Our work is divided broadly into four roles: conserving peatland habitats and their unique wildlife, informing and educating the public, fund-raising and lobbying.

Actions and achievement

The major achievements of the IPCC in each of our principal action areas are documented in the sections that follow.

Site protection, research and management

Our objectives in this area are to ensure conservation and protection, rehabilitation and management of a representative sample of peatlands in Ireland.

The Irish Peatland Conservation Council is the only organisation that maintains and makes available a list of conservation-worthy peatlands in Ireland. This list is widely endorsed by the scientific community. Our objectives are to ensure that the Government adopts and finances a peatland conservation strategy for Ireland, which takes into account acquisition, consolidation and sustainable development of the nation's peatland resource for the future.

The IPCC maintains a database with extensive information on all peatlands of European Conservation Importance in Ireland. At present, there are 802 sites on the database which are listed in the Council's fourth action plan. This document has proven invaluable as a reference source in negotiations with

Government, development and conservation agencies, the EU and with members of the public. Through official liaison with the National Parks and Wildlife Service of the Office of Public Works and continued monitoring via a system of national watch dogs, the IPCC ensures that the peatland site list remains up to date, and that the levels of various damaging activities are monitored and addressed where appropriate.

To ensure the effective conservation of a peatland area, it is essential to gain control of the hydrology of the habitat. This is usually only feasible where there is a single owner (either state or private) or where all owners are prepared to enter into a management agreement to conserve the peatland.

In either situation, the IPCC grasp any opportunity to act as a catalyst to initialise communication between landowners and conservation agencies in order to bring about conservation of further peatland sites. Out of necessity, our work has also involved us in direct site purchase (particularly with Dutch funds) and in meeting landowners to provide management solutions to further site conservation aims.

Education and public awareness

Since the formation of the IPCC, the organisation has always recognised the key role that education can play in changing attitudes and behaviour towards our natural peatland heritage (IPCC, 1986; 1989; 1992). For this reason, the IPCC made environmental education an integral part of the 'Save the Bogs Campaign'. Our aims are to increase knowledge, understanding and awareness of the peatland heritage, its conservation and management; to promote active participation by the individual in the conservation of peatlands, the protection of the environment and the prudent and rational use of natural resources.

Schools are a priority for action because they are places where all young people can be reached. Teachers are the key to getting the conservation message across to young people. Their success in changing attitudes and behaviour towards the environment, and in particular towards peatlands, is dependent on the availability of education materials and resources that span the school curriculum and that are presented in such a way as to stimulate and challenge the intellect of young people.

To assist teachers and students in this process, the IPCC have researched, developed and published a wide range of educational materials including Bogs in the Classroom (O'Connell, 1991) - an introductory pack for teachers wishing to use bogs, Peatland Education Pack for secondary school use (O'Connell, 1993), and a range of other useful publications (O'Connell, 1987; 1992b; 1994a; 1995). All IPCC education materials are produced from direct classroom work with teachers and have clear links with the school curriculum. To facilitate the uptake of our education materials in schools, we provide in-service training courses for teachers and youth leaders.

Formal education is only part of the education process. Our work in this area cannot be successful if the public at large remains ignorant of the need for, and practice of, environmental education. The IPCC education programme also uses any learning opportunities that stimulate interest amongst the general public and encourages them to learn more about peatlands. We regularly publicise current issues focusing on bog conservation in the national media and in our membership magazines *Peatland News* and *Portach*. We organise poster and photographic competitions, and have produced an exhibition 'The Wild Beauty of Bogs' which tours the country. We provide awareness walks and weekend study courses.

Fund-raising

Conservation of peatlands is an expensive undertaking. For this reason, the IPCC must devote time and energy to fund-raising. In our last ten years of existence we have achieved our funding success through maintaining a reputation for excellence and achievement which has built up the image and strength of the organisation to its current status. We have invested considerable funding in conservation projects which has paid off in terms of protecting peatlands and raising public awareness.

Our fund-raising activities are targeted at the general public, grant aiding organisations, Government bodies and the corporate sector (both in Ireland and Europe) with a view to gaining funds for peatland conservation, education and publicity programmes, for lobbying campaigns and to earn enough revenue to ensure the growth of the organisation.

To improve its performance in fund-raising the IPCC is adopting a marketing approach to its business activities which concentrates on providing services and promotions which meet the needs of members, supporters, consumers and commercial companies and hence ensures their satisfaction and support for the work of the IPCC.

Lobbying

The fourth action plan published by the IPCC devised a strategy to achieve the conservation of the 50,000 ha peatland target and published peatland site-lists for the whole of Ireland. The document - produced in 1992 - was widely circulated to those with a role to play in either the conservation, or development, of Irish peat resources. Pitfalls in legislation were highlighted with case studies.

The Action Plan suggested that a realistic annual Government budget of £850,000 needs to be assigned if peatland conservation targets are to be achieved before those bogs deserving protection become extinct.

Since the publication of the Action Plan, the IPCC has been heartened by the interest and support which international and national agencies have provided for our peatland conservation campaign and strategy.

The IPCC Action Plan calls on all interested parties to make greater efforts to implement an active and progressive peatland conservation programme, and in the meantime, the Council will continue to act as a catalyst in stimulating greater understanding, liaison and co-operation between all groups.

The future: prospects and problems

Although major success have been achieved by the IPCC in the areas of peatland conservation and awareness in Ireland over the last ten years, much work lies ahead if the peatland resource is to be fully protected, re-rehabilitated and integrated into sustainable development framework for Ireland.

Specifically, a number of hurdles and obstacles to the successful conservation of our finest bogland areas must be overcome. Principal among the issues affecting bogs are:

European Union Directives and national legislation

The Government must live up to its commitment under two European Union Directives that directly affect the ability of the State to conserve bogs, and its stated objective to conserve 50,000 ha of peatland.

In the first instance, urgent changes are required to the national legislation implementing the Environmental Impact Assessment Directive. This is wholly inadequate to address the damage being inflicted on bogs by peat extraction, afforestation, drainage and agricultural intensification.

As this document stands, it affords no protection to peatlands from the majority of developments. This is principally due to the threshold levels set for turbary (50 ha), afforestation (200 ha), and drainage (100 ha for local projects, and 1000 ha for arterial schemes). These are too high to prevent all but the largest industrial schemes from being subject to an assessment. On Natural Heritage Areas, the IPCC has called for an environmental impact assessment on any scheme affecting the site and recommends a zero hectare threshold.

Unless the thresholds and therefore the rate of damage to key conservation sites are reduced, we are in danger of losing more bogs of conservation importance through uncontrolled developments before they can be fully protected.

The Habitats Directive could provide up to 75% EU support for the purchase of raised and blanket bog listed as SACs (Special Areas for Conservation). If we are to maximise the benefit gained from this, the Natural Heritage Survey (initiated by the National Parks and Wildlife Service) must be completed and all potential conservation worthy areas be identified.

It is essential that the data from the survey is disseminated to all interested groups, and that these areas are put on a statutory footing within the Wildlife Act. An adequately funded compensation mechanism must also be put in place to allow state conservation agencies to negotiate with landowners for the purchase or sustainable management of SACs.

Peatland conservation strategy for Ireland

The most important requirement is for the Government to formulate and publish a bog conservation strategy for Ireland - using the IPCC's fourth Action Plan as a blueprint. Other governments have recognised the importance of their peatland resource and produced plans for their conservation and management e.g. the Department of the Environment for Northern Ireland (DoE-NI, 1993), the Swedish Nature Protection Agency (Löfroth, 1995) and the Swiss Federal Institute for Forestry, Snow and Landscape (Grünig, 1994). The international importance of Irish bogs, and the rate at which sites are being lost, makes the formulation of such a strategy imperative.

This strategy must allocate sufficient funds and manpower to ensure we attain our 50,000 ha conservation target before further unnecessary damage is inflicted to the remaining peatlands of conservation value.

The strategy must also address the different approach required for the conservation and management of raised versus blanket bog. The responsibility for developing the conservation strategy rests with the Department of Arts, Culture and the Gaeltacht.

Given that public support is vital for the long-term implementation of Government conservation strategies, environmental education must also feature as a central part of the strategy.

Chapter forty-two:

Peatlands Park

Mr Keith Stanfield
Environment and Heritage Service (Northern Ireland), UK.

Establishment of the Park

In the mid 1960s, the Amenity Lands Branch of the Ministry of Development was established; with two of its main aims being to declare National Nature Reserves (NNRs) and Country Parks. By 1975, over 25 NNRs and five Country Parks had been targetted, reflecting most of the natural ecosystems present in the province. Peatland sites were investigated but although a number of NNRs were declared, bogs were not included in the Country Park series.

Dr Henry Heal (of the Northern Ireland Nature Reserves Committee) and Joe Furphy (now Senior Principle Scientific Officer in the Environment and Heritage Service - EHS) recognised this anomaly. Jointly they developed the innovative idea of a site where both high scientific interest and the provision of controlled public access to a bog landscape could be accommodated; allowing bogs to be experienced, appreciated and enjoyed. It was largely due to their vision and conviction that in 1978 the Park site was acquired, the first such project of its type in the British Isles. This area was chosen for a number of reasons:

Suitability

The area was a mosaic of cutover and uncut raised bogs interspersed with a number of wooded drumlins and a series of small oligotrophic lakes. It was typical of a previously more widespread habitat and retained relict pockets of high nature conservation interest. Although 60 years of turbary had caused considerable disturbance, this was perceived as potentially advantageous, in that an influx of visitors here should not have as dramatic an impact as on an intact bog.

Accessibility

The site was only 48 km (30 miles) west of Belfast, adjacent to a motorway junction and within an hour's drive for 90% of Northern Ireland's population. This was ideal as it was anticipated the facility would be most attractive to the local population.

Availability

With the demise of their markets, the then owners, the Irish Peat Development Company were looking for a buyer. The time of purchase was most fortuitous, as within a few years the rise of a domestic machined peat and turf market in Northern Ireland sent local bogland values soaring.

Aims on establishment

By October 1979, the Department had formulated a development plan. The chief proposals were that:

1. The site was to be named The Peatlands Park.
2. A National Nature Reserve was to be declared on the most significant nature conservation areas, when practicable.
3. Management would provide for a mix of three development aims, in descending priority; conservation, education and recreation.

How were these aims to be addressed?

Conservation

The resource base, especially the bog ecosystems, was identified as the most important site asset. This ensured that the whole area would be treated as a single unit and effectively managed as a nature reserve. Other developments would only be permitted if they were not detrimental to this aim.

Using appropriate means, and as resources would permit, the Department proposed to actively enhance and restore the nature conservation interest in the now derelict industrial turbary areas.

Education

The need to promote a public awareness and appreciation strategy for this neglected habitat was seen as the secondary aim. In order to facilitate this, education was to be given a higher profile here than at any of our other sites.

Considerable thought went into devising an effective strategy compatible with our finite resources. In essence:

- the site would be used as the focus for the Government's programme of bog interpretation throughout Northern Ireland;
- specific audience groups would be targeted, to maximise the impact of the message;
- the non-commercial values of our bogs would be highlighted, such as their ecology and cultural/archaeological heritage; current issues would be presented and debate on their use and future encouraged.

It was decided that schools would be the priority target audience. They present a readily accessible market for advertising, are generally receptive to new ideas and can act as good ambassadors within their own communities.

The programme was to be carried out directly by Park staff, initially at the local level, then expanding as resources permitted.

Recreation

As with other Country Parks it was considered important to provide for passive recreation, but here developments would only take place if they were not detrimental to either the conservation or education aims of the site.

Recreational value was however judged important if the local community was to be won over, essential if the project was to succeed. It would provide a facility previously lacking in the area, would create local employment and attract revenue to the area. Concurrently it was hoped to see a rise in local self-esteem as people began to perceive themselves as the custodians of a valued resource.

It is also sound conservation management to make the most of all the site's assets!

Development of the site

In order to facilitate these aims a rolling programme of developments was undertaken as resources permitted.

Nature conservation management

Surveys and monitoring of the flora and fauna commenced in 1979. By June 1980, sufficient knowledge had been gathered to enable the declaration of the 99 ha Mullenakill and Annagarriff National Nature Reserve within the site.

Long-term control of *Rhododendron ponticum* scrub, relics from the estate which owned the land in the 19th century, were also initiated in 1980.

Hydrological management commenced in 1984 with drain infilling, drain damming, bunding, land levelling and *Sphagnum* spp. sowing all being attempted over the next few years. These programmes are all still on-going. Where necessary, specific research projects have been commissioned on-site to enhance management aims.

Access provision

Considerable planning went into the siting of access routes. Every path, road and boardwalk was designed on-site and laid by the direct labour squad.

Overall, access provision was geared to encourage visitors away from the familiar and into new experience areas. Novel means were sometimes employed; a bog garden was opened in 1987 to invite access to the open bog, while in 1988 the derelict narrow-gauge peat railway on-site was refurbished for the public. The latter in particular acted not only as a gimmick to attract the floating visitor but was used as an interpretative on-site tool for school groups.

Car parks were installed, while for recreational purposes picnic sites and a small amenity grassland facility were provided.

Educational/interpretative provision

The interpretative/education programme was rapidly developed with the first school visit taking place in 1980. The sole educational resources were the warden and the site - no packs, equipment or buildings being available. Initially promotion was done by personal approaches to local schools. Special events were attempted for the general public, but interest was disappointing and these were generally not repeated until 1990.

By 1984 we were advertising in the internal mailing system of the local education and library board. This gave us direct access to over 300 schools. Take-up levels were around 10% annually, while visits to schools accounted for a further 3%.

Experience showed that teachers and children responded best to a cross-curricular package. This combined peat ecology with historical/cultural aspects. Surprisingly the archaeological significance of our bogs was seen as of lower interest, and current threats and issues were not always fully appreciated. Practical demonstrations of traditional peat cutting or arranged visits to adjoining commercial peat works found great favour. In the absence of direction from a common curriculum, the majority of schools preferred an individually tailored programme and - where possible - this was accommodated.

From the late 1980s, as school numbers increased, Park staff were increasingly used to assist teachers in planning site visits rather than actually taking the groups directly. Good relationships were also fostered with other organisations and professionals involved in peatland education and interpretation throughout the British Isles. One outstandingly successful relationship developed with the local Education for Mutual Understanding (EMU) programme. This provided many advantages:

- access to professional educators, conversant with the new Northern Ireland Curriculum, and in tune with their professional colleagues' requirements;
- they were able to produce more relevant package material than ourselves;
- they could guarantee bookings for large numbers of schools, 12 months in advance, making planning of resource allocation easier;
- they could guarantee that all sections of our population would be equally accommodated.

In 1993 the Government's Policy Statement on Peatland Conservation acknowledged the role of the Park and promised additional funding for new developments. A full laboratory facility was provided on-site for use by visiting schools, and an education pack for Key Stage 3 was developed by a professional consultant.

The establishment of the laboratory facility was accompanied by the setting up of a small library housing reference works on peatland issues (including Government sponsored research papers). Enquiries to the Environment Service for non-policy related information on peatlands are generally directed to the Park for assistance. Development of the Park had also progressed sufficiently to facilitate unrestricted general visitor access, and with it the first series of information/interpretative leaflets on the site were produced. Without any advertising, visitor numbers were already exceeding original targets for the post-launch period.

In 1990 the Park visitor centre, railway and new interpretative leaflets were officially launched, further broadening the interpretative experience for the general visitor. The centre exhibition included living and interactive displays on all aspects of the bog issue. Self-guided, indoor and outdoor quizzes were produced for a variety of age groups. The resultant media attention encouraged further increases in visitor numbers: at over 100,000 annually, they were now five times the anticipated figures. Indeed from 1993, peak time numbers were actively restricted by limited car parking.

The Park today and in the future

Despite the large number of annual visitors, from both schools and the general public, over 80% come from within a 30 mile radius of the site. Only around 4% come from outside Northern Ireland. Considerable opportunities now exist to develop a broader marketing strategy for the site, and so enhance the spread of the peatland debate.

A review of the Park's education/interpretative strategy is underway. Despite a decade of provision, most teachers are still reluctant to deal with the peatland experience without some on-site assistance. This largely stems from unfamiliarity with the ecology, their own training having normally been in freshwater or woodland ecology. This problem is being addressed by changes in the Northern Ireland Curriculum, but until such proficiency works its way through the system, educators/naturalists will still be required on-site.

In the future visitor provision will develop along three main lines:

- for the general visitor: the use of self-guided trails and outdoor interpretative panels will expand rapidly; a re-vamped series of special events and regular walks, led by professional educators/naturalists, will be instituted;
- for the school visit: the provision of professional educators/naturalists will maximise the teacher/pupil experience value; site-specific curricular based work programmes will be expanded to fulfil all four Key Stage requirements; Key Stage 1 in particular is seen as the highest priority;
- for the professional visit: increasingly the site will be used as a demonstration venue for professional visits, workshops and seminars on all aspects of peatland conservation and management. In particular the experience and mistakes made here are seen of as special interest to a series of similar development projects started or being planned elsewhere.

In Northern Ireland, Government commitment to peatland conservation is being reaffirmed at present with resources going into the conservation of large areas under the EU Habitats Directive.

The Park's role is to assist this by winning hearts and minds, through combining education with enjoyment of our bogs, their unique flora and fauna, their traditional exploitation and their value as wilderness in an increasingly developed and managed countryside.

Chapter forty-three:

Peatland Education Initiatives

Dr Catherine O'Connell
Irish Peatland Conservation Council, Ireland.

The role of environmental education

The Irish Peatland Conservation Council (IPCC) is a non-governmental environmental organisation that aims to conserve a representative example of 50,000 ha of Ireland's bogland heritage. Their fourth action plan for peatland conservation (IPCC, 1992) recognises that conservation strategies and public support through environmental education, information and participation must go hand-in-hand if they are to be successful. Environmental education gives people the capacity to address conservation and development issues together and paves the way for a sustainable future for our peatlands.

What is an appropriate environmental education?

Educators need to address three objectives in developing an environmental education programme. These are increasing knowledge about the environment; developing the right attitudes towards the environment and providing opportunities for action to resolve environmental problems.

Environmental education involves not only imparting knowledge and information but seeks to modify behaviour and develop lifestyles that are harmonious with the environment (Keating, 1993).

Informal and formal environmental education

The IPCC's environmental education programme draws a distinction between formal in-schools education and informal out-of-school education and recognises the need to make provision for both types in planning educational activities.

Informal education includes all outside school activities, such as competitions, earth days, excursions, exhibitions etc. Many groups cater to this level of education but the effects are not lasting. The IPCC also has an on-going informal education programme. The Guide to Irish Peatlands - providing the general public with an all round introduction to bogs - was published by IPCC in 1987 (O'Connell, 1987). An exhibition 'The Wild Beauty of Bogs/De Venen Geturfd', was designed to tour through Ireland and the Netherlands for six years to highlight the lessons in bog conservation that each country could learn from the other.

On an on-going basis poster and photographic competitions are organised promoting a responsible attitude towards the use of peat resources. Awareness walks, conservation work camps and weekend study courses are provided within a 'learning-for-leisure' programme. Full details of the conservation and education campaigns over the past ten years are described in *Our Story* (Foss and O'Connell, 1992). Most informal education activities are high profile but have a limited life and hence must be supported by formal education if a change in attitude and behaviour in favour of the environment is to be achieved.

Formal education occurs in the classroom using resources and materials that are linked to the curriculum set down by the Department of Education. Within the Irish education system environmental education is not taught as a subject in its own right but is integrated into core subjects such as environmental studies, science, geography and civics at primary and secondary levels.

Teachers are the facilitators in providing an environmental education for today's young people. The IPCC's role is to act as a support group for teachers.

Resources

Over the last five years IPCC has worked with teachers to develop a whole series of curriculum-linked resources focusing on peatlands as a topic for environmental education. The resources produced are inter-disciplinary and holistic in orientation and involve the use of experiential teaching methods. They focus on a local topic as a starting point and help to develop an understanding of national and international conservation and development issues. The resources published by the IPCC include: Peatlands and the Primary School Curriculum (O'Connell, 1994a), The Peatland Education Pack (O'Connell, 1992), Discovering the Wild Boglands Video, the Bogland Wildlife Wall Chart, Bogs in the Classroom (O'Connell, 1991) and The Living Bog Ecology Slide Pack (O'Connell, 1993).

The range of resources is broad, highlighting the enormous potential of peatlands in education which, until the recent education initiatives from IPCC, were largely under utilised. One of the strengths of these resources is that the materials emerged from the classroom. For example, the production of the Peatland Education Pack took three years and involved three phases of activity. Phase one was research and development which was undertaken by the IPCC's education staff and a group of committed teachers. This resulted in the production of a draft version of the pack in six subject modules including

Gaeilge, English, geography, science, history and art. Phase two of the project involved evaluation of the materials in a network of volunteer secondary schools. The final phase involved the incorporation of the findings from the testing phase into each module and subsequent publication of the pack (O'Connell, 1994b).

There is a growing overseas interest in IPCC's educational work and a variety of packs have been published in Northern Ireland (Macartney, 1994), United Kingdom (RSPB, 1994; Scott and Agyeman, 1995; Greig, 1995; Bell *et al.*, 1994), and France (Dalbavie and Solleliet, 1994) all stimulated by IPCC's pioneering efforts in the field of peatland education.

In-service teacher training

Experience has shown that - although very important - it is not enough simply to make education materials and resources available. Teachers must be actively encouraged to make use of the materials. The IPCC organises an annual series of teacher training courses in liaison with the Department of Education's In-Career Unit and the network of teaching centres throughout the country. Courses are run during term time from April to September. Groups of teachers are shown how to use the bogs in their locality to teach environmental education.

A typical training course on peat and bogs includes the following field studies:

- a look at a peat bank
- studying bog flora using square frame quadrats
- studying differences between wet and dry parts of the bog along a transect
- pond dipping

Indoor aspects include model making of bog formation, plant identification, video presentation, discussion, distribution of resources and designing a schools environmental education programme. The course structure is similar for both primary and secondary level teachers. The field methods and the interpretation of the data is more intensive for teachers of senior students.

As bogs are well distributed in Ireland most schools are within an hour's drive from a suitable site. Only a very small portion of the bog visited needs to be used for field studies, so damage is kept to a minimum. A small bog remnant is ideal and can be very successfully used as a heritage education site. There are several examples of successfully used sites including Ardkill Bog, Co. Kildare, Lodge Bog (Peatland World), Co. Kildare and Killaun Bog, Co. Offaly. Some of these have specific resource material, study bases and education staff who promote environmental education programmes.

Killaun Bog is particularly interesting because the pupils of St Brendan's Community School act as guardians of the reserve, undertaking a range of activities such as practical conservation work, monitoring, observation and recording in addition to environmental education.

Peatland education network

Good environmental education practice tends to be localised and patchy without collaborative effort. For this reason IPCC have designed a special cross-border peatlands training programme for instructors in education centres throughout Ireland. The target audience includes full and part-time staff and contractors in privately run conservation centres, adventure centres, outdoor education centres and national parks throughout Ireland. The programme called 'Wake Up to Bogs' provides instructors with the resources, skills and equipment they need to run courses (for teachers and students) on the bogs in their locality. By 1996, IPCC hope to have in place a network of 100 centres throughout the country where peatland courses are being run on a regular basis as part of an environmental education programme. This will greatly enhance their educational work and go a long way to ensuring that as many schools as possible are utilising the bogs for environmental education. To date 600 out of the 5000 Irish schools are utilising bogs in education. The 'Wake Up to Bogs' project should bring a much greater number of schools into the network over the next few years. Two important resources have been produced from this project, an Instructors Handbook and a Training Video (O'Connell, 1995).

Communication and monitoring

Once the interest in peatlands and environmental education has been developed, it must be nurtured. IPCC provides on-going support and resource back-up for teachers. This includes a special teacher representative scheme with a wide range of benefits especially for teacher and school representatives. Schools are kept in touch with regular newsletters and receive a Catalogue of Educational Resources and Services. This is full of ideas for educational visits to heritage sites and briefs teachers on the latest publications in environmental education. This service helps to promote a wide variety of environmental education topics including habitat conservation, endangered species and development education. Through the teacher representative, IPCC can monitor the effectiveness of their environmental education programme.

Teachers can also visit the IPCC's Enviro shop and talk with the staff there who are trained to answer queries and meet their educational needs. In this way the interest of teachers is held, encouraging them to spread the message about the need to conserve our peatland heritage, so helping the save the bogs campaign to build momentum.

Environmental education in the non-school sector

One pitfall to be avoided is equating environmental education solely with schools. Environmental education must be a life long commitment and provided throughout society in all sectors: higher education, business, industry, the professions etc.

The IPCC is currently researching innovative ways of introducing environmental education to these groups. One successful on-going environmental education programme involves the use of soil improvers in gardens. This consumer led campaign seeks to educate people about the consequences of their actions - that the use of peat in the garden is causing the destruction of wildlife.

IPCC have produced a range of magazine articles, information brochures, fact sheets and public displays all of which aim to provide information on gardening skills and alternative products to members of the public wishing to garden in an environmentally friendly manner. By recycling organic kitchen and garden waste in worm bins, compost tumblers or compost heaps, people cut down on the volume of municipal organic waste being produced. Such waste is a cause for concern among many municipal authorities who cannot find sufficient new landfill sites. By using home produced compost, leaf mould, or other alternative peat free products in the garden, people cut down on the use of peat which in turn will reduce market demand for moss peat and will help promote other forms of sustainable development for bogs.

Conclusion

This chapter outlines the IPCC's activity in the areas of education, training and public awareness, and the steps taken through an environmental education plan to ensure a sustainable future for the peatland resource. Economic development cannot stop, but it must change course to become less ecologically destructive. The challenge for the remainder of this decade is to put this understanding into action, and - through environmental education - make the transition to sustainable forms of development and lifestyles.

Part Twelve

The Way Ahead

An Introduction

Dr Rob Stoneman
Scottish Wildlife Trust, UK.

The Peatlands Convention, and this volume, tell a story. It is a story that began as ice retreated back to the poles and high mountains to leave a bare and pitted landscape across the northern latitudes. Through time, aided by climatic deterioration and, in some cases, human intervention, some of the landscape became dominated by peat. In oceanic Britain and Ireland, vast blankets of bog stretched across the west and the uplands. Elsewhere in the world, a diverse range of peatlands developed creating a specialised and substantial part of global biodiversity. Within the peats, an archive of environmental and cultural information was stored. Samples of this wonder are discussed in the earlier chapters of these proceedings.

The story continues with an increasing interaction between humans and peatlands. In many cases, this proved to be a destructive partnership as shown by the '*cold, logic[al]... lists*' (Lindsay, p. 183, this volume) of peatland inventories. The partnership is changing as humankind increasingly recognises the unsustainable path human culture appears to be taking. The loss of species and their genetic makeup, global warming, the swamping of our own traditional cultures, the degradation of landscape and so on are all on an exponential rise to possibly global catastrophe. Peatlands and their degradation play a part in this picture.

So the development of sustainable uses for peatlands, largely via conservation measures, plays a significant part in sustainable development. The peatland conservation initiatives, outlined in this volume, are laudable. The challenge is working them together to form a coherent pattern.

The last part of the proceedings is concerned with this pattern in the form of a strategy to bring together the whole gamut of conservation action. Joosten provides the moral framework with his consideration of ethics. I detail an action plan for Scotland's raised bogs. Some of the detail refers specifically to UK and Scottish issues but, in general, approaches to conservation are similar for differing habitats, within different regions.

The *Way Ahead* becomes clearer as the result of gatherings such as the *Peatlands Convention*. It was an inspiring occasion - over 200 people from around the world collected together to pursue the common goal of peatland conservation. As Professor Aubrey Manning (Chairman of the Scottish Wildlife Trust) enthused,

now is the time for action.

Chapter forty-four:

Mores and Mires: Ethical Considerations on Bog Conservation

J. Hans J. Joosten
Ernst-Moritz-Arndt Universitat, Germany.

Introduction

'I point no moral, except that we seem ultimately always thrown back on individual ethics as the basis of conservation policy. It is hard to make a man, by pressure of law or money, do a thing which does not spring naturally from his own personal sense of right and wrong.'

Aldo Leopold (1937).

Nature conservationists have to operate in a society where ethical considerations are mostly regarded as essentially unrelated to political or economic criteria (Stirling, 1990). 'Food comes first, then morals' (Brecht, 1928) is a widely quoted statement. Apart, however, from our animal instinct behaviour (the 'food'), there are three types of reason why we do something: because we *like* to do so ('emotion'); because social forces *require* us to do so ('laws and customs'); and because we reach upon reflection the conclusion that we *ought* to do so ('ethics'). Subjective reasons (emotion) are dangerous, when they are universalized ('everyone must do what I like'). Law and custom will easily be infringed in the absence of safeguards. Furthermore, blind observation of laws and customs makes humans irrational slaves of external forces. Therefore ethical considerations are more important than often acknowledged.

Ethics involve the moral principles which one chooses to adopt. Morality involves the distinction between right and wrong, requiring the exercise of reason and free will (Chau and Kam-Kong, 1990). In studying the consistency and justification of moral standards (Van Hoogstraten, 1993), the science of ethics is normative: ultimately, we want to know how people *should* act, treat, or live (Callicott, 1986).

In this chapter I want to survey some ethical considerations regarding the preservation of natural entities, especially bogs and bog beings.

Ethical principles

Basic ethical approaches

Classical ethical theories can be subdivided in three main schools, respectively based on:

1. the result of the act: consequential (utilitarian) ethics
2. the act itself: deontological ethics
3. the motive of the act: virtue ethics (Beauchamp, 1991).

From a bog conservation point of view, the first approach will be especially important. These *consequential theories* judge the rightness of an act by the rightness of the result of that act. According to the *maximizing* principle, each person has the duty always to bring about the best consequences, or the most good, possible. The *egalitarian* principle cares not only about the total good but also about its distribution: a smaller quantity of good that is equally distributed may be preferred to a larger total of which some have a disproportionate share. The *satisfying* principle is less demanding than the maximizing in giving each agent the duty to bring about consequences that are reasonably good, not maximally good.

Any of these principles can be limited by a *permission* allowing agents to give some more weight to their own interests than to others'. They still have a duty to sacrifice their interests for the sake of large benefits to others (cf. Crocker, 1990, '*Live simply that others may simply live*', Salleh, 1990), but do not need to accept very great losses to secure just a small increase in the aggregate good (Hurka, 1993).

Consequentialism requires that we work to achieve the best outcomes, even when others fail to do their fair share. As a result '*few of us believe the claim, and ... none of us live in accordance with it*' (Kagan, 1989). Therefore Danielson (1993) introduced the approach of *causal responsibility*. If I am part of (the cause of) a problem, I only have to do something about my share.

An act with the best overall consequences may, nevertheless, be wrong because it violates ethical 'boundary conditions', the so-called *rights*. If the only way to save the life of five patients is to kill an innocent person and divide his/her organs among them by transplantation, most people would say that the killing is wrong even though, by saving five lives at the expense of one, it has overall good consequences (Hurka, 1993; Harris, 1975).

Prior to calculating which action leads to the greatest good, consequentialists must decide, which beings count as *moral entities*. A moral entity is one whose interests must be taken into account (Norton, 1982). The same problem applies to other ethical theories. Kant's categorical imperative states: '*One must act to treat every person as an end and never as a means only*'. Similarly, Naess (1990) formulated: '*one cannot favour a level of standard of living for oneself which depends upon others not reaching that level.'*

The central question is: who/what are the moral entities, the 'persons', the 'others'...

Moral standing and significance

'The barrier does not fall between us and the dog. It falls between you and me.'
Mary Midgley (1983)

The question of which entities do matter in ethics, is adressed in a confusing multitude of terms and concepts, including rights, interests, values, functions, standing, significance (cf. Landman, 1995). For our purpose, the concepts of 'moral standing' and 'moral significance' will suffice. If something has moral standing, its well-being does count and is the basis of a presumptive duty to that thing ('moral consideration') (Van DeVeer and Pierce, 1994). Conferring moral standing implies that self-interest is exceeded.

Moral standing is an absolute concept: something either has (+) or lacks (0) it (cf. to be or not to be pregnant). Moral significance is a comparative concept (+1, +2, +3...), which indicates the relative moral weight of one entity with regard to another (cf. 3 or 6 months pregnant).

Classically, the type entity of moral standing is the self-conscious human individual (the 'person', 'I'). Departing from 'I', the moral standing of other entities can be addressed by regarding their distance to that reference-point in a multidimensional space with the following axes:

1. the axis of space (somewhere else)
2. the axis of time (sometime later)
3. the axis of taxonomy, phylogeny, affinity (something else)
4. the axis of organisational complexity.

The least controversial moral principle says that I have to consider the interests of other human self-conscious (phylogenetical distance = 0) individuals (complexity distance = 0) in my direct surroundings (spatial distance = almost 0) at the present time (temporal distance = 0). The extension of ethical concern to humans elsewhere (spatial distance > 0) is reflected in the Universal Declaration on Human Rights (UN General Assembly, 1948).

A concern for humans in future generations (temporal distance > 0) is central to the concept of 'sustainable development': meeting 'the needs of the present without compromising the ability of future generations to meet their own needs' (World Commission on Environment and Development, 1987). With this principle, the 'reciprocity of rights and duties' (cf. Watson, 1979) is burst: present generations are able to harm future generations, but not the other way round.

A next qualitative step involves an increasing 'phylogenetical' distance, by awarding moral standing to non-person humans, non-human animals, non-animal living beings and to non-living natural elements or processes respectively. Awarding moral standing to species, communities, civilizations, ecosystems, landscapes and other environmental wholes exceeds the individualistic approach (organisational complexity distance > 0) and involves another new dimension.

Anthropocentrism

'Damage to penguins, or to sugar pines, or geological marvels is, without more, simply irrelevant... Penguins are important [only] because people enjoy seeing them walk about the rocks.'

William Baxter, 1974 (in: Warren, 1983).

Anthropocentrists believe that only human beings have moral standing and that non-human objects and processes have value only for the fulfillment of human needs and wants (Norton, 1985). This human-centred approach is the starting-point of the Brundtland report, which speaks of 'people, whose well-being is the ultimate goal of all environment and development policies' (World Commission on Environment and Development, 1987).

Functions of bogs

In anthropocentrism, the value of non-human entities is dependent on the functions these entities (may) fullfill for humankind. De Groot (1992) (cf. Naveh, 1994) distinguishes the following functions of the natural environment, that also apply to bogs:

1. *Regulation functions*, related to the capacity to regulate essential ecological processes and life-support systems, contributing to the maintenance of adequate climatic, atmospheric, hydrological, pedological, ecological and genetic conditions.
2. *Carrier functions*, related to the capacity to provide space and a suitable substrate for human habitation, cultivation (crop growing, animal husbandry, aquaculture), energy conversion, recreation and nature protection.
3. *Production functions*, related to the capacity to provide many resources, ranging from oxygen, water and food, and raw materials for industrial use to energy resources and genetic materials.
4. *Information functions*, contributing to the maintenance of mental health by providing opportunities for reflection, spiritual enrichment, cognitive development, and aesthetic experience.

Dialectics of scale

Anthropocentrism has to deal with many scale-related problems. An important issue is the (sometimes complicated) distinction between consumptional *needs* (necessities) and *wants* (amenities). According to John Maynard Keynes, the real absolute needs are those that can be satisfied. The wants based on 'keeping up with the Joneses' (mimetical desire, cf. Achterhuis, 1988) are materially insatiable (Science Action Coalitioan and Fritsch, 1980).

Time-related scale problems include immediate versus optional values, resource versus source availability, short-term versus long-term use, and reversible versus irreversible effects.

The decision not to go ahead with a 'development' project is reversible and keeps open all the options; the destruction of a bog does not.

As the 'renewability' of peat must be seriously questioned peat(land) is not (yet?) a sustainable resource (Joosten, 1995a). Because of the ongoing destruction of the natural peat producing 'sources', the living mires, and their insufficient restoration, the global peat(land) balance is shifting more and more towards increasing net peat(land) losses (Joosten, 1995d).

Economic benefits must be computed for finite times. The benefits and harms beyond the time frame chosen are necessarily ignored. Since it is virtually impossible to restore bogs, values of these ecosystems should be projected into the infinite future, while the economic activities that damage them and extinguish species normally contribute to human welfare only over a limited period of time (Randall, 1985). Furthermore, the functional value of bogs may wait upon changes in environmental conditions (climatic, atmospheric, hydrologic) and in culture (preferences, income levels, knowledge and technologies, public policies, cf. Norton, 1987a).

On the organisational level an important difference must be made between a person's position as a consumer and as a community member. The decision to have or not to have peat extracted in a country concerns the political question of what limits should be placed on the satisfaction of consumer preferences. The decision to buy a bag with peat, when it is available on the market, concerns merely the satisfaction of individual preferences within these limits (Norton, 1987a).

As individuals, we pursue self-interest, largely heedless of the cumulative effects of our individual actions. Though the maintenance of a healthy environment is in everyone's general interest, it is in no single individual's personal interest to moderate his or her consumption. Eventually, a tragedy may result (Hardin, 1968), unless the community sets limits to individual consumption.

Abundance and redundancy

Do we have to preserve all bogs, all bog species and all peat from an anthropocentric point of view? To what extent is their maintenance necessary for maximizing human happiness (or minimizing human suffering)? To what extent is their abundance redundant and useless?

The probability that a species will be directly useful for humankind in future is low, because it is the product of two low probabilities: that the species indeed is useful, and that that use will be discovered (Norton, 1987a). In a world with tens of millions of species, the loss of a single species may therefore be of negligible direct importance.

Species are also part of ecosystems. We have, however, insufficient knowledge of interdependence to judge which species are and will be redundant in ecosystem functioning (Lovejoy, 1988). A loss of one or several elements will mostly not ruin an ecosystem completely, because ecosystems are, contrary to what is often suggested, not 'wheel-works' but 'networks'. They do not collapse, they simply adjust (During and Joosten, 1992; Hargrove, 1987). Ecosystems normally contain a complex of mutually substitutional negative-feed-back mechanisms developed over time (cf. Joosten, 1993).

The same accounts for mires. Some mires are/were definitely redundant in the functioning of regional landscapes, western civilization and the global biosphere. Because there are so many bogs, bog species and processes, the likelihood that the next destruction or extinction will produce a disaster is very low. Each destruction, however, will increase the chance that a positive feed-back mechanism is initiated which ultimately may have disastrous consequences (cf. Joosten, 1993). A possible example is the increasing carbon emissions from mires as a result of the greenhouse effect resulting from increased carbon concentrations in the atmosphere. That this probability is initially very low and the effects are initially very limited encourages the fallacy that the absence of disaster so far is evidence that disaster will never come (Norton, 1987a; cf. Ehrlich and Ehrlich, 1981: 'The Rivet Poppers'). The limited direct effect may lead to a rapid destruction of the majority of bogs, especially in areas where a seeming abundance leads to the false conception that an endless resource is available (Sober, 1985; Joosten, 1995b).

Anthropocentrist strategies

How should an anthropocentrist deal with these uncertainties?

The simplest way, of course, would be not to interfere at all. But in practice, we prefer to take a small chance of great disaster in return for the high probability of a rather modest benefit (Sober, 1985): e.g. we fly in order to save some time.

If an act involves some risk of bad consequences this is a reason to avoid it. If the consequences are extremely bad, even a small risk of producing them is a reason to avoid the act and to accept some costs in doing so.

Relying on the central assumption that every species and ecosystem may have great but unquantifiable value, one could argue that species and ecosystems should be saved as long as the costs are tolerably low. In the face of high costs, society might choose a small risk of serious negative consequences (Norton, 1987a).

As bogs are self-regulating ecosystems, the costs of preservation are very low and restricted to 'external management' (Joosten, 1994). The benefits renounced by not exploiting and destroying living bogs, consist of food, electricity, oil, wood, fuel peat, agricultural and horticultural peat.

At present, bog reclamation for (unsustainable!) agricultural use is only substantial in developing countries (e.g. Indonesia). Hydro-electric project development is an important threat to bogs in Canada, where the electricity is to a largely generated for luxury consumption ('wants'). Winning and transport of oil damages large tracts of bog in Siberia. Energy in Russia is to a large extent wasted. In the west, oil products for predominantly luxury consumption create many additional environmental problems. Bog forestry in Scandinavia (for Europe) and South-East Asia (for Japan) is practised mainly for paper production and other low value uses. More than half of the annual world extraction volume of fuel peat is being used in the countries of the former Soviet Union, known for its notorious wastage of energy. Other big spenders in the west are Ireland, Finland and Sweden. Peat for fuel may also play some local economic role in developing countries (Indonesia, Malaysia, Senegal). Agricultural peat extraction is largely determined by the countries of the former Soviet Union, where enormous quantities of peat are used for restoring the fertility of mismanaged agricultural lands. Horticultural peat is used for growing flowers and luxurious vegetables (Joosten, 1996).

We may conclude that the long-term 'costs' (= renounced 'benefits') of preserving bogs are low from a global point of view. Locally and temporarily, however, peatland exploitation may have major socio-economic significance. In several areas of the world, peatland products act as a medium of unequal exchange of 'eastern needs' against 'western wants': primarily a problem of global welfare distribution. For several uses alternatives are available, that would contribute to a sustainable society, although possibly would involve somewhat lower standards (e.g. composts based on organic wastes).

Indirect anthropocentrism

From the anthropocentric standpoint, all our responsibilities with regard to non-human entities are based solely upon the realization of human happiness (Taylor, 1981). Watson (1979) presents several arguments to attribute moral standing to non-human entities in an indirect anthropocentric way, by linking them to the human moral milieu. He distinguishes:

1. The ecological argument: all livings things on Earth are interdependent parts of ecological communities. 'The land' makes possible the existence of human beings and of (human) morality, and therefore is said to earn the right to be. The women of the Chipko movement in India, who risked there lives to save their tree, clearly believe that the rights of trees are of a higher order than those of human beings because trees provide the conditions for life on Earth (Kothari, 1990).

2. The prudential argument: we should act as though non-humans also have rights to comfortable lives, to avoid the possibility that some people will treat humans in the same way as non-humans are sometimes treated. As John Locke wrote in 1693: '*People who delight in the suffering and destruction of inferior creatures, will not ... be very compassionate, or benign to those of their own kind.*'

3. The sentimental argument: in the interests of decreasing the suffering of humans who suffer when non-human entities are violated, we can assign to these entities secondary rights not to be unnecessary violated.

4. The contractual argument: nature might be said to earn a right to peaceful existence because it provides pleasure for human beings. In this argument nature is treated as something in which individual humans have a proprietary interest.

Norton (1984, 1987a) points to another important 'weak-anthropocentrist' argument: the 'transformative value' of nature. This function does not lie in the fulfillment of an existing preference, but in altering or transforming such preferences. Experience of wild species and pristine ecosystems is a necessary condition for developing a consistent and rational world view, one that fully recognizes human's place as a highly developed animal whose existence depends upon other living beings and functioning ecosystems. It provides the occasions for forming and criticizing our frame of reference: How to exist in a limited world, how to understand that world and what value to place on it?

The destruction of the biosphere lies first and foremost in the wasteful lifestyles of the world's privileged groups, and the problem of poverty emanates from this same source (Kothari, 1990). Experience of nature can promote questioning and rejection of the world views that lead to overly materialistic and consumptive felt preferences. Bogs, as extremely economical, stable, sustainably developing miniature-worlds (cf. Joosten, 1992), may play an important role in this respect.

Conclusions

In deciding on preservation or destruction of living bogs, we have to be aware of our limited knowledge, the possibly high risks, and the long-term benefits and harms. From a non-egoistic anthropocentrist point of view, an action leading to destruction of a living mire can only be justified:

1. if the action and its effects are also beneficial to humans on the long run;
2. if the beneficial action is related to genuine human needs and not luxuries;
3. if the benefits accrue to a large number of persons and not to a privileged few;
4. if alternative sustainable resources are not available;
5. if the preservation costs are too great to be borne by society;
6. if the bog type and its inhabiting species are common.

(modified after Science Action Coalition and Fritsch, 1980; cf. Joosten, 1996).

Non-anthropocentrism

Non-anthropocentrist ethics extend moral standing to parts of the environment, independent of their value for humankind. *Individualistic* environmental principles address moral standing to individual non-human entities. The rationale for this approach is ethical consistency: if we value a state such as freedom from pain in humans, it is arbitrary and 'speciesist' not to value it also in non-humans (cf. racism, sexism) (Hurka, 1993). A more radical view is the *holistic* environmental ethic, which attributes moral standing to wholes such as ecosystems, and which grants individuals moral significance only as contributing to valued properties of these wholes (Hurka, 1993). Between both ethical schools interesting debates have been going on (cf. Callicott, 1980; 1988; Warren, 1983; Hargrove, 1992).

Individualistic approaches

'But the pine is no more lumber than man is, and to be made into boards and houses is no more its true and highest use than the truest use of a man is to be cut down and made into manure.'

Henry Thoreau (1864) (in: Homan, 1991).

If all humans do matter, the characteristics necessary for 'earning' moral standing must be some common denominator, pitched so low that no human lacks them. Any such set of characteristics, however, will not be possessed only by humans (Singer, 1986). The central premise in individualistic environmental ethics is the *denial of disanalogy*. Individualistic theories attribute moral standing to different entities with different 'phylogenetic' motivations:

1. Rational beings, with a presumed morally important characteristic 'reason' or 'rationality' (cf. Immanuel Kant). This characteristic excludes some human beings (mentally disabled, severly demented, babies etc.), but does cover some animals, including great apes. Reassessment of the moral status of chimpanzees, gorillas and orang-utans is currently in discussion (Cavalieri and Singer, 1993). Conferring moral standing to great apes implies that ape-centric instrumental values of landscapes must be taken into moral consideration, e.g. the mires in Southeast Asia that are important as a habitat of orang-utans (Rieley, Chapter 3, this volume).

2. Sentient beings with a presumed morally significant characteristic 'the capacity to experience pleasure and pain' (cf. Jeremy Bentham, Peter Singer). Pain is presumed bad, since it is obvious that every creature seeks to avoid or, if unsuccessful at avoidance, to minimise it (Dizard, 1994). Pain is, both in humans and non-human animals, based on limbic-based consciousness. This parallel capacity for pain gives sentient non-human beings the right for pain not to be intentionally and needlessly inflicted upon them. Their capacity for pleasure gives them the right not to be prevented from pursuing whatever pleasures and fulfillments are natural to creatures of their kind (Warren, 1983; Singer, 1990). One important source of individual suffering among wild, non-human animals is destruction of habitat (Norton, 1982).

3. Living beings, with a presumed morally significant characteristic 'life' (cf. Arthur Schopenhauer, Albert Schweitzer, Paul Taylor). Sentiency is considered not to be the basic criterion for moral standing, since sentiency exists only as a means to another end: life (Goodpaster, 1978). In 1982 the United Nations General Assembly adopted the World Charter for Nature, which affirms that 'Every form of life is unique, warranting respect regardless of its worth to man.' This ecological egalitarianism ('the equal right to live and blossom') is the central idea of 'deep ecology' (Naess, 1973).

4. All beings, with a presumed morally significant characteristic 'being part of the (natural) whole' (cf. John Muir).

'When we try to pick out anything by itself, we find that it is bound fast by a thousand invisible cords that cannot be broken to everything in the universe. I fancy I hear a heart beating in every cristal, in every grain of sand and see a wise plan in the making and shaping and placing of every one of them.' (John Muir, 1869 in: Fox, 1981)

This approach implies, that plants, bogs, and peat all have moral standing (cf. Stone, 1972), but only in the sense that paintings, garbage dumps, and bricks do too. If everything has moral standing, ethical deliberation is empty, unless it is supplemented by some technique for weighting and comparing the moral significance of different objects (Sober, 1985). Holistic environmental ethics formulate some premises for this weighting.

Holistic approaches

'A thing is right when it tends to preserve the integrity, stability and beauty of the biotic community. It is wrong when it tends otherwise.'

Aldo Leopold (1949)

Although paradigmatic holders of moral standing are individual persons, our society also ascribes moral standing to 'superorganismic entities' - corporations and nations, for example (Callicott, 1985). Similarly, holistic environmental ethics address moral standing to a higher level of organization than the individual. These 'wholes' may be both abstract (conceptual) and concrete (tangible).

In contrast to the 'animal protection' approach, the individual is not valued as such in nature conservation. The individual is valued primarily as a representation of something beyond and distinct from it, the species. A species, however, is only a conception of a momentary group of individual entities, gradually changing in accordance with the principles of evolution. As a result, strictly speaking, we are not trying to preserve natural *patterns* so much as natural *processes* (Hargrove, 1987): the capacity of future evolution, speciation and diversification. These processes take place and are shaped in the spatio-temporal wholes, we call ecosystems.

The interests of the whole are not those of its constitutive parts summed up and averaged out. For the sake of the whole, some parts may be sacrificed (cf. the lizard's tail) (Callicott, 1980). In holistic ethics, the good of the 'whole' serves as a standard for the assessment of the moral significance of its constitutive parts. Hierarchical concepts (Klijn, 1995) may play an important role in this assessment. 'Key' *Sphagnum* species in bogs will, therefore, have a higher significance than other plant and animal species (Joosten, 1993).

Ecosystems, however, are not merely collections of interacting individuals, epiphenomenal aggregations with own characteristics. The community also shapes the morphology and behaviour of individuals within it (Rolston, 1994), especially on the long run. As such, a firm distinction between the individual and its environment cannot be made (Callicott, 1980). Gandhi stated that in extending aid to others 'altruism was unnecessary because his self embraced the whole village' (Nash, 1989).

Alternative metaphysics of conservation ethics

'All creatures are part of God's family, unfallen, indepraved, and cared for with the same species of tenderness and love as is bestowed on angels in heaven or saints on earth.'

John Muir (1916)

Apart from anthropocentrism, denial of disanalogy, and evolutionary/ecological arguments, there are various alternative metaphysics of ethics in which the moral standing of non-human entities may be grounded (Callicott, 1985):

1. **Theism**: the theological argument, that all entities are God, the image of God, or created to glorify God. Conceptions and attitudes of the major world religions regarding human responsibility for the environment are reviewed by Westhoff (1989), Engel and Engel (1990) and Coward (1993). The relation between environmental problems and Jewish/Greek/Christian philosophy is specifically addressed by White (1967), Passmore (1980), Callicott (1985), Hargrove (1989), Van der Perk (1989) and Van Hoogstraten (1993).
2. **Nature mysticism**: the intuitive feeling of human's unity with all nature, as expressed by Pythagoras, Francis of Assisi, Baruch Spinoza, Herman Hesse, Rosa Luxemburg, Guido Gezelle, John Muir, Ralph Waldo Emerson, Henri Thoreau and many others.
3. **Holistic rationalism** : the idea that this world must be the 'best' of all possible worlds, with a maximum economy of premises and fundamental laws, a maximum diversity of resulting phenomena, and its consistency, order, or 'harmony' (Callicott, 1985). In conservation literature one often finds similar concepts including 'the balance of nature' (cf. Pimm, 1991), 'nature knows best' (Commoner, 1971) and 'Gaia' (cf. Lovelock, 1991).
4. **Darwinism**: altruism is seen as a result of evolutionary selection. Human sentiments are genetically fixed reactions toward the 'given' world (Hargrove, 1989; Callicott, 1985; 1988). The survival advantages of group membership more than compensate individuals for the personal sacrifices required by morality. Since human beings are not sufficiently intelligent to make a cost-benefit analysis of their social actions, we are outfitted with 'social instincts' impelling us toward socially conducive moral behaviour (Callicott, 1988; Diamond, 1991; Maynard Smith and Szathmáry, 1995). The increasingly 'mixed community' and a changing world result in an expanding concept of moral standing (cf. Midgley, 1983; Nash, 1989). In all cultures and major religions, there is a latent premise of the worth of life, indicating an underlying core of ethical values common to all people (Skolimowski, 1990).

Altruism is thought to have evolved out of parental care in the long helpless period of childhood. Humans, furthermore, prolong infantile characteristics into maturity (neoteny) (Midgley, 1983). The most notable neotenous characteristic is the great size and long-continued growth of the brain. Humans also retain infantile behaviour, e.g. a tendency to play (cf. Huizinga, 1938), spontaniety, curiosity and the capacity for widely extended sympathy. *'It is one aspect of that openness to new impressions ... which makes us culturally malleable and enables us, through pseudo-speciation, to accept and build such varied ways of life. It carries with it, too, that still wider curiosity, that capacity for interest in other, inanimate surrounding objectives - plants and stones, stars, rocks and water -*

which extends our horizons beyond the social into the ecological, and makes us true citizens of the world' (Midgley, 1983).

This darwinistic approach leads to the conclusion, that the increasing threats to nature as a result of the enormous creativity of free-market capitalism, and the moral circles, in the course of time widening to include new groups of natural entities, are two sides of human neoteny, waiting to come into balance.

Maybe then we will understand, that a bog '... is not a mere fragment of dead history, stratum upon stratum like the leaves of a book, to be studied by geologists and antiquaries chiefly, but living poetry like the leaves of a tree, which precede flowers and fruit - not a fossil earth, but a living earth; compared with whose great central life all animal and vegetable life is merely parasitic.' (Henry David Thoreau, 1854 in Homan, 1991).

Acknowledgements

Thanks are due to my daughter Eline and the Flow Country (Scotland), whose simultaneous existence founded my interests in bog conservation ethics.

Chapter forty-five:

The Scottish Raised Bog Conservation Strategy

Dr Rob Stoneman
Scottish Wildlife Trust, UK.

Introduction

As explained below, Scotland's raised bogs have suffered extensively as a result of human interference. Whilst the data presented below is new, the plight of raised bogs has long been of concern to conservationists. The Scottish Wildlife Trust (SWT) began its involvement with this habitat as far back as the late-1960s. A strong stance to protect East Flanders Moss from afforestation (which had already led to the (temporary?) loss of Gartrenich and West Flanders Moss) both led to the protection of the moss and to the creation of a new wildlife reserve covering 40 ha of the south-western corner of the site. SWT's interest increased with further habitat loss through the 1970s and 1980s. Today, the Trust owns eight raised bog reserves and is a leading light of the Peatlands Campaign Consortium.

A clear understanding of the issue, the seriousness of the situation and a good working relationship with Scottish Natural Heritage (SNH) led to European LIFE funding to co-ordinate the Scottish Raised Bog Conservation Project. This broad multi-faceted project's chief aim was to research and produce a strategy to protect the resource. The strategy is timely since it begins the process of implementing the Habitats and Species Directive, which requires governments to bring raised bogs into favourable conservation status, and it follows in the wake of the Rio 'Earth' Conference in 1992.

At the Peatlands Convention, a draft consultative document was presented for delegates to comment on. Accordingly, it is a summary of this document (SWT, 1995b) that is presented here. It must be made clear that the strategy, at present, represents the views of the Scottish Wildlife Trust only. The role of the major partners indicated below is suggestive only. At the time of writing, the strategy is being updated for formal consultation by the major partners with the intention of integrating this strategy with habitat action plans, currently being prepared, for this habitat.

A damaged resource

The results of several differing surveys of Scottish raised bogs have been incorporated within the National Peatlands Resource Inventory (Lindsay & Immirzi, 1996). From this information, it has been possible to classify the majority of the raised bog resource on the basis of land cover and landuse characteristics. However, the different surveys (unsurveyed bogs (sites which have not been visited), the mid Strathclyde survey (McTeague and Watson, 1989) and the SRBCP survey (SWT, 1995a)) contain different types of data and are, therefore, difficult to merge together to form an overall picture. Nevertheless, careful manipulation of the data allows the cover classes listed in Table 45.1 to emerge.

Table 45.1. Cover classes.

Class	Description
Primary-natural	Uncut areas of bog with *Sphagnum* dominated vegetation.
Open degraded	Uncut or cutover (primary and secondary) areas of bog which support degraded bog flora - usually dominated by heathers, sedges and scattered scrub - but is not dominated by thick scrub or trees.
Wooded	Uncut or cutover areas of bog which support woodland. Much of this is plantation though many sites have dried out to allow tree species to colonise the bog.
Commercial workings	Areas which are being actively mined. They are usually characterised by bare peat.
Developed areas	Areas that have been built over, subject to open cast mining or otherwise developed.
Agricultural/fen	Archaic areas which have been reclaimed for agricultural or are now dominated by fen vegetation.
Unknown	This category mostly relates to unsurveyed sites to which a major land cover class has not be assigned. However, a few stray polygons have no attribute data attached.

In terms of developing strategies to conserve the resource, the main shortcoming with these classes is the inability to split open-degraded and wooded areas into uncut or cutover bog. However, the classes give a good overall impression of the state of the resource and can be used to compare types of damage across groups of raised bogs.

The present condition of the 27,000 ha of raised bog which originally existed in Scotland is shown in Table 45.2 and Fig. 45.1 (note that primary peatland is bog which has not been cut and secondary peatland is bog which has been cut; these site have often re-vegetated). The resource is highly damaged with only about 9% or 2300 ha in a 'near-natural' state. The majority of the resource has been damaged in varying ways.

Table 45.2. Condition of Scotland's Raised Bogs.

Cover Class	Totals (ha)
Primary - natural	2340.43
Open degraded (Primary/Secondary)	6208.83
Wooded (Primary/Secondary)	7945.43
Commercial workings	1088.74
Developed areas	589.61
Agricultural / fen	5746.13
Unknown	3030.15
Total:	26949.32

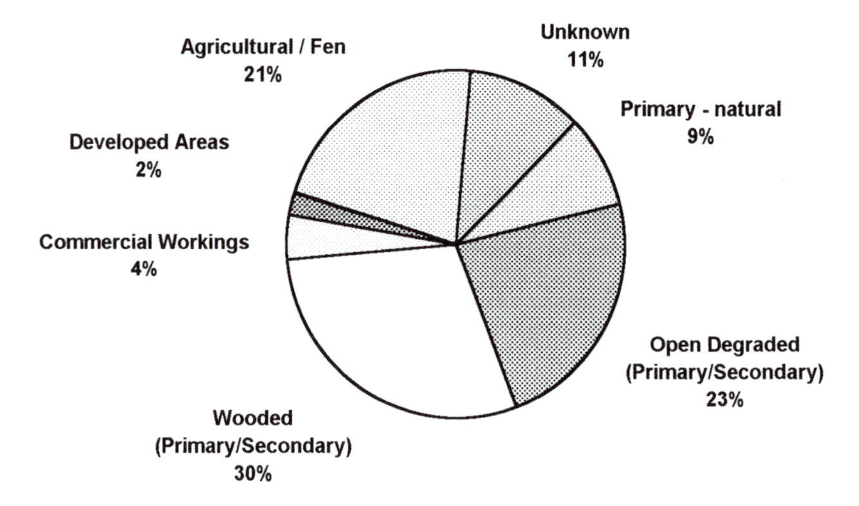

Fig. 45.1. A pie chart showing the condition of Scotland's Raised Bogs.

Overall 23% of the resource has now been effectively lost for bog wildlife as it has been developed or 'reclaimed' for agriculture. The other areas, which have little peatland nature conservation importance, are those areas which are wooded. These cover about 8000 ha; much of which relates to the commercial afforestation of uncut bog during the 1960s and 1970s. Afforestation has clearly been the main form of recent damage to raised bogs in Scotland although, today, the Forest Commission is committed to safeguarding the remaining areas of active raised bog. Whilst wooded areas have less peatland wildlife value at present, their potential value remains high (Brooks and Stoneman, Chapter 33, this volume) and these sites represent the main opportunity to dramatically expand the presently very small resource. Furthermore, many of these sites still contain full Holocene palaeoenvironmental records and are, therefore, of high heritage value (Charman, Chapter 6, this volume).

Active commercial peat extraction covers over 1000 ha which is highly significant given that commercial peat extraction currently forms the largest present direct threat. A further 6200 ha, or 23%, of the resource is still recognisable as bogland though it now exhibits degraded bog flora.

An analysis of the data on a regional basis reveals that the levels and type of damage varies considerably (Fig. 45.2). For example, the bulk of the remaining 'near-natural' areas are concentrated in the new local authorities of Stirling, Falkirk, West Lothian, North Lanarkshire (N.E. Central Belt) and South Lanarkshire. Other areas exhibit much higher levels of damage. For example, about 5000 ha of raised bog was originally found on the Grampian Plain (defined as the new local authorities of Aberdeen, Aberdeenshire and Moray and Highland). Today, only 150 ha are considered as 'near-natural'.

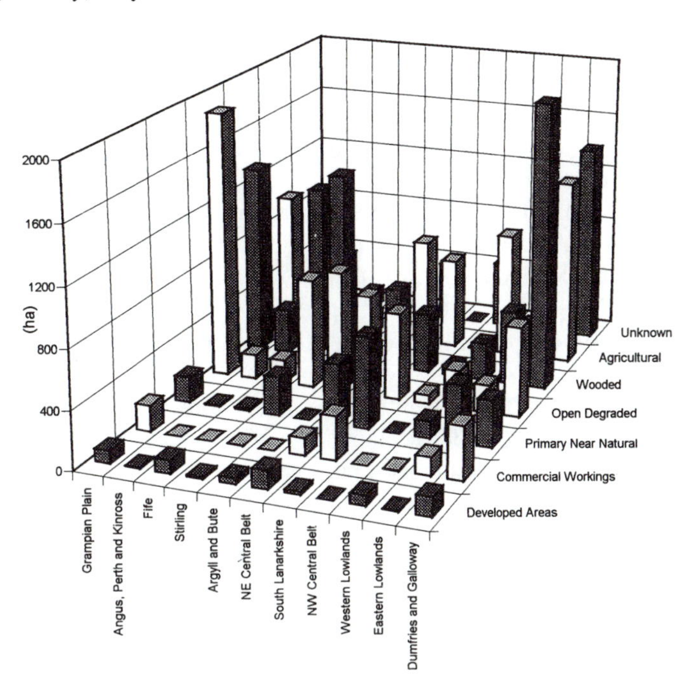

Fig. 45.2. Regional variations of raised bog condition.

This analysis confirms and reinforces the serious plight of raised bogs in Scotland already demonstrated by others (e.g. McTeague and Watson, 1989; RSNC, 1990; Barkham, 1993; Lindsay, 1993; DoE, 1994c; Lindsay, 1995; Lindsay and Immirzi, 1996). The proportion of 'near-natural' bog has fallen to a dangerously low level. Even so, these figures underplay the seriousness of this situation.

All Scottish raised bogs have been subject to a long and variable history of exploitation principally through peripheral peat extraction and drainage which are the main forms of damage on sites unaffected by forestry. Consequently, any remaining areas of 'near-natural' vegetation are always surrounded by bog (or former bog) in varying degrees of degradation. Given each bog acts as a single hydrological unit (Ingram, 1982), any damage to peripheral areas must have an effect on core 'near-natural' areas.

A further problem relates to the definition of 'near-natural' areas. The *Erica tetralix-Sphagnum papillosum* vegetation community (M18) is often considered as '*virgin ombrogenous peatland*' (Moore, 1968; Rodwell, 1991) and, therefore, 'near-natural' areas are defined (Parkyn *et al.*, 1995) as those areas where the vegetation generally fits into the National Vegetation Classification class M18. However, stratigraphic studies (Barber, 1981; Barber *et al.*, 1993) show that this community is both a transient and recent vegetation type on British raised bogs. Indeed, raised bog vegetation was formerly dominated by *Sphagnum imbricatum* (Stoneman *et al.*, 1993). Today, the species is rare; reasons for its decline are unknown but palaeoecological studies point to atmospheric nitrogen pollution as an important factor (Van Geel and Middledorp, 1988; Stoneman, 1993). 'Near-natural' vegetation may not, therefore, be as natural as it has been assumed.

The evidence suggests that Scotland's remaining raised bogs are highly degraded and continue to degrade. Without a concerted effort to halt and reverse this degradation, Scotland's raised bogs, as *sustainable*, *fully functioning* ecosystems, are likely to become extinct.

Reversing the degradation of Scotland's raised bogs

The types of damage, and, therefore, the likely threats which raised bogs are under, are examined in the forthcoming *Bog Management Handbook* (Brooks *et al.*, 1996). Briefly, damage to raised bogs relates to peat extraction, commercial afforestation, agricultural practices (drainage, grazing, 'reclamation'), burning, development (roads, housing, mining, waste disposal), recreational activities, atmospheric pollution, eutrophication, water abstraction and climate change. In the last ten years, most direct damage, in Scotland, has related to open cast mining and peat extraction. Atmospheric pollution and climate change could potentially have far-reaching consequences although quite what the effects may be have yet to be researched. Probably, the greatest threat to Scotland's raised bogs is the cumulative effect of years or centuries of small-scale damaging activities, leaving a highly degraded, fragmented and degrading resource.

This Strategy proposes a four pronged approach to halt and reverse the present degradation of the resource. These are explained in detail by SWT (1995b).

The strategy is divided into four actions:

- **Action 1**: Protecting Scotland's raised bogs.
- **Action 2**: Securing positive conservation management.
- **Action 3**: Rehabilitation of cutover or afforested bog.
- **Action 4**: Raising awareness.

Most of these approaches are based upon existing mechanisms and commitments to conserving biodiversity which were recently reviewed by the UK government as part of its response to the United Nations Conference on Environment and Development in Rio 1992 - *The Rio Earth Summit*. Four documents relating to climate change, forests, sustainability and biodiversity were produced in response to the Earth Summit (DoE, 1994d; 1995a; b; c). Each document has relevance to the protection of raised bogs - sometimes explicit or, more usually, contained within more generalised statements. The documents outline a framework for achieving the Rio agreements and detail existing and future initiatives which could contribute towards that achievement. They largely restate government policy and, therefore, act as a review of present local, national and international policies and legislation. Additionally, some new commitments are stated.

The UK's 'Rio documents' demonstrate clear government commitment to conserving raised bogs as a key component of Britain's biodiversity. This Strategy translates that commitment into a practical working plan.

An action plan to halt and reverse the present degradation of the raised bog resource is outlined below. Different partners are targeted for action and resource provision (these are shown in brackets for each action - abbreviations are shown in Table 45.3).

Table 45.3 Abbreviations used for Strategy partners.

Partner		Partner	
Forest Authority	FA	Scottish Enterprise	SE
Forest Enterprise	FE	Scottish Natural Heritage	SNH
Local authorities	LAs	Scottish Office, Environment Department	SOEND
Non-governmental organisations	NGOs	Scottish Office, Education Department	SOED
Commercial peat extractors	CPE	Scottish Office, Agriculture and Fisheries Department	SOAFD
Royal Society for the Protection of Birds	RSPB	Scottish Wildlife Trust	SWT

Effective targeting of resources can only be achieved if there is a clear nature conservation inventory of the resource. This allows bodies to enact actions in a strategic manner and also reveals the many opportunities which can arise. The raised bog inventory was due to be completed by 1995. However, this has not been fulfilled and information is lacking on 11% of the Scottish raised bog resource. Additionally, extra information is required to further refine the usefulness of the existing database (NPRI). Ideally then, the following steps are required, before proceeding with the rest of the plan, to complete the raised bog inventory:

1. Complete surveys of all completely non-forested sites or archaic sites (mainly located in Wigtownshire and north Strathclyde) (SNH).
2. Re-code mid-Strathclyde survey according to Parkyn *et al.* (1995) and re-digitise maps. All surveyed site information should be compiled into a single consistent database (SNH).
3. Survey of afforested sites to check whether trees have been planted on cut-over or uncut surfaces (FA, SNH).

Action 1: Protecting Scotland's raised bogs

The protective mechanisms which, in the main, already exist to protect raised bogs from direct development threats should be enhanced and strengthened. Note that, protective policies should consider not only the present nature conservation value but also a sites potential nature conservation value and its palaeoenvironmental and archaeological significance. The key elements to achieve this objective are as follows:

- Revise and expand the raised bog Sites of Special Scientific Interest (SSSI) network (SNH).
- Designate enough active and degraded raised bogs as Special Areas of Conservation (SACs) to ensure favourable conservation status of the resource (SOEnD) and implement wider measures to ensure protection of the whole resource (all partners).
- Include effective raised bog protection policies in local development plans (LAs).
- Ensure there is a requirement for environmental assessment for all developments likely to affect raised bogs (SOEnD, LAs).
- Widen non-statutory tools to protect raised bogs. In particular, designate wildlife sites and incorporate raised bog protection and rehabilitation policies into indicative forest strategies (SWT, LAs, FA).
- Revise development control of the peat extraction industry in Scotland (SOEnD, LAs, SNH, NGOs).
- Purchase sites as nature reserves (SWT, RSPB, SNH, LAs and others).

Action 2: Securing positive conservation management

Positive conservation management should be secured initially for the larger and less damaged sites. Management should be enacted on all existing and new SSSIs and SACs. An expansion of the present raised bog nature reserve network can be achieved by LAs declaring raised bogs as Local Nature Reserves (LNRs) and NGOs creating new raised bog wildlife reserves. This will further secure increased conservation management across the resource. Management could be resourced and enacted through the following actions:

- Form a specialist bog management team (SWT or SNH).
- Target management agreements to particular sites through a raised bog management scheme (SNH).
- Educate raised bog site owners (SNH).
- Increase the number of SWTs and other NGOs raised bog nature reserves (SWT , NGOs).
- Form new Local Nature Reserves based upon raised bogs (LAs).
- Target agri-environment schemes towards positive conservation management (SOAFD).

Action 3: Rehabilitation of cut-over or afforested raised bog

Rehabilitation of bogs seriously damaged by peat extraction and forestry remains an experimental discipline. Rehabilitation of both cut-over and afforested sites should be field tested. In tandem, rehabilitation measures should be enacted where opportunities arise. This will occur where trees have reached maturity and are felled or where peat extraction ceases at a site. All peat extraction planning permissions should have after-care policies worked into updated permissions. Summarised action are as follows:

- Implement an experimental project to field-test methods and techniques to rehabilitate commercial afforested raised bogs (FA, FE).
- Rehabilitate afforested raised bogs and cut-over bog as opportunities arise (FE, FA, SNH, private landowners, CPE).
- Change FA policy to allow felling of trees on raised bogs without requiring replanting (FA).
- Enact a policy review concerning afforested raised bog rehabilitation to consider: a presumption against replanting on raised bogs, proactive forest removal initiatives, inclusion of raised bog rehabilitation measures into forest design plans and a reworking of WGS payments to encourage raised bog rehabilitation (FA, FE, SNH).

Action 4: Raising awareness

Whilst levels of awareness of raised bogs are rising, the momentum must be maintained and increased through a variety of means:

- Create a Peatlands Park for Scotland (SE, SWT, SNH).
- Increase public access and interpretation facilities on raised bog nature reserves (SWT, other NGOs, SNH).
- Incorporate peatlands into the Scottish curriculum (SOED).
- Designate LNRs (LAs).
- Continue the Peat Campaign through the Peatlands Campaign Consortium (SWT, RSPB, FoE, WWF).

Conclusion

The land-cover survey reveals a desperately serious situation. An example of the rate of degradation is revealed from aerial photographs of Flanders Moss East. In the 1940s, the site was largely devoid of trees. Partial drainage in preparation for afforestation and peat extraction, peat cutting in the eighteenth and nineteenth centuries, and drainage of the Carse of Stirling have all contributed to the loss of 'near-natural' vegetation communities across the site. Today, trees and scrub are rapidly colonising the site (see Parkyn and Stoneman, Chapter 21, this volume). Fully, two-thirds of the Moss is now classified as supporting degraded bog vegetation.

Flanders Moss East, is often considered to be one of the biggest and 'best' remaining raised bogs, yet its future remains in doubt.

More positively, the threat of peat extraction has now been removed and a programme of positive management is being implemented across the bog. Perhaps then, Flanders Moss East will be conserved for future generations.

However, it is not enough to conserve a few isolated sites. The raised bog resource has degraded to such an extent that we need to widen our goals to encompass the whole of the remaining resource. This Strategy is one way forwards. Whatever way is adopted, the time for action is now.

References

Åberg, E. (1992) Tree colonisation of three mires in southern Sweden. In: Bragg, O.M. *et al.* (eds.) *Peatland Ecosystems and Man: An Impact Assessment.* International Peat Society, Helsinki pp. 268-270.

Aaron, J. R. (1982) *Conifer bark: its properties and uses.* Forestry Commission Forest Record 110. HMSO, London.

Achterhuis, H. (1988) *Het rijk van de schaarste.* Van Thomas Hobbes tot Michel Foucault. Ambo, Baarn.

Adamus, P.R. (1992) Rapid methods for evaluating ranking or categorising wetlands. In: World Wildlife Fund *Statewide Wetland Strategies: a Guide to Protecting and Managing the Resource.* Island Press, Washington DC, World Wildlife Fund. 268 pp.

Aerts, R., Wallén, B. and Malmer, N. (1992) Growth-limiting nutrients in *Sphagnum*-dominated bogs subject to low and high atmospheric nitrogen supply. *Journal of Ecology* **80**, 131-140.

Ahmad-Shah, A. and Soepadmo, E. (1992) Current status of exploitation and utilisation of peat soil in Peninsular Malaysia. In: *Proceedings of the International Symposium on Peat/Peatland Characteristics and Uses.* Bemidji, USA, 1989, pp. 57-72.

Ahmad-Shah, A., Radzi-Abas, M. and Mohd-Jamil, A.S. (1992) Characteristics of tropical peat under a secondary forest and an oil palm plantation in Selangor, Malaysia. In: *Proceedings of the 9th International Peat Congress,* Uppsala, Sweden, Vol. 1 (Special Edition of the International Peat Journal) pp. 256-269.

Ainslie, W.B. (1994) Rapid wetland functional assessment: its role and utility in the regulatory arena. *Water, Air and Soil Pollution* **77**, 433-444

Aiton, W. (1811) *A Treatise on the Origin, Qualities and Cultivation of Moss-Earth, with Directions for Converting it into Manure.* Ayr.

Anderson, A.R., Pyatt, D.G. and White, I.M.S. (1995) Impacts of conifer plantations on blanket bogs and prospects of restoration. In: Wheeler, B.D.,

Shaw, S.C., Fojt, W.J. and Robertson, R.A. (eds) *Restoration of Temperate Wetlands,* Wiley, Chichester, pp. 533-548.

Anderson, J.A.R. (1963) The flora of the peat swamp forests of Sarawak and Brunei including a catalogue of all recorded species of flowering plants, ferns and fern allies. *Gardens Bulletin Singapore,* **20**, 131-228.

Anderson, J.A.R. (1964) The structure and development of the peat swamps of Sarawak and Brunei. *Journal of Tropical Geography*, **18**, 7-16.

Anderson, J.A.R. (1983) The tropical peat swamps of western Malaysia. In: Gore, A.J.P. (ed.) *Ecosystems of the World. Vol. 4B. Mires, swamp, bog, fen and moor. Regional studies.* Elsevier, Amsterdam and Oxford, pp. 181-199.

Anderson, M.L. (1967) *A History of Scottish Forestry I - From the Ice Age to the French Revolution* (ed. C.J.Taylor) Nelson and Sons Ltd, London, Edinburgh.

Anderson, P. and Radford, E. (1994) Changes in vegetation following reduction in grazing pressure on the National Trust's Kinder Estate, Peak District, Derbyshire, England. *Biological Conservation* **69**, 55-63.

Andriesse, J.P. (1974) *Tropical Lowland Peats in South-east Asia.* Royal Tropical Institute, Amsterdam, Communication 63.

Andriesse, J.P. (1988) *Nature and Management of Tropical Peat Soils.* F.A.O. Soils Bulletin 59. Food and Agriculture Organization of the United Nations, Rome, Italy.

Anon. (1894) *Mountain, Moor and Loch, Illustrated by Pen and Pencil on the Route of the West Highland Railway.* London.

Anon. (1987) *Malaysian Wetland Directory.* Department of Wildlife and National Parks, Kuala Lumpur, Malaysia, 316 pp.

Armentano, T.V and Verhoeven, J.T.A. (1988) The contribution of freshwater wetlands to the global biogeochemical cycles of carbon, nitrogen and sulphur. In: Patter, B. (ed.) *Wetlands and Shallow Water Bodies 1.* SPB Academic Publishing, The Hague, Netherlands.

Arnott, D.A., Whitfield, D.P., Clark, I.R. and Bates, M.A. (1994) *A survey of moorland breeding birds in Caithness and Sutherland.* Research and Advisory Services Directorate Commissioned Research Report (Confidential), Scottish Natural Heritage, Edinburgh.

Askew, R.R. (1988) *The Dragonflies of Europe*. Harley Books, Colchester.

Aspinwall and Company (1994) *Peat-based and Alternative Products in the Gardening and Landscape Markets.* Department of the Environment, London.

Atkinson, T.C., Briffa, K.R. and Coope, G.R. (1987) Seasonal temperatures in Britain during the past 22,000 years, reconstructed using beetle remains. *Nature (London)* **325**, 587-592.

Avery, M. and Leslie, R. (1990) *Birds and Forestry.* T. and A.D. Poyser, London.

Avery, M.I. and Haines-Young, R.H. (1990) Population estimates for the dunlin *Calidris alpina* derived from remotely-sensed satellite imagery of the Flow Country of northern Scotland. *Nature* **344**, 860-862.

Backeus, I. (1972) Bog vegetation re-mapped after sixty years. Studies on Skagerhultsmossen, central Sweden. *Oikos* **23**, 384-393

Backeus, I. (1978) Skyddsvärda myrar i Oerebro län. *Medd. Växtbiol. Inst. Upps.* **1978** (5), 217.

Backeus, I. (1990) The cyclic regeneration on bogs - a hypothesis that became an established truth. *Striae* **31**, 33-35.

Bailey, M.P. (1992) The white-faced dragonfly *Leucorrhinia dubia* (Vander Linden) at Chartley Moss National Nature Reserve, Staffordshire. *Journal of the British Dragonfly Society* **8** (1) 1-3.

Baillie, M.G.L. (1991) Do Irish bog oaks date the Shang dynasty? *Current Archaeology* **117**, 310-313.

Baillie, M.G.L. (1992) Dendrochronology and Past Environmental Change. *Proceedings of the British Academy* 77, 5-23.

Baillie, M.G.L. and Brown, D.M. (1988) An overview of oak chronologies. In: Slater, E.A. and Tate, J.O. (eds) *Science and Archaeology, Glasgow 1987.* British Archaeological Reports 196, Oxford, pp. 543-548.

Baillie, M.G.L. and Munro, M.A.R. (1988) Irish tree rings, Santorini and volcanic dust veils. *Nature* **332**, 344-346.

Baines, C. (1991) Expert evidence for Peat Inquiry In: *Plantlife Commission of Inquiry into Peat and Peatlands. Public Hearing No. 2: Evidence*. October 23rd 1991. Plantlife, London.

Ball, M.E. (1974) Floristic changes on grasslands and heaths on the Isle of Rhum after a reduction or exclusion of grazing. *Journal of Environmental Management* **2**, 299-318.

Bannister, P. (1966) Biological Flora of the British Isles: *Erica tetralix*. *Journal of Ecology* **54**, 795-813.

Barber, J.W. and Crone, B.A. (1993) Crannogs: a diminishing resource? A survey of the crannogs of South-west Scotland and excavations at Buiston Crannog. *Antiquity* **67**(1993), 520-533.

Barber, K.E. (1981) *Peat Stratigraphy and Climatic Change - a Palaeoecological Test of the Theory of Cyclic Bog Regeneration.* A.A. Balkema, Rotterdam.

Barber, K.E. (1982) Peat-bog stratigraphy as a proxy climate record. In: Harding, A.F. (ed.) *Climatic Change in Later Prehistory.* Edinburgh University Press, Edinburgh pp. 103-133.

Barber, K.E. (1985) Peat stratigraphy and climatic change: some speculations. In: Tooley, M.J. and Twigger, S.N. (eds) *The Climatic Scene*. Allen and Unwin, London, pp. 175-185.

Barber, K.E. (1993) Peatlands as scientific archives of past biodiversity. *Biodiversity and Conservation* **2**, 474-489.

Barber, K.E. (1994) Deriving Holocene Palaeoclimates from peat stratigraphy - some misconceptions regarding the sensitivity and continuity of the record. *Quaternary Newsletter* **72**, 1-9.

Barber, K.E. (1995) Peat stratigraphy and the Lindow Bog Body: a reconsideration of the evidence. In: Turner, R.C. and Scaife, R.G. (eds) *Bog Bodies. New Discoveries and New Perspectives.* British Museum Press, London pp. 50-51.

Barber, K.E., Dumayne, L. and Stoneman, R. (1993) Climatic change and human impact during the late Holocene in northern Britain. In: Chambers, F.M. (ed.) *Climate Change and Human Impact on the Landscape*. Chapman and Hall, London.

Barber, K.E., Chambers, F.M. and Maddy, D. (1994a) Sensitive high-resolution records of Holocene palaeoclimate from ombrotrophic bogs. In: Funnell, B.M. and Kay R.L.F. (eds) *Palaeoclimate of the Last Glacial/Interglacial Cycle.* NERC, Swindon, pp. 57-60.

Barber, K.E., Chambers, F.M., Maddy, D., Stoneman, R.E. and Brew, J.S. (1994b) A sensitive high-resolution record of late Holocene climatic change from a raised bog in northern England. *The Holocene* **4**, 198-205.

Barber, K.E., Chambers, F.M., Dumayne, L., Haslam, C.J., Maddy, D., and Stoneman, R.E. (1994c) Climate change and human impact in North Cumbria: peat stratigraphic evidence from Bolton Fell Moss and Walton Moss. In: Boardman, J. and Walden, J. (eds) *The Quaternary of Cumbria: Field Guide*. Quaternary Research Association, Oxford.

Barbier, E.B., Burgess, J.C. and Folke, C. (1994) *Paradise Lost, the Ecological Economics of Biodiversity.* Earthscan, London.

Barkham, J.P. (1993) For peat's sake: conservation or exploitation? *Biodiversity and Conservation* **2** (5), 556-566.

Barkman, J.J., Moravec, J. and Rauschert, S. (1986) Code of the phytosociological nomenclature. Second edition. *Vegetatio* **67** (3), 145-195.

Bartels, R. (1977) Die landwirtschaftliche Nutzung von Moorboden. *Geologisches Jahrbuch* **F4**, 141-174.

Bateman, I. and Bryan, F. (1994) Recent advances in monetary evaluation of environmental preferences. In: Wood, R. (ed.) *Environmental Economics, Sustainable Management and the Countryside.* Proceedings of a workshop held on 25 April 1994. University of Wales.

Bates, M.A., Shepherd, K.B., Whitfield, P. and Arnott, D.A. (1994a) *A breeding wader and upland bird survey of selected sites in Caithness and Sutherland.* Research and Advisory Services Directorate Commissioned Research Report (Confidential), Scottish Natural Heritage, Edinburgh.

Bates, M.A., Shepherd, K.B., Whitfield, P. and Arnott, D.A. (1994b) *A breeding wader and upland bird survey of selected sites in Lewis and Harris.* Research and Advisory Services Directorate Commissioned Research Report (Confidential), Scottish Natural Heritage, Edinburgh.

Bather, D.M. and Miller, F.R. (1991) *Peatland Utilisation in the British Isles.* Paper No. 21, Centre for Agricultural Statistics, University of Reading.

Bayfield, N.G. (1979) Recovery of four montane heath communities on Cairngorm, Scotland, from disturbance by trampling. *Biological Conservation* **15**, 165-179.

Beauchamp, T.L. (1991). *Philosophical Ethics. An Introduction to Moral Philosophy,* 2nd edn. McGraw-Hill, New York.

Beaumont, D., Dugan, D., Evans, G. and Taylor, S. (1995) Deer management and tree regeneration in the RSPB Reserve at Abernethy Forest. *Scottish Forestry* **49**, 155-161.

Bell, S., Dunlop, J. and Watson, R. (1994) *CADISPA Primary Fish Farming, Tourism and Boglands.* World Wide Fund for Nature, United Kingdom, 70 pp.

Bellamy, D.J. (1986) *Bellamy's Ireland: the Wild Boglands.* Country House, Dublin.

Bennett, K.D. (1994) Post-glacial dynamics of pine (*Pinus sylvestris* L) and pinewoods in Scotland. In: Aldhous, J.R. (ed.) *Our Pinewood Heritage.* Forestry Commission/Royal Society for the Protection of Birds/Scottish Natural Heritage, pp. 23-39.

Bennett, K.D., Boreham, S., Sharp, M.J., and Switsur, V.R. (1992) Holocene history of the environment, vegetation and human settlement on Catta Ness, Lunnasting, Shetland. *Journal of Ecology* **80** (2), 241-273.

Bennetts, D.A. (1995) Modelling climate change, 1860-2050. *The Globe* **24**, 2-4.

Bennington, R. (1996) *Heaps of Success? - An Evaluation of the Wildlife Trusts' Guide to Community Composting.* The Wildlife Trusts, Lincoln.

Bennington, R. and Steel, C. (1994) *Growing Wiser - Case Studies on the successful use of Peat-Free Products.* The Wildlife Trusts, Lincoln.

Betts, N.L. (1982) Climate. *Northern Ireland Environment and Natural Resources.* Queen's University of Belfast and New University of Ulster pp. 9-42.

Beynon, T.G. (1995) *Leucorrhinia dubia* (Vander Linden) at Shooters Pool, Chartley Moss, Staffordshire, in 1994. *Journal of the British Dragonfly Society* **11** (1), 1-9.

Billamboz, A., Dieckmann, B., Maier, U. and Vogt, R. (1992) Exploitation du sol et de la forêt à Hornstaad-Hörnle I. In: Bodensee RFA *Archéologie et Environnement des Milieux Aquatiques. Actes du 116e Congrès Nationale des Sociétés Savantes. Commission de Pré- et Proto-histoire.* Editions du CTHS, Paris, pp. 119-148.

Biodiversity Challenge Group (1995) *Biodiversity Challenge: an agenda for conservation in the UK.* 2nd edn. Royal Society for the Protection of Birds, Sandy, UK.

Birks, H.H. (1975) Studies in the vegetation history of Scotland. IV. Pine stumps in Scottish blanket peats. *Philosophical Transactions of the Royal Society of London* **B270,** 181-226.

Birks, H.J.B. (1988) Long-term ecological change in the British uplands. In: Usher, M.B. and Thompson D.B.A. (eds) *Ecological Change in the Uplands.* Blackwell Scientific, Oxford, pp. 37-56,

Birks, H.J.B. (1989) Holocene isochrone maps and patterns of tree spreading in the British Isles. *Journal of Biogeography* **16**, 503-540.

Birks, H.J.B. (1993) Quaternary palaeoecology and vegetation science - current contributions and possible future developments. *Review of Palaeobotany and Palynology* **79**, 153-77.

Birks, H.J.B. (1994) Did Icelandic volcanic eruptions influence the post-glacial vegetational history of the British Isles? *Trends in Ecology and Evolution* **9**, 312-314.

Birse, E.L. and Robertson, J.S. (1976) *Plant Communities and Soils of the Lowland and Southern Upland Regions of Scotland.* Macauley Institute for Soil Research, Aberdeen.

Blackford, J.J. (1993) Peat bogs as sources of proxy climate data: past approaches and future research. In: Chambers F.M. (ed.) *Climate Change and Human Impact on the Landscape.* Chapman and Hall, London, pp. 47-56.

Blackford, J.J. and Chambers, F.M. (1991) Proxy records of climate from blanket mires: evidence for a Dark Age (1400BP) climatic deterioration in the British Isles. *The Holocene* **1**, 63-67.

Blackford, J.J. and Chambers, F.M. (1995) Proxy-climate record for the last 1,000 years from Irish blanket peat and a possible link to solar variability. *Earth and Planetary Science Letters* **130**, 145-150.

Blackford, J.J., Edwards, K.J., Dugmore, A.J., Cook, G.J. and Buckland, P.C. (1992) Icelandic volcanic ash and the mid-Holocene Scots pine (*Pinus sylvestris*) pollen decline in northern Scotland. *The Holocene* **2**, 260-265.

Blackman, J. (1993) Peat use as seen by ADAS. *The Grower* August 19th, 12-13.

Bliss, L.C., Heal, O.W. and Moore, J.J. (eds) (1981) *Tundra Ecosystems: a Comparative Analysis.* Cambridge University Press, Cambridge.

Blytt, A. (1882) Die Theorie der wechselnden kontinentaren und insularen Klimate. *Botanische fur Systematik, Pflanzengeschicte und Planzen-geographie* **2**, 1-50, 177-184.

Boelter, D.H. (1964) Water storage characteristics of several peats *in situ*. *Soil Science Society of America Proceedings.* **28**, 433-435.

Bohn, H.L. (1978) On organic soil carbon and carbon dioxide. *Tellus* **30**, 472-475.

Bostock, J.L. (1980) The history of the vegetation of the Berwyn Mountains, North Wales, with emphasis on the development of the blanket mire. PhD thesis (2 vols), University of Manchester.

Boswijk, G. (1996) Tree ring analysis and Thorne Moors. *Thorne and Hatfield Papers* (in press).

Botch, M.S. and Masing, V.V. (1983) Mire ecosystems in the U.S.S.R. In: Gore, A.J.P. (ed.) *Ecosystems of the World. Vol. 4B. Mires, swamp, bog, fen and moor. Regional studies.* Elsevier, Amsterdam, Oxford, New York, pp. 95-152.

Boutron, C.F. and Görlach, U. (1990) The occurrence of heavy metals in Antarctic and Greenland ancient ice and recent snow. In: Broekart, J.A.C., Gücer, S. and Adams, F. (eds) *Metal Speciation in the Environment.* Springer-Verlag, New York, pp. 137-151.

Bower, M.M. (1959) An investigation of erosion in blanket peat. MSc thesis, University of London.

Bower, M.M. (1961) The distribution of erosion in blanket peat bogs in the Pennines. *Transactions of the Institute of British Geographers* **29**, 17-30.

Bradshaw, R. (1993) Forest response to Holocene climatic change: equilibrium or non-equilibrium. In: Chambers, F.M. (ed.) *Climatic Change and Human Impact on the Landscape.* Chapman and Hall, London, pp. 57-66.

Bradshaw, R. and McGee, E. (1988) The extent and time course of mountain blanket peat erosion in Ireland. *New Phytologist* **108**, 219-224.

Bragg, O. (1988) The importance of water in mire ecosystems. In: Fojt, W. and Meade, R. (eds) *Cut-over lowland raised mires*, Research and Survey in Nature Conservation 24. Nature Conservancy Council, Peterborough, pp. 61-82.

Bragg, O. (1992) *Wedholme Flow, Cumbria: An Ecohydrological Study.* Report to the Nature Conservancy Council, North West England.

Bragg, O.M. and Brown, J.A.B. (1996) *A Handbook for Ground Water Mound Modelling.* Scottish Natural Heritage Research, Battleby (in press).

Bragg, O.M. (1995) Towards an ecohydrological basis for raised mire restoration. In: Wheeler, B.D., Shaw, S.C., Fojt, W.J. and Robertson, R.A. (eds) *Restoration of Temperate Wetlands.* Wiley, Chichester, pp. 305-314.

Bragg, O.M., Hulme, P.D., Ingram, H.A.P., Johnston, J.P. and Wilson, A.I.A. (1994) A maximum-minimum recorder for shallow water-tables, developed for ecological studies on mires. *Journal of Applied Ecology* **31**, 589-592.

Bragg, O., Moldaschl, E., Reiter, K. and Steiner, G.M. (1993). *Erste Expertise zum Schutz und Management des Puergschachenmooses und seiner näheren Umgebung im steirischen Ennstal, Gemeinde Ardning, Bezirk Liezen.* Unpublished Report to WWF.

Bragg, O. and Steiner, G.M. (1996). Applying groundwater mound theory to bog management on Puergschachenmoos in Austria. In: *Proceedings. of the 6th IMCG conference in Norway 1994* (in press).

Braun-Blanquet, J. and Tüxen, R. (1943) *Übersicht der Höheren Vegetationseinheiten Mitteleuropas.* Communication SIGMA **84**, 11 pp.

Braun-Blanquet, J. and Tüxen, R. (1952) Irische Pflanzengesellschaften. *Veröffentlichungen Geoboanischen Institutes Rübel in Zürich.* **25,** 222-421.

Brecht, B. (1928) Die Dreigroschenoper. In: *Die Stücke von Bertolt Brecht in einem Band.* Suhrkamp, Frankfurt, 1978.

Bridge, M.C., Haggart, B.A. and Lowe, J.J. (1990) The history and palaeoclimatic significance of subfossil remains of *Pinus Sylvestris* in blanket peats from Scotland *Journal of Ecology* **78**, 77-99.

Briffa, K.R. (1994) Mid and Late Holocene climate change: evidence from tree growth in northern Fennoscandia. In: Funnell B.M. and Kay R.L.F. (eds) *Palaeoclimate of the Last Glacial/Interglacial Cycle.* NERC, Swindon, pp. 61-65.

Broecker, W.S. (1992) *The Glacial World According to Wally (draft), III. Records.* Columbia University, Palisades.

Broggi, M.F. (1990) *Inventaire des Bas-marais d'Importance Nationale. Projet Mis en Consultation.* Berne, DFI (Département Fédéral de l'Intérieur) 75 pp.

Brooks, S.J. (1994) How much does acidity affect the distribution of acidophilic dragonflies? *Journal of the British Dragonfly Society* **10** (1), 16-18.

Brooks, S.J. (1995) Langlands Moss monitoring proposals In: Anon. (1994) *Monitoring Raised Bogs: Workshop Report.* Unpublished Report, English Nature and Scottish Wildlife Trust, pp. 59-66

Brooks, S.J., Parkyn, E.L., Stoneman, R.E., and Wilson, E.V. (1994) *Peatland Conservation Management in Switzerland, Germany and the Netherlands.* Unpublished Report. Scottish Wildlife Trust, Cramond House, Edinburgh.

Brooks, S.J., Parkyn, E.L., Stoneman, R.E., and Wilson, E.V. (1996) *Conserving Bogs: A Management Handbook.* Scottish Natural Heritage, Battleby (in press).

Brown, A. (1985) Review of lignin in biomass. *Journal of Applied Biochemistry* **7,** 371-387.

Brown, A., Mathur, S.P. and Kushner., D.J. (1989) An ombrotrophic bog as a methane reservoir. *Global Biogeochemical Cycles* **3,** 205-213.

Brown, A.F. and Shepherd, K.B. (1993) A method for censusing upland breeding waders. *Bird Study* **40**, 189-195.

Brown, D.A. (1995) Carbon cycling in peat and the implications for rehabilitation of bogs. In: Cox, M., Starker, V. and Taylor, D. (eds) *Wetlands: Archaeology and Nature.*HMSO, Norwich. pp. 116-124.

Brown, D.A. and Overend, R.P. (1993) Methane metabolism in raised bogs of northern wetlands. *Geomicrobiology Journal* **11,** 35-48.

Brown, D.M. (1991) Studies on *Pinus sylvestris* L. from Garry Bog, County Antrim. MSc thesis, Queen's University, Belfast.

Brown, L.T. (1968) A survey of turf-working in County Down. Unpublished MSc Thesis, Queen's University, Belfast.

Buckland, P.C. (1979) *Thorne Moor: a palaeoecological study of a Bronze Age Site.* University of Birmingham, Department of Geography Occasional Paper 8.

Buckland, P.C. (1993) Peatland archaeology: a conservation resource on the edge of extinction. *Biodiversity and Conservation* **2**, 513-27.

Buckland, P.C. (1995) Peat stratigraphy and the age of the Lindow bodies. In: Turner, R.C. and Scaife, R.G. (eds) *Bog Bodies. New Discoveries and New Perspectives.* British Museum Press, London, pp. 47-50.

Buckland, P.C. and Coope, G.R. (1991) *A Bibliography and Literature Review of Quaternary Entomology.* J.R. Collis Publications, Department of Archaeology and Prehistory, University of Sheffield.

Buckland, P.C. and Dinnin, M.H. (1993) Holocene woodlands: the fossil insect evidence. In: Kirby, K. and Drake, C.M. (eds) *Dead Wood Matters: The Ecology and Conservation of Saproxylic Invertebrates in Britain* English Nature Science 7, English Nature. Peterborough, pp. 6-20.

Buckland, P.C. and Dinnin, M.H. (1994) Peatlands and floodplains: the loss of a major palaeontological resource. In: Stevens, C., Gordon, J.E., Green, C.P. and Macklin, M.G. (eds) *Conserving our Landscape (Proceedings of the Conference: Conserving our Landscape: Evolving Landforms and Ice-Age Heritage, Crewe, UK, May 1992).* English Nature, Peterborough, pp. 145-149.

Buckland, P.C. and Johnson, C. (1983) *Curimopsis nigrita* (Palm) (Coleoptera: Byrrhidae) from Thorne Moors, South Yorkshire. *Naturalist (Hull)* **108**, 153-154.

Buckland, P.C. and Kenward, H.K. (1973) Thorne Moors: a Palaeoecological Study of a Bronze Age Site. *Nature* **241**, 405-408.

Buckland, P.C., Eversham, B.C. and Dinnin, M.H. (1994) Conserving the Holocene record: a challenge for geomorphology, archaeology and biological conservation. In: O'Halloran, D., Green, C., Harley, M., Stanley, M. and Knill, J. (eds) *Geological and Landscape Conservation.* Geological Society, London, pp. 201-204.

Buckley, P. (1996) Critical natural capital: operational flaws in a valid concept. *Ecos* **16**, 13-18

Burke, W. (1969) Drainage of blanket peat at Glenamoy. *Second International Peat Congress, Leningrad 1963*, 809-817.

Burley, T.M. (1961) Land use or land utilisation? *Professional Geographer* **13** (6), 18-20.

Butcher, B. (1995) *Wetland Information and Evaluation Systems.* Somerset Environmental Records Centre

Butterfly Conservation, FoE, Plantlife, RSPB, The Wildlife Trusts, WWF (1995) *Biodiversity Challenge,* 2nd edn. RSPB, Sandy.

Caithness and Sutherland HRC Working Party (1989) *Caithness and Sutherland HRC Working Party Summary Report and Land Use Strategy.* Highland Regional Council, Inverness.

Callicott, J.B. (1980) Animal liberation: a triangular affair. *Environmental Ethics* **2**, 311-338.

Callicott, J.B. (1985) On the intrinsic value of nonhuman species. The preservation of species. In: Norton, B.G. (ed.) *The Value of Biological Diversity.* Princeton University Press, pp. 138-172.

Callicott, J.B. (1986) Moral considerability and extraterrestrial life. In: Hargrove, E.C. (ed.) *Beyond Spaceship Earth: Environmental Ethics and the Solar System*. Sierra Club Books, San Francisco, pp. 238-259.

Callicott, J.B. (1988). Animal liberation and environmental ethics: back together again. *Between the Species* **5**, 163-169.

Campbell, E.O. (1983) Mires of Australia. In: Gore, A.J.P. (ed.) *Ecosystems of the World. Vol. 4B. Mires, swamp, bog, fen and moor. Regional studies.* Elsevier, Amsterdam and Oxford, pp. 153-180.

Campbell, M. (1965) Goat-keeping in the Old Highland economy. *Scottish Studies* **9**, 182-186.

Campbell, S. and Bowen, D.Q. (1989) *Geological Conservation Review: Quaternary of Wales*. Nature Conservancy Council, Peterborough.

Campeau, S and Rochefort, L. (1996) *Sphagnum* regeneration on bare peat surfaces: Field and greenhouse experiments. *Journal of Applied Ecology* (in press).

Cannell, M.G.R., Dewar, R.C. and Pyatt, D.G. (1993) Conifer plantations of drained peatlands in Britain: a net gain or loss of carbon? *Forestry* **66**, 353-369.

Casparie, W. (1993) The Bourtanger Moor: endurance and vulnerability of a raised bog system. *Hydrobiologia* **265**, 203-215.

Casparie, W.A. (1972) *Bog Development in South-eastern Drench*. Junk, The Hague.

Caulfield, S. (1978) Neolithic fields: the Irish evidence. In: Bowen, H.C. and Fowler, P.J. (eds) *Early Land Allotment in the British Isles. A survey of recent work.* BAR British Series, 48, Oxford, pp. 137-43.

Caulfield, S. (1983) The Neolithic settlement of N. Connaught. In: Reeves-Smith, T. and Hamond, F. (eds) *Landscape Archaeology in Ireland.* BAR British Series, 116, Oxford, pp. 195-215.

Cavalieri, P. and Singer, P. (eds) (1993). *The Great Ape Project. Equality beyond humanity*. Fourth Estate, London.

Central Statistical Office (CSO) (1995) *Annual Abstract of Statistics.* HMSO, London.

Chambers, F.M. (1994) Peatlands - evolution, heritage and conservation. In: Stevens, C., Gordon, J.E., Green, C.P. and Macklin, M.G. (eds) *Conserving our Landscape (Proceedings of the Conference: Conserving our Landscape: Evolving Landforms and Ice-Age Heritage, Crewe, UK, May 1992).* English Nature, Peterborough, pp. 163-167.

Chambers, F.M., Grant, M.E. Lageard, J.G.A., Roberts, L.J. and Thomas, P.A. (1996) The Palaeoenvironmental Record. In: Berry, A., Gale, F. and Allmark B. (eds) *Fenn's and Whixall Moss.* Clwyd Archaeological Service, Mold, (in press).

Chambers, W. (1864) *The History of Peebleshire*. Chambers, Edinburgh.

Chapman, S.B. and Rose, R.J. (1991) Changes in the vegetation at Coom Rigg Moss National Nature Reserve within the period 1958-86. *Journal of Applied Ecology* **28**, 140-153.

Charman, D.J. (1994a) Holocene ombrotrophic peats: conservation and value in Quaternary research. In: Stevens, C., Gordon, J.E., Green, C.P. and Macklin, M.G. (eds) *Conserving our Landscape (Proceedings of the Conference: Conserving our Landscape: Evolving Landforms and Ice-Age Heritage, Crewe, UK, May 1992).* English Nature, Peterborough. pp. 158-162.

Charman, D.J. (1994b) Geological conservation of Holocene peatlands in the National Parks of England and Wales. In: O'Halloran, D., Green, C., Harley, M., Stanley, M. and Knill, J. (eds) *Geological and Landscape Conservation.* Geological Society, London. pp. 197-200.

Charman, D.J. (1994c) Patterned fen development in northern Scotland: developing a hypothesis from palaeoecological data. *Journal of Quaternary Science* **9**, 285-297.

Charman, D.J. (1995) Patterned fen development in northern Scotland: hypothesis testing and comparison with ombrotrophic blanket mire. *Journal of Quaternary Science* **10**, 327-342.

Charman, D.J. and Smith, R.S. (1992) Forestry and blanket mires of Kielder Forest, Northern England: long term effects on vegetation. In: Bragg, O.M., Hulme, P.D., Robertson, R.A. (eds) *Peatland Ecosystems and Man: An Impact Assessment*. University of Dundee.

Chau, S.S.C. and Kam-Kong, F. (1990). Ancient wisdom and sustainable development from a Chinese perspective. In: Engel, J.R. and Engel, J.G. (eds) *Ethics of Environment and Development. Global Challenge and International Response*. Belhaven Press, London, pp. 222-231.

Chester, D. (1993) *Volcanoes and Society*. Edward Arnold, London.

Childs, E.C. (1969) *An Introduction to the Physical Basis of Soil Water Phenomena.* Wiley, London.

Clarke, D.J. (1994) Notes on the larvae and generation time of *Aeshna caerulea* (Ström) in Scotland, with particular reference to the south-west. *Journal of the British Dragonfly Society* **10** (2), 29-36.

Clarke, D.J., Hewitt, S.M., Smith, E.M. and Smith, R.W.J. (1990). Observations on the breeding habits and habitat of *Aeshna caerulea* (Ström) in Scotland. *Journal of the British Dragonfly Society* **6** (2), 24-29.

Clarke, J.L., Welch, D. and Gordon, I.J. (1995) The influence of vegetation pattern on the grazing of heather moorland by red deer and sheep. I. The location of animals on grass/heather mosaics. *Journal of Applied Ecology* **32**, 166-176.

Clymo, R.S. (1963) Ion exchange in *Sphagnum* and its relation to bog ecology. *Annals of Botany, London* NS **27**, 309-324.

Clymo, R.S. (1967) Control of cation concentrations, and in particular of pH in *Sphagnum*-dominated communities. In: Golterman H.L. and Clymo, R.S.

(eds) *Chemical Environment in Aquatic Habitat.* North Holland, Amsterdam, pp. 273-284

Clymo, R.S. (1973) The growth of *Sphagnum*: some effects of environment. *Journal of Ecology* **61**, 849-869.

Clymo, R.S. (1983) Peat. In: Gore, A.J.P. (ed.) *Mires: Swamp, Bog, Fen and Moor, Ecosystems of the World, Vol. 4A.* Elsevier Scientific Publishing Company, pp. 159-224.

Clymo, R.S. (1984a) The limits to peat bog growth. *Philosophical Transactions of the Royal Society of London B* **303**, 605-654.

Clymo, R.S. (1984b) *Sphagnum*-dominated peat bog: a naturally acid ecosystem. *Philosophical Transactions of the Royal Society of London B* **305**, 487-499.

Clymo, R.S. (1987) Interactions of *Sphagnum* with water and air. In: Hutchinson, T.C. and Meema, K.M. (eds) *Effects of Atmospheric Pollutants on Forests, Wetlands and Agricultural Ecosystems.* Springer-Verlag, Berlin, pp. 513-529.

Clymo, R.S. (1990) The record of atmospheric deposition on a rainwater-dependent peatland. *Philosophical Transactions of the Royal Society of London B* **327**, 331-338.

Clymo, R.S. (1991) Peat growth. In: Shane, L.C.K. and Cushing, E.J. (eds) *Quaternary Landscapes.* University of Minnesota Press, Minneapolis, pp. 76-112.

Clymo, R.S. (1992) Models of peat growth. *Suo* **43**, 127-136.

Clymo, R.S. and Duckett, J.G. (1986) Regeneration of *Sphagnum. New Phytologist* **102**, 589-614.

Clymo, R.S. and Hayward, P.M. (1982) The ecology of *Sphagnum.* In: Smith, A.J.E. (ed.) *Bryophyte Ecology.* Chapman and Hall, London, pp. 229-289.

Clymo, R.S. and Lewis Smith, R.I. (1984) An extraordinary peat-forming community on the Falkland Islands *Nature* **309**, 617-620.

Clymo, R.S. and Reddaway, E.J.F. (1972) Productivity of *Sphagnum* (bog-moss) and peat accumulation. *Hidrobiologia (Bucharest)* **12**, 181-192.

Coles, B. (1995a) *Wetland Management. A Survey for English Heritage.* WARP Occasional Paper 9, Exeter.

Coles, B. (1995b) Archaeology and wetland restoration. In: Wheeler, B.D., Shaw, S.C., Fojt, W.J. and Robertson, R.A. (eds) *Restoration of Temperate Wetlands.* John Wiley, Chichester, pp. 19-32.

Coles, B. and Coles, J. (1989) *People of the Wetlands.* Thames and Hudson, London.

Coles, B. and Coles, J. (1996) *Enlarging the Past.* Society of Antiquaries of Scotland Monograph Series No. 11 and WARP Occasional Paper No. 10., Edinburgh.

Colquhoun, I.R. (1971) The grazing ecology of red deer and blackface sheep in Perthshire, Scotland. PhD thesis, University of Edinburgh.

Commoner, B. (1971) *The Closing Circle: Nature, Man, and Technology.* Knopf, New York.

Condry, W.M. (1981) *The Natural History of Wales*. Collins, London.

Conrad, R. (1989) Control of methane production in terrestrial ecosystems. In: Andreae, M.O. and Schimel, D.S. (eds) *Exchange of Trace Gases between Terrestrial Ecosystems and the Atmosphere* John Wiley, Konferensen, pp. 39-58.

Consumers Association (1992) Mulch trials. *Gardening from 'Which?'* April 1992. Consumers Association, London.

Consumers Association (1995) Compost Directory. *Gardening from 'Which?'* January/February 1995. Consumers Association, London.

Conway, V.M. (1949) Ringinglow Bog, Sheffield. II. The present surface. *Journal of Ecology* **37**, 148-170.

Conway, V.M. (1954) Stratigraphy and pollen analysis of southern Pennine blanket peats. *Journal of Ecology*. **42,** 117-147.

Corbet, P.S. (1957) The life-histories of two summer species of dragonfly (Odonata: Zygoptera). *Proceedings of the Zoological Society of London* **128**, 403-418.

Corbet, P.S. (1962) *A Biology of Dragonflies*. Witherby, London.

Council for Nature Conservation and the Countryside (CNCC) (1992) *Peatland Conservation Strategy*. Belfast.

Council of the European Commission (CEC) (1989) Council Regulation (EEC) No 3906/89 of 18 December 1989 on economic aid to the Republic of Hungary and the Polish Peoples Republic. *Official Journal of the European Communities* **L375**, 11.

Council of the European Commission (CEC) (1991a) Council Regulation (EEC, EURATOM) No 2157/91 of 15 July 1991 concerning the provision of technical assistance to economic reform and recovery in the Union of Soviet Socialistic Republics. *Official Journal of the European Communities* **L 201**, 2-4.

Council of the European Commission (CEC) (1991b) *CORINE Biotopes Manual*. Habitats of the European Community. Data Specifications part 2: 300.

Council of the European Commission (CEC) (1992) Council Regulation (EEC) No 1973/92 of 21 May 1992 establishing a financial instrument for the Environment (LIFE). *Official Journal of the European Communities* **L 206**, 1-6.

Council of the European Commission (CEC) (1993) *Community Structural Funds 1994-1999. Revised regulations and comments*. Luxembourg, Office for official publications of the European Communities.

Council of the European Commission (CEC) (1995) Wise use and conservation of wetlands. *Communication from the Commission to the Council and the European Parliament*. Com (95) 189 final. Brussels 29.05.95

Coward, H. (1993). Religious responsibility. In: Coward, H. and Hurka, T. (eds) *The Greenhouse Effect. Ethics and Climate Change*. Wilfrid Laurier University Press, Waterloo, pp. 39-60.

Cramp, S. and Simmons, K.E.L. (eds) (1983) *The Birds of the Western Palearctic*. Volume III. Oxford University Press, Oxford.

Crocker, D.A. (1990). The hope for just, participatory ecodevelopment in Costa Rica. In: Engel, J.R. and Engel, J.G. (eds) *Ethics of Environment and Development. Global Challenge and International Response* Belhaven Press, London, pp. 150-163.

Crofts, A. (1992) *Guide to Community Composting*. RSNC, The Wildlife Trusts Partnership, Lincoln.

Cronberg, N. (1991) Reproductive biology of *Sphagnum. Lindbergia* **17**, 69-82.

Crozier, J.D. (1910) The Sitka spruce as a tree for hill planting. *Transactions of the Royal Scottish Arboricultural Society* **23**, 7-16.

Cruickshank, M.M. and Tomlinson, R.W (1988) *Northern Ireland Peatland Survey*. Report to Countryside and Wildlife Branch, Department of the Environment (NI).

Cummins, R.D. and Miller, G.R. (1982) Damage by red deer (*Cervus elaphus*) enclosed in planted woodland. *Scottish Forestry* **36**, 1-8.

Cunliffe, B. (1984) *Danebury. An Iron Age hillfort in Hampshire. Vol.I. The excavations 1969 - 1978: the site.* CBA Research Report No.52. Council for British Archaeology, London.

Dalbavie, T. and Solleliet, J.P. (1994) *Les Tourbieres au Pays des Plantes Carnivores.* Espaces et Recherches, France, 24 pp.

Damman, A.W.H. (1978) Distribution and movement of elements in ombrotrophic peat bogs. *Oikos* **30**, 480-495.

Damman, A.W.H. (1986) Hydrology, development and biogeochemistry of ombrogenous peat bogs with special reference to nutrient relocation in a western Newfoundland bog. *Canadian Journal of Botany*. **64**, 384-394.

Daniels, R.E. and Eddy, A. (1990) *Handbook of European Sphagna*. Natural Environment Research Council, HSMO, London.

Danielson, P. (1993) Personal responsibility. In: Coward, H. and Hurka, T. (eds) *The Greenhouse Effect. Ethics and Climate Change* Wilfrid Laurier University Press, Waterloo. pp. 81-98.

Dargie, T.C.D. and Tantrum, D.A.S. (1995) *Thorne Moors: Monitoring the Distribution of Plant Species Indicative of Successful Rewetting*. Report to English Nature, Wakefield.

Darling, F.F. (1945) *Crofting Agriculture: its Practice in the West Highlands and Islands.* Oliver and Boyd, Edinburgh.

Davis, F.W., Stoms, D.M., Estes, J.E., Scepan, J., Scott, J.M. (1990) An information systems approach to the preservation of biological diversity. *International Journal of Geographic Information Systems* **4**, 55-78.

Davis, M.B. (1994) Ecology and palaeoecology begin to merge. *Trends in Ecology and Evolution* **9**, 357-358.

De Groot, R.S. (1992) *Functions of Nature. Evaluation of nature in environmental planning, management and decision making.* Wolters-Noordhoff, Groningen.

Decker, R. and Decker, B. (1989) *Volcanoes*. Freeman, New York.

Department of the Environment (DoE)(1990) Archaeology and Planning. *Planning Policy Guidance Note 16*. HMSO, London..

Department of the Environment (DoE)(1994a) *Peat-based and Alternative Products in the Gardening and Landscape Markets*. Department of the Environment, London.

Department of the Environment (DoE)(1994b) *Report of the Working Group on Peat Extraction and Related Matters*. Department of the Environment, London.

Department of the Environment (DoE)(1994c) *Peat Extraction and Related Matters - Report of the Working Group on Peat Extraction*. Department of the Environment, London.

Department of the Environment (DoE)(1994d) *Climate Change: The UK Programme*. HMSO, London.

Department of the Environment (DoE)(1994e) Nature Conservation. *Planning Policy Guidance Note 9*. HMSO, London.

Department of the Environment (DoE)(1995a) *Biodiversity: The UK Action Plan*. HMSO, London.

Department of the Environment (DoE)(1995b) *Sustainable Development: The UK Strategy*. HMSO, London.

Department of the Environment (DoE)(1995c) *Sustainable Forestry. The UK Programme*. HMSO, London.

Department of the Environment (DoE)(1995d) *MPG13. Mineral Planning Guidance: Guidelines for Peat Provision in England, including the Place of Alternative Materials*. Department of the Environment, London.

Department of the Environment, Northern Ireland (1993) *Conserving Peatland in Northern Ireland - A Statement of Policy*. Department of the Environment, Northern Ireland.

Département Fédéral de l'Intérieur (DFI) (1990) *Inventaire des Bas Marais d'Importance Nationale - Projet Mis en Consultation*. Berne, OFEFP (Office Fédéral de l'Environnement, des Forêts et du Paysage). 24 volumes.

Département Fédéral de l'Intérieur (DFI) (1991) *Inventaire Fédéral des Hauts-marais et des Marais de Transition d'Importance Nationale*. Berne, OCFIM (Office Central Fédéral des Imprimés et du Matériel). 23 volumes.

Diamond, J. (1991). *The Rise and Fall of the Third Chimpanzee*. Vintage edition 1992, Random House, London.

Diemont, W.H. and Supardi (1987) Accumulation of organic matter and inorganic constituents in a peat dome in Sumatra, Indonesia. *Proceedings of the International Symposium on Peat and Peatlands for Development*, Yogyakarta, Indonesia (unpublished).

Dierssen, K. (1982) *Die wichtigsten Pflanzengesellschaften der Moore N.W.-Europas*. Conservatoire et Jardin Botaniques Geneve, pp. 382.

Dinel, H.S., Mathur, S.P., Brown, A. and Lévesque, M. (1988) A field study of the effect of depth on methane production in peatland waters: equipment and preliminary results. *Journal of Ecology* **76,** 1083-1091.

Dizard, J.E. (1994). *Going Wild. Hunting, Animal Rights and the Contested Meaning of Nature*. University of Massachusetts Press, Amherst.

Doar, N. (1995) *Peatlands Campaigning Spring Promotion of Peat-free Gardening Products: Joint Project Between B& Q and Scottish Wildlife Trust, Easter weekend (14 + 17 April) 1995. Market Research Project Report.* Unpublished Report. Scottish Wildlife Trust, Cramond House, Edinburgh.

Donkin, R.A. (1962) Cattle on the estates of medieval Cistercian monasteries in England and Wales. *Economic History Review, 2nd Series* **15**, 31-53.

Doyle, G.J. (1982) The vegetation, ecology and productivity of Atlantic blanket bog in Mayo and Galway, western Ireland. *Journal of Life Sciences of the Royal Dublin Society* **3**, 147-164.

Doyle, G.J. (1990a) Bog Conservation in Ireland. In: Schouten M.G.C. and Nooren M.J. (eds) *Peatlands, Economy and Conservation* SPB Academic Publishing, The Netherlands, pp. 45-58.

Doyle, G.J. (1990b) Phytosociology of Atlantic blanket bog complexes in north-west Mayo. In: Doyle, G.J. (ed.) *Ecology and Conservation of Irish Peatlands.* Royal Irish Academy, Dublin, pp. 75-90.

Doyle, G.J. and Foss, P.J. (1984) *Vaccinium oxycoccus* L. growing in the blanket bog area of west Mayo (H27). *Irish Naturalists' Journal* **22** 101-104.

Doyle, G.J. and Moore, J.J. (1980) Western blanket bog (Pleurozio purpureae-Ericetum tetralicis) in Ireland and Great Britain. *Colloque Phyto-sociologique* **7** 213-223.

Doyle, G.J., McNally, A. and O'Connell, C.A. (1989) Subfossil pine woodlands on Irish raised bogs. In: Spigarelli, S.A. (ed.) *Proceedings of the International Symposium on Peat/Peatland Characteristics and Uses.* Centre for Environmental Studies, Bemidji State University, Bemidji, Minnesota, pp. 85-95.

Doyle, G.J., O'Connell, C.A. and Foss, P.J. (1987) The vegetation of peat islands in bog lakes in County Mayo, western Ireland. *Glasra* **10,** 23-35.

Driessen, P.M. (1977) Peat soils. In: *Soils and Rice.* IRRI, Los Banos, Philippines, pp. 763-779

Dubois, A.D. and Ferguson, D.K. (1985) The climatic history of pine in the Cairngorms based on radiocarbon dates and stable isotope analysis, with an account of events leading up to its colonisation. *Review of Palaeobotany and Palynology* **46**, 55-80.

Dugan, P.J. (ed.)(1990) *Wetland Conservation: a Review of Current Issues and Required Action*, IUCN, Gland, Switzerland, 96 pp.

Dugmore, A.J. (1989) Icelandic volcanic ash in Scotland. *Scottish Geographical Magazine* **105**, 168-172.

Dugmore, A.J. (1991) Tephrochronology and UK archaeology. In: Budd, P., Chapman B., Jackson, C., Janaway, R. and Ottaway, B. (eds) *Archaeological Sciences 1989*. Oxbow Books, Oxford, pp. 242-250.

Dugmore, A.J. and Newton, A.J. (1992) Thin tephra layers in peat revealed by X-Radiography. *Journal of Archaeological Science* **19**, 163-171.

Dugmore, A.J., Larsen, G. and Newton, A.J. (1995) Seven tephra isochrones in Scotland. *The Holocene* **5**, 257-266.

Dugmore, A.J., Larsen, G., Newton, A.J., and Sugden, D.E. (1992) Geochemical stability of fine-grained silicic Holocene tephra in Iceland and Scotland. *Journal of Quaternary Science* **7**, 173-183.

Dundee District Council (1994) *Discovery Compost*. Unpublished Report. City of Dundee District Council, Cleansing Department, Dundee.

During, R. and Joosten, J.H.J. (1992). Referentiebeelden en duurzaamheid: tijd voor beleid. *Landschap* **9**, 285-295.

Dutton, E.J. and Christy J.R. (1992) Solar radiative forcing at selected locations and evidence for global lower tropospheric cooling following the eruptions of El Chich∴n and Pinatubo. *Geophysical Research Letters* **19**, 2313-2316.

Edwards, K.J., Buckland, P.C., Blackford, J.J., Dugmore, A.J. and Sadler, J.P. (1994) The impact of tephra: proximal and distal studies of Icelandic eruptions. *Münchener Geographische Abhandlungen* **B12**, 79-99.

Edwards, K.J., Dugmore, A.J., Buckland, P.C., Blackford, J.J. and Cook, G.T. (1996) Hekla-4 ash, the pine decline in Northern Ireland and the effective use of tephra isochrones: a comment on Hall, Pilcher and McCormac. *The Holocene*, (in press).

Eggelsmann, R. (1987) Hochmoor-Regeneration verlangt nahezue horizontale Mooroberfläche. *Natur und Landschaft* **62**, 241-246.

Eggelsmann, R. (1988) Ökotechnische Aspekte der Hochmoor-Regeneration. *Telma* **17**, 59-74.

Egloff, M. (1989) *Des Premiers Chasseurs au Début du Christianisme*. Editions Gilles Attinger B Hauterive/Suisse.

Ehrlich, P.R. and Ehrlich, A. (1981) *Extinction: the Causes and Consequences of the Disappearance of Species*. Random House, New York.

Eigner, J. and Schmatzler, E. (1991) *Handbuch des Hochmoorschutzes - Bedeutung, Pflege und Entwicklung*. Kilda, Greven.

Einarsson, T. (1986) Tephrochronology. In: Berglund, B.E. (ed.) *Handbook of Holocene Palaeoecology and Palaeohydrology*. John Wiley, Chichester, pp. 329-342.

Ellmauer, Th. and Steiner, G.M. (1992) Vegetationsökologische Untersuchungen an einem Kondenswassermoor in Tragöss (Steiermark). *Berichte Naturwissenschaftlich-Medizinischen Vereins in Innsbruck* **79**, 37-47.

Engel, J.R. and Engel, J.G. (1990). *Ethics of Environment and Development. Global Challenge and International Response*. Belhaven Press, London.

English Nature and Scottish Wildlife Trust (1995) *Monitoring Raised Bogs, Workshop Report.* English Nature, Peterborough.

Esterle, J.S., Calvert, G., Durig, D, Tie, Y.L. and Supardi (1992) Characterization and classification of tropical wood peats from Baram River, Sarawak and Jambi, Sumatra. *Proceedings of the International Symposium on Tropical Peatland.* Kuching, Sarawak, Malaysia, pp. 33-48.

Eurola, S. (1962) Über die regionale Einteilung der südfinnischen Moore. *Annales Botanicae Societatis Fennicae Vanamo.* **33** (2), 243.

European Commission (EC) (1994) *Manual for the Interpretation of Annex 1 Priority Habitat Types of the Directive 92/43/EEC.* European Commission, Directorate-General Environment, Nuclear Safety and Civil Protection. Nature Protection and Soil Conservation (Habitats 94/3 FINAL), Brussels.

Eversham B.C. (1991) Thorne and Hatfield Moors: Implications of Land use Change for Nature Conservation. In: Bain, C.G. and Eversham, B.C. (eds) *Thorne and Hatfield Moors Papers*, Vol. 2. Thorne and Hatfield Moors Conservation Forum, Doncaster, pp. 3-18.

Fenton, A. (1970) Paring and burning. In: Fenton, A. and Gailey, A. (eds) *The Spade in Northern and Atlantic Europe.* Fallon, Belfast, pp. 155-193.

Fenton, A. and Walker, B. (1981) *The Rural Architecture of Scotland.* John Donald, Edinburgh.

Fenton, J. (1985) Regeneration of native pine in Glen Affric. *Scottish Forestry* **39**, 104-116.

Ferguson, P. and Lee, J.A. (1983) Past and present sulphur pollution in the southern Pennines. *Atmospheric Environment* **17**, 1131-1137.

Fojt, W., (1992) *Rehabilitation of Cut-over Mires in Holland and Germany* English Nature Foreign Travel Report, English Nature, Peterborough.

Foody, G.M. and Cox, D.P. (1991) Estimation of sub-pixel land cover composition from spectral mixture models. *Proceedings of a Symposium Spatial Data 2000 Conference, Christ Church, Oxford University, September 1991.*

Forestry Authority (1995) The management of semi-natural woodlands - wet woodlands. *Forestry Practice Guide 8.* The Forestry Authority, Edinburgh

Forestry Commission (1990) *Forest Nature Conservation Guidelines.* HMSO, London.

Forestry Commission (1993) *Forests and Water Guidelines, 3rd edn.* HMSO, London.

Forestry Commission (1994) *Forestry Commission Facts and Figures 1993-94.* Forestry Commission, Edinburgh.

Forestry Commission (1995) *Forests and Archaeology Guidelines.* HMSO, London.

Foss, P.F. and O'Connell, C. (eds) (1992) *Our Story: 10 Years of the Save the Bogs Campaign.* Unpubl. report. Irish Peatland Conservation Council, Dublin, 47pp.

Foss, P.J. (ed.) (1991) *Irish Peatlands, the Critical Decade.* Irish Peatland Conservation Council, Dublin.

Foss, P.J., Rushe, E.M. and Doyle, G.J. (1984) *Andromeda polifolia* L. in West Mayo (H27). *Irish Naturalists' Journal* **21**, 343.

Foster, D.R. and Wright, H.E. Jr (1990) Role of ecosystem development and climate in bog formation in central Sweden. *Ecology* **71**, 450-463.

Foster, G.N. (1995) Evidence for pH insensitivity in some insects inhabiting peat pools in the Loch Fleet catchment. *Chemistry and Ecology* **9**, 207-215.

Fox, S. (1981) *John Muir and his Legacy.* The American Conservation Movement. Little, Brown, Boston.

Francis, I.S. (1990) Blanket peat erosion in a mid-Wales catchment during two drought years. *Earth Surface Processes and Landforms* **15**, 445-456.

Francis, P. (1993) *Volcanoes, a Planetary Perspective.* Oxford University Press, Oxford.

Freeman, C., Lock, M.A. and Reynolds, B. (1993) Fluxes of CO_2, CH_4 and N_2O from a Welsh peatland following simulation of water table draw-down: potential feedback to climatic change. *Biogeochemistry* **19**, 51-60.

Friis-Christensen, E. and Lassen, K. (1991) Length of the solar cycle: an indicator of solar activity closely associated with climate. *Science* **254**, 698-700.

Fritts, A.C. and Swetnam, T.W. (1989) Dendroecology: a tool for evaluating variations in past and present forest environments. *Advances in Ecological Research* **19**, 111-118.

Fullen, M.A. (1983) Some changes in air temperature and wind velocity after burning of heather *(Calluna vulgaris)* moor and their relations to moorland surface processes. *Naturalist Hull* **108**, 19-24.

Fuller, R.J., Reed, T.M., Buxton, N.E., Webb, A., Williams, T.D. and Pienkowski, M.W. (1986) Populations of breeding waders Charadrii and their habitats on the crofting lands of the Outer Hebrides, Scotland. *Biological Conservation* **37**, 353-361.

Fuller, R.M., Groom, G.B. and Jones, A.R. (1994) The land cover map of Great Britain: An automated classification of landsat thematic mapper data. *PERS* **60** (5), 553-562.

Gadjewski, K. (1993) The role of palaeoecology in the study of global climatic change. *Review of Palaeobotany and Palynology* **79**, 141-151.

Gailey, F.A.Y. and Lloyd, O.L. (1993) Spatial and temporal patterns of airborne metal pollution: the value of low technology sampling to an environmental epidemiology study. *The Science of the Total Environment* **133**, 201-219.

Galloway, J.N., Thornton, J.D., Norton, S.A., Volchok, H.L. and McLean, R.A.N. (1982) Trace metals in atmospheric deposition: a review and assessment. *Atmospheric Environment* **16**, 1677-1700.

Gams, H. and Ruoff, S. (1929) Geschichte, Aufbau und Pflanzendecke des Zehlaubruches. *Schr.Phys.Ök.Ges.Königsberg/Pr.* **46** (1), 192.

Gardner, A.E. (1953). The life-history of *Leucorrhinia dubia* (Vander Linden) (Odonata). *Entomologists' Gazette* **4**, 175-201.

Gash, J.H.C., Wright, I.R. and Lloyd, C.R. (1980) Comparative estimates of interception loss from three coniferous forests in Great Britain. *Journal of Hydrology* **48**, 89-105.

Gear, A.J. and Huntley, B. (1991) Rapid changes in the range limits of Scots pine 4,000 years ago. *Science* **251**, 544-547.

Gibbons, D.W., Reid, J.B. and Chapman, R.A. (comps.) (1993) *The New Atlas of Breeding Birds in Britain and Ireland: 1988-1991.* Poyser, London.

Gimingham, C.H. (1960) Biological Flora of the British Isles: *Calluna vulgaris* (L.) Hull. *Journal of Ecology* **48**, 455-483.

Gipp, W. (1986) Geschichte der Moor- und Torfnutzung in Bayern. *Telma* **16**, 305-317.

Girling, M. (1984) A Little Ice Age extinction of a water beetle from Britain *Boreas* **13**, 1-4.

Girling, M. A. (1985) An Old-Forest Beetle Fauna from a Neolithic and Bronze Age Peat Deposit at Stileway. *Somerset Levels Papers* **11**, 80-83.

Godwin, H. (1931). Studies in the ecology of Wicken Fen - I. The ground water level of the fen, *Journal of Ecology* **19**, 449-473.

Godwin, H. (1978) *Fenland: Its Ancient Past and Uncertain Future.* University Press, Cambridge.

Godwin, H. (1981) *The Archives of the Peat Bogs.* Cambridge University Press, Cambridge.

Goodpaster, K. (1978). On being morally considerable. *Journal of Philosophy* **22**, 308-325.

Gordon, I.J. (1989) Vegetation community selection by ungulates on the Isle of Rhum II. Vegetation community selection. *Journal of Applied Ecology* **26**, 53-64.

Gordon, J.E. and Sutherland, D.G. (1993) *The Quaternary of Scotland.* Chapman and Hall, London.

Gore, A.J.P. (1983) Introduction. In: Gore, A. (ed.) *Ecosystems of the World 4B. Mires: Swamp, Bog, Fen and Moor. Regional studies.* Elsevier Scientific, Amsterdam, pp. 1-34.

Gorham, E. (1991) Northern peatlands: rôle in the carbon cycle and probably responses to climatic warning. *Ecological Applications* **1**, 182-195.

Gramsch, B. (1992) Friesack mesolithic wetlands. In: Coles, B. (ed.) *The Wetland Revolution in Prehistory.* Occasional Paper 6, WARP, Exeter, pp. 65-72.

Granlund, E. (1932) De svenska högmossarnas geologi. *Svensk. Geol. Unders. Ser. C.* **373**, 193.

Grant, M.E. (1995) The Dating and Significance of *Pinus sylvestris* remains at Fenn's/Whixall Moss: Palaeoecological and Comparative Modern Analyses. PhD thesis, Keele University.

Grant, S.A., Bolton, G.R. and Torvell, L. (1985) The responses of blanket bog vegetation to controlled grazing by hill sheep. *Journal of Applied Ecology* **22**, 739-751.

Grant, S.A., Lamb, W.I.C., Kerr, C.D. and Bolton, G.R. (1976) The utilization of blanket bog vegetation by grazing sheep. *Journal of Applied Ecology* **13**, 857-869.

Grattan, J. and Charman, D.J. (1994) Non-climatic factors and the environmental impact of volcanic volatiles: implications of the Laki fissure eruption of AD 1783. *The Holocene* **4**, 101-106.

Greig, J.R.A. (1982) Past and present limewoods of Britain. In: Bell, M. and Limbrey, S. (eds) *Archaeological Aspects of Woodland Ecology.* British Archaeological Reports S146, Oxford, pp 23-56.

Greig, S. (1995) *Wildlife Habitats: An Environmental Education Resource for Key Stage 2.* Friends of the Earth, London, 40 pp.

Grenon, A. (1989) *Peatlands and Remote Sensing - Inventory of Peatlands in Quebec.* Service de la Cartographie Centre, Quebec de Co-ordination de la Teledetection, Ministrie de l'Energie et des Resources, Sector Tries, Sainte-Foy, Quebec.

Grieve, I.C., Hipkin, J.A. and Davidson, D.A. (1994) Soil erosion sensitivity in upland Scotland. *Scottish Natural Heritage Research Survey and Monitoring Report* No. 24, Perth.

Grimmet, R.F.A. and Jones, T.A. (eds)(1989) *Important Bird Areas in Europe. International Council for Bird Preservation*, Technical Publication 9. International Waterfowl and Wetlands Research Bureau, Cambridge.

Grosvernier, Ph., Matthey, Y, and Buttler, A. (1995) Microclimate and physical properties of peat: new clues to the understanding of bog restoration. In: Wheeler, B., Shaw, S., Fojt, W. and Robertson, A. (eds) *The Restoration of Temperate Wetlands.* John Wiley, Chichester. pp. 435-450.

Grünig, A. (1994) Findings from the two national inventories of mire habitats. In: Grünig, A. (ed.) *Mires and Man. Mire Conservation in a Densely Populated Country - the Swiss Experience. Excursion Guide and Symposium Proceedings of the 5th Field Symposium of the International Mire Conservation Group (IMCG) to Switzerland 1992.* Swiss Federal Institute for Forest, Snow and Landscape Research, Birmensdorf. pp. 89-95.

Grünig, A., Vetterli, L. and Wildi, O. (1986) *Les Hauts-Marais et Marais de Transition de Suisse. Rapp. 281.* Birmensdorf, Swiss Federal Institute for Forest, Snow and Landscape Research, 58 pp.

Haines-Young, R. and Bunce, B. (1995) Remote Sensing: Land Classification and Environmental Survey. In: Delbaere, B. and Gulinck, H. (eds) *European Collaborative Programme Workshop on Remote Sensing in Landscape Ecological Mapping.* Leuven 1994 Joint Research Center of the European Commission, European Commission, Brussels, pp. 51-62.

Hall, D. and Coles, J.M. (1994) *Fenland Survey. An Essay in Landscape and Persistence.* Archaeological Report 1, English Heritage, London.

Hall, D. and Wells, C. (1995) *The Wetlands of Greater Manchester.* North West Wetland Survey 2, Lancaster Imprints 3, Lancaster.

Hall, D.A. and Smith, C.A. (1994) An evaluation of perlite-based substrates for ornamental pot plant production. *Acta Horticulturae* **361**, 486-490.

Hall, V.A., Pilcher, J. and McCormac, F.G. (1994) Icelandic volcanic ash and the mid-Holocene Scots pine (*Pinus sylvestris*) decline in the north of Ireland: no correlation. *The Holocene* **4**, 79-83.

Halliday, S.P. (1993) Marginal agriculture in Scotland. In: Smout, T.C. (ed.) *Scotland Since Prehistory: Natural Change and Human Impact.* Scottish Cultural Press, Aberdeen, pp. 64-78.

Hamilton, G. (1991) Plantlife Peat Enquiry. Evidence from Geoff Hamilton. In: Plantlife *Commission of Inquiry into Peat and Peatlands.* Public Hearing No. 2: Evidence October 23rd 1991. Plantlife, London.

Hammond, C.O. (1983) *The Dragonflies of Britain and Ireland.* [2nd edn revised by R. Merritt]. Harley Books, Colchester

Hammond, P.M. (1974) Changes in the British coleopterous fauna In: Hawksworth, D.L. (ed.) *The Changing Flora and Fauna of Britain* Systematics Association Special Volume 6. Academic Press, London, pp. 323-369.

Hammond, R.F. (1979) The peatlands of Ireland. *Soil Survey Bulletin No. 35.* An foras Taluntais, Dublin.

Handler, P. (1989) The effect of volcanic aerosols on global climate. *Journal of Volcanology and Geothermal Research* **37**, 233-249.

Handreck, K.A. (1992) Rapid assessment of the rate of nitrogen immobilisation in organic components of potting media: I. Method development. *Communications in Soil Science and Plant Analysis* **223,** 201-215.

Hardin, G. (1968) The tragedy of the commons. *Science* **162**, 1243-1248.

Hargrove, E.C. (1989) *Foundations of Environmental Ethics.* Prentice Hall, Englewood Cliffs, New Jersey.

Hargrove, E.C. (ed.) (1992) *The Animal Rights/Environmental Ethics Debate. The Environmental Perspective.* State University of New York Press, Albany.

Hargrove, E.G. (1987) The foundations of wildlife protection attitudes. *Inquiry* **30**, 3-31.

Harris, J. (1975) The survival lottery. *Philosophy* **50**, 81-87.

Haslam, C.J. (1987) Late Holocene Peat Stratigraphy and Climatic Change - a Macrofossil Investigation from the Raised Mires of North Western Europe. PhD thesis, University of Southampton.

Havas, P. (1961) Vegetation und Ökologie der ostfinnischen Hangmoore. *Annales Botanicae Societatis Fennicae Vanamo* **31**(2), 188.

Hayward, P.M. and Clymo, R.S. (1982) Profiles of water content and pore size in *Sphagnum* and peat, and their relation to bog ecology. *Proceedings of the Royal Society of London* B **215**, 299-325.

Headley, A.D., Wheeler, B.D. and Baker, A.J.M. (1992) The impact of man on the vegetation of Crymlyn Bog. In: Bragg, O.M., Hulme, P.D., Ingram, H.A.P. and Robertson, R.A. (eds) *Peatland Ecosystems and Man: An Impact Assessment*. Dept Biological Sciences, University of Dundee, pp. 257-261.

Heathwaite, A.L. (1993) Disappearing peat - regenerating peat? The impact of climate change on British peatlands. *The Geographical Journal* **159**, 203-208.

Heaver, D. and Eversham, B. (1991) *Thorne and Hatfield Moors Invertebrate Survey. Final Report.* Thorne and Hatfield Moors Conservation Forum.

Heiken, G. (1974) An atlas of volcanic ash. *Smithsonian Contributions to the Earth Sciences.* **12**, 1-99.

Hester, A.J. and Miller, G.R. (1995) Scrub and woodland regeneration: prospects for the future. In: Thompson, D.B.A., Hester A.J. and Usher, M.B. (eds) *Heaths and Moorland: Cultural Landscapes* HMSO, Edinburgh, pp. 140-153.

Hintermann, U. (1992) *Inventaire des sites marécageux d'une beauté particulière et d'importance nationale.* Cahier de l'Environnement no 168 Nature et Paysage. Berne, OFEFP (Office Fédéral de l'Environnement, des Forêts et du Paysage) 214 pp.

Hintermann, U. (1994) The inventory of the Swiss mire landscapes of particular beauty and national importance. In: Grünig, A. (ed.) *Mires and Man. Mire Conservation in a Densely Populated Country - The Swiss experience. Excursion Guide and Symposium Proceedings of the 5th field Symposium of the International Mire Conservation Group (IMCG) to Switzerland 1992* Birmensdorf, Swiss Federal Institute for Forest, Snow and Landscape Research, pp. 95-103.

Hirons, G., Goldsmith, B., and Thomas, G. (1995) Site management planning. In: Sutherland, W.J. and Hill, D.A., (eds) *Managing Habitats for Nature Conservation.* Cambridge University Press, pp. 22-41.

Hobbs, N.B. (1986) Mire morphology and the properties and behaviour of some British and foreign peats. *Quarterly Journal of Engineering Geology, London* **19**, 7-80.

Hodge, S. (1993) *Trials of Organic Backfill Amendments on Trunk Road sites.* ARB 114/93. Arboricultural Advisory and Information Service, Farnham.

Holdgate, (1994) Ecology, development and global policy. *Journal of Ecology* **31**, 201-211

Holloway, C.W. (1967) The effect of red deer and other animals on naturally regenerated Scots pine. PhD thesis, University of Aberdeen.

Holmes, P. (1992) *Border Mires Invertebrate Project.* English Nature, Peterborough.

Homan, T. (ed.) (1991) *A Yearning Toward Wildness. Environmental Quotations from the Writings of Henry David Thoreau*. Peachtree Publishers, Atlanta.

Hopkins, J.J. (1995) The Habitats Directive - Selecting the UK sites. *British Wildlife* **6** (5), 297-306.

Houghton, J.T., Callander, B.A. and Varney, S.K. (1992) *Climate change 1992: The Supplementary Report to the IPCC Scientific Assessment.* Cambridge University Press, Cambridge.

Houghton, J.T., Jenkins, G.J. and Ephraums, J.J. (eds) (1990) *Climate Change - The IPCC Assessment.* Cambridge University Press, Cambridge.

Hove, T.T. (1983) *Tørvegravning i Danmark.* Poul Kristensens Forlag, Herning.

Hughes, R.E., Dale, J., Ellis-Williams, I. and Rees, D.I. (1973) Studies in sheep population and environment in the mountains of North-west Wales. I. The status of the sheep in the mountains of North Wales since medieval times. *Journal of Applied Ecology* **10,** 113-132.

Huizinga, J. (1938). *Homo ludens. Proeve ener bepaling van het spelelement der cultuur.* 8th printing/1985. Wolters-Noordhof, Groningen.

Hulme, P.D. (1994) A palaeobotanical study of paludifying pine forest on the island of Hailuoto, Northern Finland. *New Phytologist* **126**, 153-162.

Hunt, J.B. and Hill, P.G. (1993) Tephra geochemistry: a discussion of some persistent analytical problems. *The Holocene* **3**, 271-287.

Hunter, R.F. (1962) Hill sheep and their pasture: a study of sheep-grazing in South-east Scotland. *Journal of Ecology* **50**, 651-680.

Huntley, B. (1979) The past and present vegetation of the Caenlochan National Nature Reserve, Scotland. *New Phytologist* **83**, 215-283.

Huntley, B. and Birks, H.J.B. (1983) *An Atlas of Past and Present Pollen Maps for Europe 0-13000 years ago.* Cambridge University Press, Cambridge.

Hurka, T. (1993). Ethical principles. In: Coward, H. and Hurka, T. (eds) *The Greenhouse Effect. Ethics and Climate Change* Wilfrid Laurier University Press, Waterloo, pp. 23-38.

Immirzi, C.P. and Maltby, E. (1992) *The Global Status of Peatlands and their Role in Carbon Cycling.* A Report for Friends of the Earth by the Wetland Ecosystems Research Group, Department of Geography, University of Exeter. Friends of the Earth, London.

Ingram, H.A.P. (1978) Soil layers in mires: function and terminology. *Journal of Soil Science* **29**, 224-227.

Ingram, H.A.P. (1982) Size and shape in raised mire ecosystems: a geophysical model. *Nature* **297**, 300-303.

Ingram, H.A.P. (1983). Hydrology In: Gore, A.J.P. (ed.) *Mires: Swamp, Bog, Fen and Moor (Ecosystems of the World* 4a*)*, Elsevier. Amsterdam, pp. 67-158.

Ingram, H.A.P. (1992) Introduction to the ecohydrology of mires in the context of cultural perturbation. In: Bragg, O.M., Hulme, P.D., Ingram, H.A.P. and Robertson, R.A. (eds) *Peatland Ecosystems and Man: An Impact Assessment* University of Dundee, Dundee, pp. 67-93.

Ingram, H.A.P. (1994) *Conserving the cultural landscape?* (unpublished comment, doc. no. I-0273. MRG, circulated to members of the Mires Research Group).

Irish Peatland Conservation Council (IPCC) (1986) *The Irish Peatland Conservation Council Action Plan 1986-1989*. Irish Peatland Conservation Council, Dublin.

Irish Peatland Conservation Council (IPCC) (1989) *The IPCC Action Plan 1989-1992*. Irish Peatland Conservation Council, Dublin.

Irish Peatland Conservation Council (IPCC) (1992) *Irish Peatland Conservation Council Policy Statement and Action Plan 1992-1997*. Unpubl. report. Irish Peatland Conservation Council, Dublin. 100 pp.

Ivanov, K.E. (1981) *Water Movement in Mirelands* (transl. Thomson, A. and Ingram, H.A.P.). Academic Press, London.

Jacobson, G.L. and Bradshaw, R.H.W. (1981) The selection of sites for palaeovegetational studies. *Quaternary Research* **16**, 80-96.

Jessen, K. (1949) Studies in the late Quaternary deposits and flora-history of Ireland. *Proceedings of the Royal Irish Academy* **52B**, 85-290.

Jewell, N. and Brown, R.W. (1988) *The Use of Landsat TM Data for Vegetation Mapping in the North York Moors National Park*. National Remote Sensing Centre, Farnborough.

Johnson, B.R., and Rowarth, P.C. (1994) The Revival of Thorne Moors, a unique lowland wilderness in Yorkshire. *Annual Report of the Yorkshire Philosophical Society*, York.

Johnson, G.A.L. and Dunham, K.C. (1963) *The Geology of Moor House*. HMSO, London.

Johnson, R.H. (1957) Observations on the stream patterns of some peat moorlands in the southern Pennines. *Memoirs of the Manchester Literature and Philosophical Society* **99**, 1-18.

Joint Nature Conservation Committee (JNCC) (1994) *Supplement to The Guidelines for Selection of Biological Sites of Special Scientific Interest: Bogs, Nature Conservancy Council (1989)*. Joint Nature Conservation Committee, Peterborough.

Jones, J.M. (1985) Heavy metals and magnetic minerals in ombrotrophic peat. PhD thesis, University of Liverpool.

Jones, J.M. (1987) Chemical fractionation of copper, lead and zinc in ombrotrophic peat. *Environmental Pollution* **48**, 131-144.

Jones, J.M. and Hao, J. (1993) Ombrotrophic peat as a medium for historical monitoring of heavy metal pollution. *Environmental Geochemistry and Health* **15**, 67-74.

Joosten, H. (1992) Hoogveen: het Madurodam van Gaia. In: Schrier, A., van de Merwe, A., Savelsberg, B., Koster, L., Goumans, M.-J. and Stassar, M. (eds) *Of zoals de fransen zeggen H_2O*. Utrechtse Biologen Vereniging, Utrecht, pp. 20-25.

Joosten, H. (1993) Denken wie ein Hochmoor: Hydrologische Selbstregulation von Hochmooren und deren bedeutung für Wiedervernässung und Restauration. *Telma* **23**, 95-115.

Joosten, H. (1994) Turning the tides: experiences and perspectives of mire conservation in the Netherlands. In: Grünig, A. (ed.) *Mires and Man, Mire Conservation in a Densely Populated Country - The Swiss Experience* Swiss Federal Institute for Forest, Snow and Landscape Research, Birmensdorf pp. 300-310.

Joosten, J.H.J. (1985) A 130 year micro- and macrofossil record from regeneration peat in former peasant peat pits in the Peel, the Netherlands: a palaeoecological study with agricultural and climatological implications. *Palaeogeography, Palaeoclimatology, Palaeoecology* **49**, 277-312.

Joosten, J.H.J. (1995a) Time to regenerate: long-term perspectives of raised bog regeneration with special emphasis on palaeoecological studies. In: Wheeler, B.D., Shaw, S.C., Fojt, W.J. and Robertson, R.A. (eds) *Restoration of Temperate Wetlands.* John Wiley, Chichester, pp 379-404.

Joosten, J.H.J. (1995b) Mire development and peatland utilization in Estonia and the Netherlands: a comparison. In: *Peat Industry and Environment.* Environment Information Centre, Ministry of Environment, Tallinn, pp. 24-25.

Joosten, J.H.J. (1995c) Time to Regenerate: long-term perspectives of raised bog regeneration with special evidence on palaeoecological studies. In: Wheeler, B.D., Shaw, S.C., Fojt, W.J. and Robertson, R.A. (eds) *Restoration of Temperate Wetlands.* John Wiley, Chichester, pp. 379-404.

Joosten, J.H.J. (1995d) *The Golden Flow: The Changing World of International Peat Trade*. Gunneria **70**, 269-292 (in press).

Joosten, J.H.J. (1996) A world of mires: criteria for identifying mires of global conservation significance. In: Lüttig, G.W. (ed.) *Peatlands Use - Present, Past and Future.* Vol 13. Schweizerbart, Stuttgart, pp. 18-25.

Kagan, S. (1989) *The Limits of Morality*. Oxford University Press, Oxford.

Kästner, M. and Flössner, W. (1933) Die Planzengesellschaften der erzebirgishen moore. In: Kästner, M., Flössner, W. and Uhlig, J. (eds) *Die Planzengesellschaften des Westsächsischen Berg- und Hugellandes* Vol. II. Dresden.

Kaufmann, A. (1948) Notes on the distribution of the British Longicorn Coleoptera. *Entomologist's Monthly Magazine* **84**, 66-85.

Kazda, J. (1977) The importance of *Sphagnum* bogs in the ecology of mycobacteria. *Zentralblatt fur Bakteriologie Mikrobiologie und Hygiene* **165**, 323-334.

Keating, M. (1993) *The Earth Summit's Agenda for Change. A Plain Language Version of Agenda 21 and the Other Rio Agreements.* The Centre for Our Common Future, Switzerland, 70 pp.

Kerr, R.A. (1991) Could the sun be warming the climate? *Science* **254**, 652-653.

Keys, D. (1992) *Canadian Peat Harvesting and the Environment.* Sustaining wetlands Issues. Paper No 1992-3, North American Wetlands Conservation Council, Ottawa.

Kinnaird, J.W., Welch, D. and Cummins, C. (1979) Selective stripping of rowan (*Sorbus aucuparia* L.) bark by cattle in north-east Scotland. *Transactions of the Botanical Society of Edinburgh* **43**, 115-125.

Klijn, J.A. (1995) *Hierarchical Concepts in Landscape Ecology and its Underlying Disciplines.* Staring Centrum, Wageningen.

Knight, A., Coutts, L., Jamison, L., Cox, C., Ball, A. and Ball, C. (1995) *How Green is My Front Door? B & Q's second Environmental Review.* B & Q, Eastleigh.

Kohli, E. (1994) The legal basis for mire conservation in Switzerland and its implementation. In: Grünig, A. (ed.) *Mires and Man. Mire Conservation in a Densely Populated Country - The Swiss Experience. Excursion guide and symposium proceedings of the 5th field symposium of the International Mire Conservation Group (IMCG) to Switzerland 1992.* Birmensdorf, Swiss Federal Institute for Forest, Snow and Landscape Research, pp. 85-89.

Korhola, A. (1992) Mire induction, ecosystem development and lateral extension on raised bogs in the southern coastal area of Finland. *Fennia* **170**, 25-92.

Kothari, R. (1990) Environment, technology, and ethics. In: Engel, J.R. and Engel, J.G. (eds) *Ethics of Environment and Development. Global Challenge and International Response.* Belhaven Press, London, pp. 27-35.

Krisai, R. (1966) Pflanzensoziologische Untersuchungen in Lungauer Mooren. *Verh. Zool. Bot. Ges. Wien* **105/106**, 94-136.

Kuhry, P. (1994) The role of fire in the development of *Sphagnum*-dominated peatlands in western boreal Canada. *Journal of Ecology* **82**, 899-910.

Kuntze, H. and Eggelsmann, R. (1981) Zur Schutzfähigkeit nordwestdeutscher Moore. *Telma* **11**, 197-212.

Kyuma, K., Kaneko, N., Zahari, A.B. and Ambak, K. (1992) Swamp forest and tropical peat in Johore, Malaysia. *Proceedings of the International Symposium on Tropical Peatland.* Kuching, Sarawak, Malaysia, 1991, pp. 300-306.

Labadz, J.C., Burt, T.P. and Potter, A.W.R. (1991) Sediment yields and delivery in the blanket peat moorlands of the southern Pennines. *Earth Surface Processes and Landforms* **16**, 255-271.

Lageard, J.G.A. (1992) Vegetational history and palaeoforest reconstruction at White Moss, South Cheshire, UK. PhD thesis, University of Keele.

Laine, J. (1984). Estimation of evapotranspiration from peatlands by means of daily water table hydrographs, *Publication Department of Peatland Forestry (Univ. Helsinki),* **5**.

Lamb, H.H. (1977) *Climate: Past, Present and Future. Vol. 2: Climatic History and the Future.* Methuen, London.

Lamb, H.H., Lewis, R.P.W. and Woodruffe, A. (1966) Atmospheric circulation and the main climatic variables between 8000 and 0 B.C.: meteorological evidence. In: Sawyer, J.S. (ed.) *World Climate from 8000 to 0 B.C.* Royal Meteorological Society, London pp. 174-217.

Lancashire County Council (1990) *Lancashire - a green audit.* Lancashire County Council, Preston.
Landman, W.A., (1995) Moral standing, value, and environmental ethics. *South African Journal of Philosophy* **14**, 9-18.
Lavoie, C. and Rochefort, L. (1996) The natural regeneration of a harvested peatland in southern Quebec: a spatial and dendro-ecological analysis. *Ecoscience*, **3**, 101-111.
Lawrence, W.J.C. and Newell, J. (1939) *Seed and Potting Composts.* George Allen and Unwin, London.
le Maitre, D.C., van Wilgen, B.W., Chapman, R.A. and McKelly, D.H. (1996) Invasive plants and water resources in the western Cape province, South Africa: modelling the consequences of a lack of management *Journal of Applied Ecology* **33** (1), 161-172.
Leach, S.J. and Corbett, P.McM. (1987) A preliminary survey of raised bogs in Northern Ireland. *Glasra* **10**, 57-73.
Lee, J.A. and Tallis, J.H. (1973) Regional and historical aspects of lead pollution in Britain. *Nature* **245**, 216-218.
Lehman, S.J. and Keigwin, L.D. (1992) Sudden changes in North Atlantic circulation during the last deglaciation. *Nature* **356**, 757-762.
Leopold, A. (1937) Conservationist in Mexico. *American Forests* **43**, 118-120.
Leopold, A. (1949) *A Sand County Almanac and Sketches Here and There.* Oxford University Press, New York.
Lewis, F.J. (1905) The plant remains in the Scottish peat mosses. Part I. The Scottish Southern Uplands. *Transactions of the Royal Society of Edinburgh* **41**, 699-723.
Lewis, F.J. (1906a) The history of the Scottish peat mosses and their relation to the glacial period. *Scottish Geographical Magazine* **22**, 235-48.
Lewis, F.J. (1906b) The plant remains in the Scottish peat mosses. Part II. The Scottish Highlands. *Transactions of the Royal Society of Edinburgh* **45**, 335-360.
Lewis, F.J. (1907) The plant remains in the Scottish peat mosses. Part III. The Scottish Highlands and the Shetland Islands. *Transactions of the Royal Society of Edinburgh* **46**, 33-70.
Lewis, F.J. (1911) The plant remains in the Scottish peat mosses. Part IV. The Scottish Highlands and Shetland, with an appendix on the Icelandic peat deposits. *Transactions of the Royal Society of Edinburgh* **47**, 793-833.
Lindsay, R.A. (1977) *Monitoring the effects of the 1976 fire at Glasson Moss.* Unpublished report. Nature Conservancy Council. Peterborough.
Lindsay, R.A. (1993) Peatland conservation - from cinders to Cinderella. *Biodiversity and Conservation* **2** (5), 528-540.
Lindsay, R.A. (1995) *Bogs: the Ecology, Classification and Conservation of Ombrotrophic Mires.* Scottish Natural Heritage, Edinburgh.
Lindsay, R.A., Charman, D.J., Everingham, F., O'Reilly, R.M., Palmer, M.A., Rowell, T.A. and Stroud, D.A. (1988) *The Flow Country. The peatlands of*

Caithness and Sutherland. Nature Conservancy Council, Peterborough, pp. 148
Lindsay, R.A. and Immirzi, C.P. (1996) *An Inventory of Lowland Raised Bogs in Great Britain.* Scottish Natural Heritage Research, Battleby, (in press).
Livett, E.A., Lee, J.A. and Tallis, J.H. (1979) Lead, zinc and copper analysis of British blanket peats. *Journal of Ecology* **67**, 865-891.
Löfroth, M. (1995) *The Swedish Mire Protection Programme.* Swedish Nature Protection Agency, Stockholm.
Lonard, R.J. and Clairain, E.J. (1983) Identification of methodologies and the assessment of wetland functions and values. In: Kusler, J.A. and Riexinger, P. (eds) *Proceedings of the National Wetlands Assessment Symposium.* Portland, ME pp. 66-72
Louwe Koojimans, L.P. (1996) Prehistory or paradise. In: Cox, M., Straker, V. and Taylor, D. (eds) *Wetlands: Nature Conservation and Archaeolog,y* HMSO, Norwich. pp. 000-000.
Lovejoy, T.E. (1988). Diverse considerations. In: Wilson, E.O. (ed.) *BioDiversity.* National Academy Press, Washington D.C., pp. 421-427.
Lovelock, J. (1991) *Gaia. The Practical Science of Planetary Medicine.* Gaia Books, London.
Lowe, J.J., Coope, G.R., Keen, D. and Walker, M.J.C. (1994) High resolution stratigraphy of the last glacial-interglacial transition (LGIT) and inferred climatic gradients. In: Funnell B.M. and Kay R.L.F. (eds) *Palaeoclimate of the Last Glacial/Interglacial Cycle.* NERC, Swindon, pp. 47-52.
Lowe, S. (1993) *Border Mires Redefinition of Hydrological Boundaries.* Unpublished report. English Nature North East Region.
Lu, J. and Zhang, X. (1986) Discovery of Tertiary *Sphagnum* coal in Jinsuo Basin, Yunnan Province and its significance. *Kexue Tongbao* **31**, 1556-1559.
Lunn, A.G. (1979) *The Vegetation of Northumberland. A 1:200,000 map.* University of Newcastle upon Tyne. Dept of Geography.
Luthin, J.N. and Kirkham, D. (1949). A piezometer method for measuring permeability of soils in situ below a water table. *Soil Scientist* **68**, 349-358.
Maas, D. and Poschlod, P. (1991) Restoration of exploited peat areas in raised bogs - technical management and vegetation development. In: Ravera, O. (ed.) *Terrestrial and Aquatic Ecosystems: Perturbation and Recovery.* Ellis Horwood, Chichester, pp. 379-386.
Macartney G. (1994) *Peatlands Forever? An Education Resource Pack.* Department of the Environment for Northern Ireland, Belfast, 58 pp.
Macaulay Land Use Research Institute (MLURI) (1993) *The Land Cover of Scotland 1988 Final Report.* The Macaulay Land Use Research Institute, Aberdeen.
MacCulloch, J. (1824) *The Highlands and Western Isles of Scotland - Vol. I.* London.

Macdonald, J.A.B. (1928) A seventy-year-old plantation on peat. *Forestry* **2**, 85-92.

Mackay, A.W. (1993) The recent vegetational history of the Forest of Bowland, Lancashire. PhD thesis, University of Manchester.

Mackay, A.W. and Tallis J.H. (1996) Summit-type blanket mire erosion in the Forest of Bowland, Lancashire, UK: predisposing factors and implications for conservation. *Biological Conservation*, **76**, 31-44.

Mackey, E.C. and Tudor, G.T. (1996). Estimating land cover change in Scotland over the past 50 years. *Proceedings from Biogeography Research Group Annual Conference. Vegetation Mapping: From Patch to Planet.* May 1995, Chester College. John Wiley, Chichester. (in press).

Maher, M.J. and Thomson, D. (1991) Growth and manganese content of tomato (*Lycopersicon esculentum*) seedlings grown in Sitka spruce (*Picea sitchensis* (Bong.) Carr.) bark substrates. *Scientia Horticulturae* **48**, 223-231.

Malmer, N. (1962) Studies on mire vegetation in the Archaean area of south-western Götaland (south Sweden). II. Distribution and seasonal variation in elementary constituents on some mire sites. *Opera Botanica* **7**(2), 67.

Malmer, N. (1965) The southern mires. In: *The Plant Cover of Sweden. Acta Phytogeography Suecica* **50**, 149-158.

Malmer, N. and Wallén, B. (1993) Accumulation and release of organic matter in ombrotrophic bog hummocks - processes and regional variation. *Ecography* **16**, 193-211.

Maltby, E. (1986) *Waterlogged Wealth.* An Earthscan Paperback, pp. 183.

Maltby, E. and Caseldine, C.J. (1982) Prehistoric soil and vegetation development on Bodmin Moor. *Nature* **297**, 397-400.

Maltby, E. and Crabtree, K. (1976) Soil organic matter and peat accumulation on Exmoor: A contemporary and palaeoenvironmental evaluation. *Transactions Institute of British Geographers, New Series* **1** (3) 259-278.

Maltby, E. and Gabriel, K.A.S. (1986) Experimental revegetation of severely eroded and burnt moorland at Cock Heads, Glaisdale Moor. In: *Moorland Managemen*, North York Moors National Park Authority, Helmsley, pp. 96-101.

Maltby, E. and Immirzi, C.P. (1993) Carbon dynamics in peatlands and other wetland soils: regional and global perspectives. *Chemosphere* **27**, 999-1023.

Maltby, E., Legg, C.J. and Proctor, M.C.F. (1990) Ecology of severe moorland fire on the North York Moors: effects of the 1976 fires, and subsequent surface and vegetation development. *Journal of Ecology* **78**, 490-518.

Maltby, E.M., D.V. Hogan, D.V., Immirzi, C.P., Tellam, J.H. and van der Peijl, M. (1994) Building a new approach to the investigation and assessment of wetland ecosystem functioning. In: Mitsch, W.J. (ed.) *Global Wetlands: Old and New.* Elsevier Science, Amsterdam.

Marrs, R.H. and Welch, D. (1991) *Moorland Wilderness: the Potential Effects of Removing Domestic Livestock, Particularly Sheep.* Department of the Environment, London.

Marrs, R.H., Bravington, M. and Rawes, M. (1988) Long-term vegetation change in the *Juncus squarrosus* grassland at Moor House, northern England. *Vegetatio* **76** 179-187.

Matthews, E. (1984) Global inventory of pre-agricultural and present biomass. *Progress in Biometereology* **3**, 237-246.

Maynard-Smith, J. and Szathmáry, E. (1995) *The Major Transitions in Evolution.* Freeman, Oxford.

McCormick, M.P., Thomason, L.W. and Trepte, C.R. (1995) Atmospheric effects of the Mt. Pinatubo eruption. *Nature* **373**, 399-404.

McEvoy, J. (1802) *Statistical Survey of the County of Tyrone*. Dublin.

McGreal, W.S. and Larmour, R.A. (1979) Blanket peat erosion: theoretical considerations and observations from selected conservation sites in Slievenorra Forest National Nature Reserve, County Antrim. *Irish Geography* **12**, 57-67.

McNally, A. (1990) Dendrochronology of subfossil pine as evidence of environmental change. In: Doyle, G.J. (ed.) *Ecology and Conservation of Irish Peatlands.* Royal Irish Academy, Dublin, pp. 15-22.

McNally, A. and Doyle, G.J. (1984) A study of subfossil pine layers in a raised bog complex in the Irish Midlands - I and II. *Proceedings of the Royal Irish Academy* Vol. **84**B, 57-70 and 71-81.

McTeague, E. and Watson, K. (1989) *A Peatland Survey of Mid-Strathclyde.* Nature Conservancy Council Contract Survey No. 136. NCC, Peterborough.

McVean, D.A. and Ratcliffe, D.A. (1962) *Plant Communities of the Scottish Highlands. Nature Conservancy Monograph No. 1.* HMSO, London.

Meade, R. (1992). Some early changes following the rewetting of a vegetated cutover peatland surface at Danes Moss, Cheshire, UK, and their relevance to conservation management. *Biological Conservation* **61**, 31-40.

Mendel, H. (1992). *Suffolk dragonflies.* Suffolk Naturalists' Society, Ipswich.

Middleton, R., Wells, C. and Huckerby, E. (1995) *The Wetlands of North Lancashire.* North West Wetland Survey 3, Lancaster Imprints 4, Lancaster.

Midgley, M. (1983) *Animals and Why They Matter*. University of Georgia Press, Athens.

Miles, J. (1978) The influence of trees on soil properties. *ITE Annual Report 1977.* Institute of Terrestrial Ecology, Cambridge, pp. 7-11.

Miles, J. and Kinnaird, J.W. (1979a) Grazing: with particular reference to birch, juniper and Scots pine in the Scottish Highlands. *Scottish Forestry* **33**, 280-289.

Miles, J. and Kinnaird, J.W. (1979b) The establishment and regeneration of birch, juniper and Scots pine in the Scottish Highlands. *Scottish Forestry* **33**, 102-119.

Miller, R.I. (1994) Setting the scene. In: Miller, R.I. (ed.) *Mapping the Diversity of Nature*. Chapman and Hall, London, pp. 3-7.

Mitchell, A. (ed.) (1906) *Geographical Collections Relating to Scotland made by Walter Macfarlane - Vol.I.* Scottish History Society, Edinburgh.

Mitchell, B., McCowan, D. and Willcox, N.A. (1982) Effects of deer in a woodland restoration enclosure. *Scottish Forestry* **36**, 102-112.

Mitchell, F.J.G. and Kirby, K.J. (1990) The impact of large herbivores on the conservation of semi-natural woods in the British Uplands. *Forestry* **63**, 333-353.

Mitchell, J. (1984) A Scottish bog-hay meadow. *Scottish Wildlife* **20**, 15-17.

Moen, A. (1985) Classification of mires for conservation purposes in Norway. *Aquila, Seria Botanica* **21**, 95-100.

Money, R.P. (1995) Re-establishment of a *Sphagnum*-dominated flora on cut-over lowland raised bogs. Regeneration with special evidence on Palaeoecological Studies. In: Wheeler, B.D., Shaw, S.C., Fojt, W.J. and Robertson, R.A. (eds) *Restoration of Temperate Wetlands.* J. Wiley, Chichester, pp. 405-422.

Moore, J.J. (1962) The Braun-Blanquet system: a reassessment. *Journal of Ecology* **50**, 761-769.

Moore, J.J. (1968) A classification of the bogs and wet heaths of northern Europe. In: Tüxen, R. (ed.) *Pflanzensoziologische Systematik.* Den Haag,Junk N.V., pp. 306-320.

Moore, J.J., Dowding, P. and Healy, B. (1975) Glenamoy, Ireland. *Ecological Bulletin (Stockholm)* **20**, 321-343.

Moore, N.W. (1986) Acid water dragonflies in eastern England - their decline, isolation and conservation. *Odonatologica* **15** (4), 377-385.

Moore, P. (1987) A thousand years of death. *New Scientist* **113,**,46-48.

Moore, P.D. (1975) Origin of blanket mires. *Nature* **256**, 267-269.

Moore, P.D. (1993). The origin of blanket mire, revisited. In: Chambers, F.M. (ed.) *Climate Change and Human Impact on the Landscape*. Chapman and Hall, London, pp. 217-224.

Moore, P.D. and Beckett, P.J. (1971) Vegetation and development of Llyn, a Welsh mire, *Nature,* **231**, 363-365.

Moore, P.D. and Bellamy, D.J. (1974) *Peatlands.* Elek Scientific Books Ltd., London.

Moore, T.R. and Knowles, R. (1989) The influence of water table levels on methane and carbon dioxide emissions from peatland soils. *Canadian Journal of Soil Science* **69**, 33-38.

Moseley, F. and Walker, D. (1952) Some aspects of the Quaternary period in North Lancashire. *Naturalist Hull* **841**, 41-54.

Munaut, A.V. (1966) Recherches dendrochronologiques sur *Pinus sylvestris.* 2. Première application des méthodes dendrochronologiques à l'étude pins sylvestres subfossils (Terneuzen, Pays-Bas). *Agricultura* **14**, 361-89.

Munaut, A.V. and Casparie, W.A. (1971) Étude dendrochronologique de *Pinus sylvestris* L. subfossils provenant de la tourbière d'Emmen (Drenthe, Pays-Bas). *Reviews of Palaeobotany and Palynology* **11**, 201-26.

Naess, A. (1973). The shallow and the deep, long-range ecology movement: a summary. *Inquiry* **16**, 95-100.

Naess, A. (1990) Sustainable development and deep ecology. In: Engel, J.R. and Engel, J.G. (eds). *Ethics of Environment and Development. Global Challenge and InternationalRresponse.* Belhaven Press, London, pp. 87-96.

Nash, R.F. (1989) *The Rights of Nature. A History of Environmental Ethics.* University of Wisconsin Press, Madison.

National Research Council (NRC) (1995) *Wetland Characteristics and Boundaries.* National Academy Press, Washington DC.

Naveh, Z. (1994). Functions of Nature. What is the conservation and restoration of nature worth for human society? A book review and some comments based on Naveh and Lieberman 1993. *Restoration Ecology* **2**, 71 - 74.

NCC (1983) *Offerance Moss: SSSI notification document.* Nature Conservancy Council, Edinburgh.

NCC (1989) *Guidelines for Selection of Biological SSSIs.* NCC, Peterborough (and subsequent amendments on peat bog guidelines by JNCC 1994).

Neuberg, M.F. (1960) Leafy mosses from the Permian deposits of Angarida. *Trudy Geologischeskogo instituta. Akademiya nauk SSSR (Leningrad)* **19**, 1-104.

Newson, M. (1992) Land and water. Convergence, divergence and progress in UK policy. *Land Use Policy* **9**, 111-121

Newson, M. and Rumsby, B. (1991) *Border Mires Hydrological Boundaries - a Reconnaissance Stud.y* University of Newcastle upon Tyne for NCC.

Newson, M. and White. A. (1993) *Hydrology and Management Options for a Selection of Border Mires Conservation Units.* University of Newcastle upon Tyne for EN.

Ng, F.S.P. and Low, C.M. (1982) *Check List of Trees of the Malay Peninsula.* Malaysian Forestry Department, Research Pamphlet 88.

Nicholson, I.A. and Robertson, R.A. (1958) Some observations on the ecology of upland grazing in north-east Scotland with special reference to Callunetum. *Journal of Ecology* **46**, 239-270.

Nolan, A.J., Henderson, D.J. and Merrell, B.G. (1995) The vegetation dynamics of wet heaths in relation to sheep grazing intensity. In: Thompson, D.B.A., Hester, A.J. and Usher, M.B. (eds) *Heaths and Moorlands: Cultural Landscapes.* HMSO, Edinburgh, pp. 174-179.

Norton, B.G. (1982) Environmental ethics and nonhuman rights. *Environmental Ethics* **4**, 17-36.

Norton, B.G. (1984) Environmental ethics and weak anthropocentrism. *Environmental Ethics* **6**, 131-148.

Norton, B.G. (1985) Introduction to part II. The preservation of species. In: Norton, B.G. (ed.) *The Value of Biological Diversity* Princeton University Press, Princeton, pp. 77-78.

Norton, B.G. (1987a) *Why Preserve Natural Variety?* Princeton University Press, Princeton.

Norton, S.A. (1987b) The stratigraphic record of atmospheric loading of metals at the ombrotrophic Big Heath Bog, Mt. Desert Island, Maine, USA. In: Hutchinson, T.C. and Meema, K.M. (eds) *Effects of Atmospheric Pollutants on Forests, Wetlands and Agricultural Ecosystems.* NATO ASI Series, Vol. 16, Springer-Verlag, Berlin, pp. 561-576.

Norton, S.A. and Kahl, J.S. (1987) A comparison of lake sediments and ombrotrophic peat deposits as long-term monitors of atmospheric pollution. In: Boyle T.P. (ed.) *New Approaches to Monitoring Aquatic Ecosystems, ASTM STP 940.* American Society for Testing and Materials, Philadelphia, pp. 40-57.

O'Críodáin, C. and Doyle, G.J. (1994) An overview of Irish small-sedge vegetation: syntaxonomy and a key to communities belonging to the Scheuchzerio-Caricetea nigrae (Nordh. 1936) Tx. 1937. *Biology and Environment: Proceedings of the Royal Irish Academy* **94B,** 127-144.

O'Connell, C. (1987) *The IPCC Guide to Irish Peatlands.* Unpubl. report. Irish Peatland Conservation Council, Dublin, 102 pp.

O'Connell, C. (1991) *Bogs in the Classroom.* Unpubl. report. Irish Peatland Conservation Council, Dublin, 55 pp.

O'Connell, C. (1992) *Peatland Education Pack: Science, Geography, English, Gaeilge, Art and History Modules.* Unpubl. report. Irish Peatland Conservation Council, Dublin, 626 pp.

O'Connell, C. (1993) *The Living Bog Ecology Slide Pack.* Unpubl. report. Irish Peatland Conservation Council, Dublin.

O'Connell, C. (1994a) *Peatlands and the Primary School Curriculum.* Irish Peatland Conservation Council, Dublin, 144 pp.

O'Connell, C. (1994b) The IPCC Peatland Education Programme. In: Grünig, A. (ed.) *Mires and Man. Mire Conservation in a densely Populated Country - the Swiss Experience.* Swiss Federal Institute for Forest, Snow and Landscape Research, Birmensdorf, Switzerland. pp. 336-340.

O'Connell, C. (1995) *Wake Up to Bogs Instructor's Hand Book and Video.* Unpubl. report. Irish Peatland Conservation Council, Dublin, 48 pp.

O'Connell, M. (1990) Origins of Irish lowland blanket bog. In: Doyle, G.J. (ed.) *Ecology and Conservation of Irish Peatlands* Royal Irish Academy, Dublin. pp. 49-71.

Osborne, P.J. (1965) The effect of forest clearance on the distribution of the British insect fauna. *Proceedings XII International Congress of Entomology,* London, 1964, pp. 556-557.

Osborne, P.J. (1972) Insect faunas of Late Devensian and Flandrian age from Church Stretton, Shropshire. *Philosophical Transactions of the Royal Society of London* **B263**, 327-367.

Osborne, P.J. (1980) The Late Devensian-Flandrian transition depicted by serial insect faunas from West Bromwich, Staffordshire, England. *Boreas* **9**, 139-147.

O'Sullivan, A. (1974) Possible mapping schemes for the heath and bog vegetation of north-west Ireland. In: Sommer, W.H. and Tüxen, R. (eds) *Tatsachen und Probleme der grenzen in der Vegetation.* Cramer, Vaduz, pp. 327-340.

Osvald, H. (1923) Die Vegetation des Hochmoores Komosse. *Sv. Vxtsoc. SllSk. Handl.* Uppsala, **1**, 434.

Osvald, H. (1925) Die Hochmoortypen Europas *Veröff. Geobot. Inst. Eidg. Tech. Hochsch. Stift. Rübel. Zürich 3*, 707-723.

Osvald, H. (1949) Notes on the vegetation of British and Irish Mosses. *Acta Phytogeography Suecica.* **26**, 1-62.

Overbeck F. (1975) *Botanisch - geologische Moorkunde.* Neumünster, 719 pp.

Paasio, I. (1940) Zur Kenntnis der Waldhochmoore Mittelfinnlands. *Acta Forestalia Fennica* **39** (3) 1-41.

Pabsch, E. (1989) Planung und Ausführung von Baumassnahmen für das Renaturierungsprojekt Rotes Moor. *Telma* **2**, 67-75.

Painter, T.J. (1991a) Lindow Man, Tollund Man and other peat-bog bodies: the preservative and anti-microbial action of sphagnan, a reactive glycuronoglycan with tanning and sequestering properties. *Carbohydrate Polymers* **15**, 123-142.

Painter, T.J. (1991b) Preservation in peat. *Chemistry and Industry* **17/6**, 421-424.

Painter, T.J. (1995) Chemical and micro-biological aspects of the preservation process in *Sphagnum* peat. In: Turner, R.C. and Scaife, R.G. (eds) *Bog bodies - new discoveries and new perspectives* British Museum Press, London, pp. 88-99.

Päivänen, J. (1973) Hydraulic conductivity and water retention in peat soils. *Acta Forestalia Fennica* **129,** 1-70.

Pajunen, V.I. (1962). Studies on the population ecology of *Leucorrhinia dubia* (Vander Linden) (Odon. Libellulidae). *Annales Zoologici Societatis Zoologici Societatis Zoologicae-Botanicae Fennicae Vanamo* **24**, 1-79.

Palm. T. (1951) Die Holz und Rindenkäfer der nordschwedischen Laubbäume. *Meddelanden från Statens Skogsforskningsinstitut* **40**, 2.

Palm. T. (1959) Die Holz und Rindenkäfer der süd- und mittelschwedischen Laubbäume. *Opuscula Entomologica Supplement* **16**.

Parkyn, L., Stoneman, R., Everingham, F., Hughes, J., and Tanner, I. (1995) *Lowland raised bog land cover survey - Scottish Raised Bog Conservation Project.* Unpublished report, Scottish Wildlife Trust, Edinburgh.

Parr, M.J. and Parr, M. (1979). Some observations on *Ceriagrion tenellum* (De Villers) in southern England (Zygoptera: Coenagrionidae). *Odonatologica* **8** (3), 171-194.

Parr, R. (1992) The decline to extinction of a population of Golden Plover in north-east Scotland. *Ornis Scandinavica* **23**, 152-158.

Parr, R. (1993) *Moorland Birds and their Predators in Relation to Afforestation.* PhD thesis, University of Aberdeen.

Passmore, J. (1980). *Man's Responsibility for Nature.* 2nd edn. Duckworth, London.

Pearsall, W.H. (1950) *Mountains and Moorlands.* Collins, London.

Pfadenhauer, J. (1989) Renaturierung von Torfabbauflächen in Hochmooren des Alpemnvorlands. *Telma,* Beiheft **2**, 313-332.

Pfadenhauer, J. and Kinberger, M. (1985) Torfabbau und Vegetationsentwicklung im Kulbinger Filz (Region Südostbayern). *Berichte ANL Laufen* **9**, 37-44.

Pfadenhauer, J. and Klötzli, F. (1996) Restoration experiments in middle European wet terrestrial ecosystems; an overview. *Vegetation* (in press).

Pfadenhauer, J., Siuda, C. and Krinner, C. (1990) Ökologisches Entwicklungskonzept Kendlmühlfilzen. *Schriftenreihe Bayer. Landesamt fur Umweltschutz*, 91, München

Phillips, J. (1981) Moorburning. In: Phillips, J., Yalden, D. and Tallis J.H. (eds) *Peak District Moorland Erosion Study, Phase I Report.* Peak Park Joint Planning Board, Bakewell, pp. 171-175.

Phillips, J., Yalden, D. and Tallis J.H. (1981) *Peak District Moorland Erosion Study, Phase I Report.* Peak Park Joint Planning Board, Bakewell.

Pilcher, J.R. (1993) Radiocarbon dating and the palynologist: a realistic approach to precision and accuracy. In: Chambers, F.M. (ed.) *Climate Change and Human Impact on the Landscape.* Chapman and Hall, London, pp. 23-32.

Pilcher, J.R. and Hall, V.A. (1992) Towards a tephrochronology for the Holocene of the north of Ireland. *The Holocene* **2**, 255-259.

Pilcher, J.R., Bailllie, M.G.L., Brown, D.M., McCormac, P.B. MacSweeney, P.B. and McLawrence, A.S. (1995) Dendrochronology of subfossil pine in the north of Ireland. *Journal of Ecology* **83**, 665-671.

Pimm, S.L. (1991). *The Balance of Nature? Ecological Issues in the Conservation of Species and Communities.* University of Chicago Press, Chicago.

Pétrequin, P. (1996) Lac de Chalain (Fontenu, Jura, France). Protection des sites lacustres et contexte sociale d'application. In: Ramseyer, D. and Lambert, M.J. (eds). *Erosion et Protection. Sauvegarde des Sites Archéologiques et des Zones Naturelles en Milieux Humides* Centre Jurassien du Patrimoine, Musée d'Archéologie, Lons-le-Saunier, (in press).

Plantlife (1992) *Commission of Enquiry into Peat and Peatlands: Commissioners Report.* Plantlife, London.

Pooley, M.R. and Jones, M.M. (1995) *Application of Remote Sensing to Habitat Mapping and Monitoring*. Scottish Natural Heritage Review, Perth.

Poschlod, P. (1988) Vegetationsentwicklung ehemaliger Torfabbaugebiete in Hochmooren des bayerischen Alpenvorlandes. *Tuexenia* **8**, 31-55.

Poschlod, P. (1990) *Vegetationsentwicklung in abgetorften Hochmooren des bayerischen Alpenvorlandes unter besonderer Berücksichtigung standortkundlicher und populationsbiologischer Faktoren*. Dissertationes Botanicae 152.

Poschlod, P. and Pfadenhauer, J. (1989) Regeneration of vegetative parts of peat mosses - a comparative study of nine *Sphagnum* species. *Telma* **19**, 77-88.

Prentice, H.C. and Prentice, J.C. (1975) The hill vegetation of North Hoy, Orkney. *New Phytologist* **75**, 313-367.

Pritchard, D.E., Housden, S.D., Mudge, G.P., Galbraith, C.A. and Pienkowski, M.W. (1992) *Important Bird Areas in the UK Including the Channel Islands and the Isle of Man*. Royal Society for the Protection of Birds, Sandy.

Pryce, S. (1991) *The Peat Alternatives Manual - A Guide for the Professional Horticulturist and Landscaper*. Friends of the Earth, London.

Pyatt, D.G. (1990) Long term prospects for forests on peatland. *Scottish Forestry* **44**, 19-25.

Pyatt, D.G. and John, A.L. (1989) Modelling volume changes in peat under conifer plantations. *Journal of Soil Science* **40**, 695-706.

Quinty, R. and Rochefort, L. (1997) Plant reintroduction on a harvested peat bog. In: Trettin, C.C., Jurgensen, M., Grigal, D., Gale, M., and Jeglem, J. (eds) *Northern Forested Wetlands: Ecology and Management*. Lewis Publishers, New York, pp. 137-150 (in press).

Rackham, O. (1976) *Trees and Woodland in the British Landscape*. Dent, London.

Radjagukguk, B. (1992) Utilisation and management of peatlands in Indonesia for agriculture and forestry. *Proceedings of the Symposium on Tropical Peatland*. Kuching, Sarawak, Malaysia, 1991, pp. 21-27.

Radley, J. (1962) Peat erosion of high moors of Derbyshire and West Yorkshire. *East Midland Geographer* **3**, 40-50.

Radley, J. (1965) Significance of major moorland fires. *Nature London* **205**, 1254-9.

Rampino, M.R. and Self, S. (1982) Historic eruptions of Tambora (1815), Krakatau (1883), and Agung (1963), their stratospheric aerosols and climatic impact. *Quaternary Research* **18**, 127-143.

Randall, A. (1985). Human preferences, economics, and the preservation of species. In: Norton, B.G. (ed.) *The Preservation of Species. The Value of Biological Diversity*. Princeton University Press, Princeton, pp. 79-109.

Ratcliffe, D. (1995) The Governments response to the European Union Habitats and Species Directive. *British Wildlife* **6** (nE5) 307-310

Ratcliffe, D.A. (1959) The vegetation of the Carneddau, North Wales. I. Grasslands, heaths and bogs. *Journal of Ecology* **47**, 371-413.

Ratcliffe, D.A. (1976) Observations on the breeding of the Golden Plover in Great Britain. *Bird Study* **23**, 63-116.

Ratcliffe, D.A. (1977) *A Nature Conservation Review.* Cambridge University Press, Cambridge.

Ratcliffe, D.A. (1991) *Bird Life of Mountain and Upland.* Cambridge University Press, Cambridge.

Ratcliffe, D.A. (1964) In: Burnett, J.H. (ed.) *The Vegetation of Scotland.* Oliver and Boyd, Edinburgh.

Ratcliffe, D.A. and Thompson, D.B.A. (1988) The British uplands: their ecological character and international significance. In: Usher, M.B. and Thompson, D.B.A. (eds) *Ecological Change in the Uplands.* Blackwell Scientific Publications, Oxford, pp. 9-36.

Ratcliffe, D.A. and Walker, D. (1958) The Silver Flowe, Galloway. *Journal of Ecology* **46**, 407-445.

Rawes, M and Welch, D. (1969) Upland productivity of vegetation and sheep at Moor House National Nature Reserve, Westmorland, England. *Oikos Supplement* **11**, 72 pp.

Rawes, M. and Hobbs, R. (1979) Management of semi-natural blanket bog in the northern Pennines. *Journal of Ecology* **67**, 789-807.

RCAHMS (1990) *North-east Perth: an Archaeological Landscape.* Royal Commission on the Ancient and Historical Monuments of Scotland, Edinburgh

Regez, J.-L. and Tobler, H.-J. (1977) Gespräche mit seltsamen *Maschinen. Sandoz-Bulletin* **13** (44), 10-16.

Reid E., Mortimer, G.N., Thompson, D.B.A., Lindsay, R.A. (1994) Blanket Bogs in Great Britain: an assessment of large-scale pattern and distribution using remote sensing and GIS. In: Edwards, P.J., May, R.M. and Webb, N.R. (eds) *Large-Scale Ecology and Conservation Biology.* 35th Symposium of the British Ecological Society. Blackwell Scientific Publications, Oxford, pp. 229-246.

Reitter, E. (1911) *Fauna Germanica. Die Käfer der Deutches Reiches.* Stuttgart.

Renberg, I. and Sergerstrom, V. (1981) Application of varied lake sediments, in palaeoenvironmental studies. *Wahlenbergia* 7, 125-133.

Rennie, R. (1807) *Essays on the Natural History and Origin of Peat Moss.* Edinburgh.

Rennie, R. (1810) *Essays on the Natural History and Origin of Peat Moss.* Edinburgh.

Reynolds W.D., Brown, D.A., Mathur, S.P. and Overend, R.P. (1992) Effect of in-situ gas accumulation on the hydraulic conductivity of peat. *Soil Science* **153,** 397-408.

Rieley, J.O., Page, S.E., Limin, S.H. and Winarti, S. (1996). The peatland resource of Indonesia and the Kalimantan peat swamp forest research project. *Proceedings of the International Symposium on the Biodiversity,*

Environmental Importance and Sustainability of Tropical Peat and Peatlands. (in press).

Rieley, J.O., Sieffermann, R.G., Fournier, M. and Soubies, F. (1992) The peat swamp forests of Borneo: their origin, development, past and present vegetation and importance in regional and global environmental processes. *Proceedings of the 9th International Peat Congress,* (Special edition of the International Peat Journal) Uppsala, Sweden, vol. 1, pp. 78-95.

Rippey, B., Murphy, R.J. and Kyle, S.W. (1982) Anthropogenically derived changes in the sedimentary flux of Mg, Cr, Ni, Cu, Zn, Hg, Pb and P in Lough Neagh, Northern Ireland. *Environmental Science and Technology* **16**, 23-30.

Ritchie, J. (1920) *The Influence of Man on Animal Life in Scotland.* Cambridge University Press, Cambridge.

Robertson, R.A. (1993) Peat, horticulture and the environment. *Biological Conservation* **2**, 541-547.

Robinson, M. and Newson, M.D. (1986) Comparison of forest and moorland hydrology in an upland area with peat soils. *International Peat Journal* **1**, 49-68.

Rochefort, L., Gautheir, R. and Lequéré, D. (1995) *Sphagnum* regeneration - Toward and optimisation of bog restoration. In: Wheeler, B., Shaw, S., Fojt, W. and Robertson, A. (eds) *The Restoration of Temperate Wetlands.* John Wiley, Chichester pp. 423-434.

Rodwell, J.S. (ed.) (1991) *British Plant Communities. Vol. 2. Mires and Heaths.* Cambridge University Press,Cambridge.

Rodwell, J.S. (ed.) (1992) *British plant communities. Vol. 3. Grasslands and Montane Communities.* Cambridge University Press, Cambridge.

Rolston III, H. (1994) *Conserving Natural Value.* Columbia University Press, New York.

Roper, T. (1996) Fossil insect faunas and the development of raised mire at Thorne Moors, South Yorkshire. *Biodiversity and Conservation* (in press).

Rothmaler, W. (1988) *Exkursionsflora für die Gebiete der DDR und der BRD.* Band 4 - Kritischer band. Volk und Wissen Volkseigener Verlag, Berlin.

Roulet, N.R., Ash, R. and Moore, T.R. (1992) Low boreal wetlands as a source of atmospheric methane. *Journal of Geophysical Research* **97,** 3739-3749.

Roulet, N.T., Jano, A., Kelly, C.A., Klinger, L.F., Moore, T.R., Protz, R., Ritter, J.A. and Rouse, W.R. (1994) Role of the Hudson Bay lowland as a source of atmospheric methane. *Journal of Geophysical Research* **99,** 1439-1454.

Rowell, T.A. (1991) *SSSIs: A Health Check* Wildlife. Link, London.

Royal Society for Nature Conservation (1990) *The Peat Report.* The Royal Society for Nature Conservation (RSNC) The Wildlife Trusts Partnership, Lincoln.

Royal Society for the Protection of Birds (RSPB) (1985) *Forestry in the Flow Country - the Threat to Birds.* RSPB, Sandy, UK.

Royal Society for the Protection of Birds (RSPB) (1994) *Environmental Issues: Peat and Peatlands, Estuaries, Farmland and Countryside. Studies for Science and Geography 14 - 16 years.* Unpubl. report., Royal Society for the Protection of Birds, Sandy 104 pp.

Royal Society for the Protection of Birds (1995) *Possible Special Areas of Conservation (SACs) in the UK. RSPB submission to DoE proposed list of SACs.* RSPB, Sandy.

Royal Society for the Protection of Birds and Plantlife (1993) *Out of the Mire - a Future for Lowland Peat Bogs.* RSPB, Sandy.

Rudd, J.W.D., Harris, R., Kelly, C.A. and Hecky, R.E. (1993) Are hydroelectric reservoirs significant sources of green house gases? *Ambio* **22,** 246-248.

Ruuhijaervi, R. (1963) Zur Entwicklungsgeschichte der nordfinnischen Hochmoore. *Annales Botanicae Societatis Fennicae Vanamo* **34** (2), 1-40.

Ruuhijaervi, R. (1983) The Finnish mire types and their regional distribution. In: Gore A.J.P. (ed.) Mires: Swamp, Bog, Fen and Moor. Ecosystems of the World. Vol. 4B. Elsevier, Amsterdam, pp. 47-67.

Ryan, J.B. and Cross, J.C. (1984) Conservation of Peatlands in Ireland. *Proceedings of the 7th International Peat Congress, Dublin.*

Rydin, H. and Clymo, R.S. (1989) Transport of carbon and phosphorous compounds about *Sphagnum. Proceedings of the Royal Society of London B* **237**, 63-84.

Salleh, A., (1990) Living with nature: reciprocity or control? In: Engel, J.R. and Engel, J.G. (eds) *Ethics of Environment and Development. Global Challenge and International Response.* Belhaven Press, London, pp. 245-253.

Salonen, V. (1987) Relationship between the seed rain and the establishment of vegetation in two areas abandoned after peat harvesting. *Holarctic Ecology* **10**, 171-174.

Salonen, V. (1992) Effects of artificial plant cover on plant colonisation of a bare peat surface. *Journal of Vegetation Science* **3**, 109-112.

Samuelsson, G. (1910) Scottish peat mosses. A contribution to the knowledge of the late Quaternary vegetation and climate of north-western Europe. *Bulletin of the Geological Institute of the University of Upsala* **10**, 197-260.

Schiemenz, H. (1954) Über die angebliche Bindung der Libelle *Leucorrhinia dubia* v.d.L. an das Hochmoor. *Zoologische Jahrbücher (Systematik)* **82**, 473-480.

Schlichtherle, H. (1990) Aspekte der siedlungsarchäologischen Erforschung von Neolithikum und Bronzezeit im südwestdeutschen Alpenvorland. *Siedlungsarchäologische Untersuchungen im Alpenvorland,* **5**, 208-244.

Schmeidl, H. (1965) Oberflächentemperaturen in Hochmooren. *Wetter und Leben* **14**, 77-82.

Schouwenaars, J.M. (1988) The impact of water management upon groundwater fluctuation in a disturbed bog relict. *Agricultural Water Management* **14**, 439-449.

Science Action Coalition and Fritsch, A.J. (1980) *Environmental Ethics. Choices for Concerned Citizens.* Anchor Press, New York.

Scott, A. and Agyeman, J. (1995) *Wildlife Habitats: An Environmental Education Resource for Key Stage 3.* Friends of the Earth, London, 44 pp.

Scottish Natural Heritage (1995) *Boglands.* Scottish Natural Heritage, Perth.

Scottish Office (1991) *Scottish Office Circular 13/1991.* Nature Conservation, Scottish Office, Edinburgh.

Scottish Wildlife Trust (1994a) *Tree Removal from Langlands Moss.* Unpublished report. Scottish Wildlife Trust, Cramond House, Edinburgh.

Scottish Wildlife Trust (1994b) *A Monitoring Programme for Langlands Moss.* Unpublished report. Scottish Wildlife Trust, Cramond House, Edinburgh.

Scottish Wildlife Trust (1995a) *Scottish Raised Bog Land Cover Survey 1995 - Site Reports.* Unpublished Report. Scottish Wildlife Trust, Edinburgh.

Scottish Wildlife Trust (1995b) *A Conservation Strategy for Scotland's Raised Bogs - Draft Consultative Document.* Unpublished Report. Scottish Wildlife Trust, Cramond House, Edinburgh.

Sear, C.B., Kelly, P.M., Jones P.D. and Goodess, C.M. (1987) Global surface-temperature responses to major volcanic eruptions. *Nature* **330**, 365-367.

Segerstrom, U., Bradshaw, R., Hornberg, G. and Bohlin, E. (1994). Disturbance history of a swamp forest refuge in northern Sweden. *Biological Conservation* **68**, 189-196.

Self, S. and Sparks, R.S.J. (eds) (1981) *Tephra studies.* Reidel, Dordrecht.

Sernander, R. (1908) On the evidences of post-glacial change of climate furnished by the peat mosses of Northern Europe. *Geologiska Foreningens I Stockholm Forhandlingar* **30**, 70-103.

Sheail, J. and Wells, T.C.E. (1983) The fenlands of Huntingdonshire, England: a case study in catastrophic change. In: Gore, A.J.P. (ed.) *Ecosystems of the World. Vol. 4B. Mires, Swamp, Bog, Fen and Moor. Regional Studies.* Elsevier, Amsterdam and Oxford, pp. 375-393.

Shepherd, K.B., Batchelor, P.K., Hulka, S., Stirling, J.P., Watson, D. and Whitfield, P. (1995) *A Survey of Moorland Breeding Birds on the Isle of Lewis, Outer Hebrides, Scotland in 1995.* Research and Advisory Services Directorate Commissioned Research Report (Confidential), Scottish Natural Heritage, Edinburgh.

Shimwell, D. (1974) Sheep grazing intensity in Edale 1692-1747 and its effect on blanket peat erosion. *Derbyshire Archaeological Journal* **94**, 35-40.

Shotyk, W. (1988) Review of the inorganic geochemistry of peats and peatland waters. *Earth-Science Reviews* **25**, 95-176.

Shotyk, W., Nesbitt, H.W. and Fyfe, W.S. (1990) The behaviour of major and trace elements in complete vertical peat profiles from three *Sphagnum* bogs. *International Journal of Coal Geology* **15**, 163-190.

Shotyk, W., Nesbitt, H.W. and Fyfe, W.S. (1992) Natural and anthropogenic enrichments of trace metals in peat profiles. *International Journal of Coal Geology* **20**, 49-84.

Sieffermann, R.G., Fournier, M., Truitomo, S., Sadelman, M.T. and Semah, A.M. (1988) Velocity of tropical peat accumulation in Central Kalimantan province, Indonesia (Borneo). *Proceedings of the 8th International Peat Congress: Volume 1* Leningrad, pp. 90-98.

Sieffermann, R.G., Rieley, J.O. and Fournier, M. (1992) The lowland peat swamps of Central Kalimantan (Borneo); a complex and vulnerable ecosystem. *Proceedings of the International Conference on Geography in the Asean Region.* Yogyakarta, Indonesia, 26 pp.

Silvius, M.J., Simons, H.W. and Verheught, W.J.M. (1984) *Soils, Vegetation and Nature Conservation of the Berbak Game Reserve, Sumatra, Indonesia.* RIN Contributions to Research on Management, Arnhem Research Institute for Nature Management, The Netherlands.

Simmons, I.G. (1974) The ecological setting of Mesolithic man in the Highland zone. In: Evans, J.G. (ed.) *The Effect of Man on the Landscape: The Highland Zone.* The Council for British Archaeology, London, pp. 57-63.

Simola, H. and Lodenius, M. (1982) Recent increase in mercury sedimentation in a forest lake attributable to peatland drainage. *Bulletin of Environmental Contamination and Toxicology*, **29**, 298-305.

Singer, P. (1986). All animals are equal. In: Singer, P. (ed.) *Applied Ethics* Oxford University Press, Oxford, pp. 215-228.

Singer, P. (1990) *Animal Liberation* (revised edn). The New York Review, New York.

Sjörs, H. (1948) Myrvegetation i Bergslagen. *Acta Phytogeographica. Suecica.* **21,** 299.

Skolimowski, H. (1990) Reverence for life. In: Engel, J.R. and Engel, J.G. (eds) *Ethics of Environment and Development. Global challenge and InternationalResponse* Belhaven Press, London pp. 97-103.

Smart, P.J., Wheeler, B.D. and Willis, A.J. (1986) Plants and peat cuttings: historical ecology of a much exploited peatland - Thorne Waste, Yorkshire, UK. *New Phytologist* **104**, 731-748.

Smart, P.J., Wheeler, B.D. and Willis, A.J. (1989) Revegetation of peat excavations in a derelict raised bog. *New Phytologist* **111**, 733-748.

Smith, C. (1994) *Coir - A Viable Alternative to Peat for Potting.* Unpublished report. Scottish Agricultural College, Ayr.

Smith, C.A. (1995) Coir: a viable alternative to peat for potting. *The Horticulturist* **4,** 12 and 25-28.

Smith, C.A. and Hall, D.A. (1994) The development of perlite as a potting substrate for ornamental plants. *Acta Horticulturae* **361**, 159-166.

Smith, C.A., MacDonald, P. and Sword, A.M. (1995) Culture of hardy woody ornamentals in pure-mineral container substrates. *Acta Horticulturae* **401**, 161-167.

Smith, R.D. (1992) A conceptual framework for assessing the functions and values of wetlands. *Wetlands Research Program Technical Report.* TR-

WRP-DE-3. US Army Corps of Engineers Waterways Experiment Station, Vicksburgh, MS.

Smith, R.S. and Charman, D.J. (1987) The vegetation of upland mires within conifer plantations in Northumberland, Northern England. *Journal of Applied Ecology* **25**, 579-594.

Smyth, P. (1991) *Osier Culture and Basket-Making: A Study of the Basket-Making Craft in South West County Antrim.* Lurgan, Co. Armagh.

Sober, E. (1985) Philosophical problems for environmentalism. In: Norton, B.G. (ed.) *The Preservation of species. The Value of Biological Diversity* Princeton University Press, Princeton, pp. 173-194.

Soeffing, K. (1986) Ecological studies on eggs and larvae of *Leucorrhinia rubicunda* (L.) (Odonata: Libellulidae). *Jber. Forsch Inst. Borstel.* **1986**, 234-237.

Soeffing, K. (1988) The importance of mycobacteria for the nutrition of larvae of *Leucorrhinia rubicunda* (L.) in bog water (Anisoptera: Libellulidae). *Odonatologica* **17** (3), 227-233.

Solem, T. (1986) Age, origin and development of blanket mires in Sr-Trρndelag, Central Norway. *Boreas* **15**, 101-115.

Solem, T. (1994) Age, origin and development of blanket mires in Central Norway. *Norsk Geografisk Tidsskrift* **48**, 93-8.

Sorensen, K.W. (1993) Indonesian peat swamp forests and their role as a carbon sink. *Chemosphere* **27**, 1065-1082.

Sparling, J. (1967) The occurrence of *Schoenus nigricans* L. in blanket bogs. I. Environmental conditions affecting the growth of *Schoenus nigricans* in blanket bog. *Journal of Ecology* **55**, 1-13.

Steele, A. (1826) *The Natural and Agricultural History of Peat-moss or Turf-bog.* Edinburgh.

Steiner, G.M. (1992) *Oesterreichischer Moorschutzkatalog,* 4th edn.. Grüne Reihe des BMUJF **1**, pp. 509.

Steiner, H. (1948) Die bindung der Hochmoorlibele *Leucorrhinma dubia* Vand. an ihren Biotop. *Zoologische Jahrbücher Jena (Systematik)* **78**, 65-96.

Sternberg, K. (1990) Autökologie von sechs Libellenarten der Moore und Hochmoore des Schwarzaldes und Ursachen ihrer Moorbindung. PhD thesis, Univ. Freiburg i. Br.

Stevenson, A.C., Jones, V.J. and Battarbee, R.W. (1990) The cause of peat erosion: a palaeolimnological approach. *New Phytologist* **114**, 727-35.

Stirling, S.R. (1990). Towards an ecological world view. In: Engel, J.R. and Engel, J.G. (eds) *Ethics of Environment and Development. Global Challenge and International Response.* Belhaven Press, London, pp. 77-86.

Stirling-Maxwell, J. (1907) The planting of high moorland. *Transactions of the Royal Scottish Arboricultural Society* **20**, 1-7.

Stirling-Maxwell, J. (1936) Turf nurseries. *The Scottish Forestry Journal* **50**, 18-22.

Stokes, J.S. (1990) Developments in wetlands regulations. *Proceedings of the Environmental Law Conference - 20 July 1990.*

Stoms, D.M. and Estes, J.E. (1993) A remote sensing research agenda for mapping and monitoring biodiversity. *International Journal of Remote Sensing* **4** (10), 1839-1860.

Stone, C.D. (1972). *Should Trees have Standing? Toward Legal Rights for Natural Objects.* Kauffman, Los Altos.

Stoneman, R. (1993) Holocene palaeoclimates form peat stratigraphy: extending and refining the model. PhD thesis, University of Southampton.

Stoneman, R., Barber, K.E. and Maddy, D. (1993) Present and past ecology of *Sphagnum imbricatum* and its significance in raised peat - climate modelling. *Quaternary Newsletter* **70**,14-22.

Street, M.W. and McClees, J.D. (1981) North Carolina's Coastal Fishing Industry and the Influence of Coastal Alterations. In: Richardson, C.J. (ed.) *Pocosin Wetlands.* Hutchinson Ross Publishing Company, USA, pp. 238-250.

Stroud, D.A., Condie, M., Holloway, S.J., Rothwell, A.J., Shepherd, K.B., Simons, J.R. and Turner, J. (1988) *A Survey of Moorland Birds on the Isle of Lewis in 1987.* Chief Scientist's Directorate Report 776, Nature Conservancy Council, Peterborough.

Stroud, D.A., Reed, T.M. and Harding, N.J. (1990) Do moorland breeding waders avoid plantation edges ? *Bird Study* **37**, 177-186.

Stroud, D.A., Reed, T.M., Pienkowski, M.W. and Lindsay, R.A. (1987) *Birds, Bogs and Forestry. The Peatlands of Caithness and Sutherland.* Nature Conservancy Council, Peterborough.

Swanson, T.M. and Barbier, E.B. (1992) *Economics for the Wilds. Wildlife, wildlands, diversity and development.* Earthscan, London.

Sweatman, T.R. and Long, J.V.P. (1969) Quantitative electron microprobe analysis of rock forming minerals. *Journal of Petrology* **7**, 332-379.

Sykes, J.M. (1992) Caledonian pinewood regeneration: progress after sixteen years of exclosure at Coille Coire Chuilc, Perthshire. *Arboricultural Journal* **16**, 61-67.

Symon, J.A. (1959) *Scottish Farming Past and Present.* Oliver and Boyd, Edinburgh.

Takahashi, H. and Yonetani, Y. (1996) Studies on microclimate and hydrology of peat swamp forest in Central Kalimantan, Indonesia. *Proceedings of the International Symposium on the Biodiversity, Environmental Importance and Sustainability of Tropical Peat and Peatlands* (in press).

Tallis, J.H. (1964a) Studies on southern Pennine peats, II. The pattern of erosion. *Journal of Ecology* **52**, 333-44.

Tallis, J.H. (1964b) Studies on southern Pennine peats, III. The behaviour of *Sphagnum*. *Journal of Ecology* **52**, 345-53.

Tallis, J.H. (1965) Studies on southern Pennine peats, IV. Evidence of recent erosion. *Journal of Ecology* **53**, 509-20.

Tallis, J.H. (1981a) The nature and extent of soil and peat erosion in the Peak District - literature survey. In: Phillips, J., Yalden, D. and Tallis, J.H. (eds) *Peak District Moorland Erosion Study, Phase I Report.* Peak Park Joint Planning Board, Bakewell, pp. 65-73.

Tallis, J.H. (1981b) Rates of erosion. In: Phillips, J., Yalden, D. and Tallis, J.H. (eds) *Peak District Moorland Erosion Study, Phase I Report* Peak Park Joint Planning Board, Bakewell pp. 74-83.

Tallis, J.H. (1981c) Uncontrolled burns. In: Phillips, J., Yalden, D. and Tallis, J.H. (eds) *Peak District Moorland Erosion Study, Phase I Report.* Peak Park Joint Planning Board, Bakewell, pp. 176-177.

Tallis, J.H. (1985a) Mass movement and erosion of a southern Pennine blanket peat. *Journal of Ecology* **73**, 283-315.

Tallis, J.H. (1985b) Erosion of blanket peat in the southern Pennines: new light on an old problem. In: Johnson, R.H (ed.) *The Geomorphology of Northwest England.* Manchester University Press, Manchester, pp. 313-336.

Tallis, J.H. (1987) Fire and flood at Holme Moss: erosion processes in an upland blanket mire. *Journal of Ecology* **75**, 1099-1129.

Tallis, J.H. (1995) Blanket mire in the upland landscape. In: Wheeler, B.D., Shaw, S.C., Fojt, W.J. and Robertson, R.A. (eds) *Restoration of temperate wetlands.* John Wiley and Sons, Chichester, pp. 495-508.

Tallis, J.H. and Livett, E.A. (1994) Pool-and-hummock patterning in a southern Pennine blanket mire, I. Stratigraphic profiles for the last 2800 years. *Journal of Ecology* **82**, 775-88.

Tallis, J.H. and Yalden, D.W. (1983) *Peak District Moorland Restoration Project Phase 2 Report: Revegetation Trials.* Peak Park Joint Planning Board, Bakewell.

Tansley, A.G. (1939) *The British Islands and Their Vegetation.* Cambridge University Press, London.

Tansley, A.G. (1949) *Britain's Green Mantle.* Allen and Unwin, London.

Tansley, A.G. (1953). *The British Isles and their Vegetation.* Cambridge University. Press, Cambridge.

Taylor, J.A. (1983) The Peatlands of Great Britain and Ireland. In: Gore, A.J.P. (ed.) *Ecosystems of the World. Vol. 4B. Mires, Swamp, Bog, Fen and Moor. Regional Studies.* Elsevier, Amsterdam and Oxford, pp. 1-46.

Taylor, K.C., Lamorey, G.W., Doyle, G.A., Alley, R.B., Grootes, P.M., Mayewski, P.A., White, J.W.C. and Barlow, L.K. (1993) The 'flickering switch' of late Pleistocene climate change. *Nature* **361**, 432-436.

Taylor, P.W. (1981). The ethics of respect for nature. *Environmental Ethics* **3**, 197-218.

Thompson, D.B.A. and Miles, J. (1995) Heaths and moorlands: some conclusions and questions about environmental change. In: Thompson, D.B.A., Hester, A.J. and Usher, M.B. (eds) *Heaths and Moorlands: CulturalLandscapes.* HMSO, Edinburgh. pp. 362-387

Thompson, D.B.A., MacDonald, A.J., Marsden, J.H., Galbraith, C.A. (1995) Upland Heather Moorland In Great Britain: A Review Of International Importance, Vegetation Change and Some Objectives For Nature Conservation. *Biological Conservation* **71**, 163-178.

Thompson, D.B.A., Stroud, D.A. and Pienkowski, M.W. (1988) Afforestation and upland birds: consequences for population ecology. In: Usher, M.B. and Thompson, D.B.A. (eds) *Ecological Change in the Uplands.* Blackwell Scientific Publications, Oxford, pp. 237-259.

Thorarinsson, S. (1981) Greetings from Iceland: ashfalls and volcanic aerosols in Scandinavia. *Geografiska Annaler* **63A**, 109-118.

Thornthwaite, C.W. (1948) An Approach Toward a Rational Classification of Climate. *Geographical Review* **38**, 55-94.

Tivy, J. (ed.) (1973) *The Organic Resources of Scotland: Their Nature and Evaluation.* Oliver and Boyd, Edinburgh.

Tudor, G.J., Mackey, E.C., and Underwood, F.M. (1994) *The National Countryside Monitoring Scheme: The Changing Face of Scotland, 1940s to 1970s. Main Report.* Scottish Natural Heritage, Perth.

Turtle, C.E. (1984) Peat erosion and reclamation in the southern Pennines. PhD thesis, University of Manchester.

UK Government (1994) *Climate Change. The UK Programme.* HMSO, London.

United Nations General Assembly (1948) Universal declaration on human rights. In: *UN General Assembly, third session, first part, Official Records, 'Resolutions'*, pp. 71-77.

Urapeepatanapong, C. and Pitayakajornwute, P. (1992) The peat swamp forests of Thailand. *Workshop on Integrating, Planning and Management of Tropical Lowland Peatlands,* Cisarua, Indonesia.

Urban, N.R., Eisenreich, S.J., Grigal, D.F. and Schurr, K.T. (1990) Mobility and diagenesis of Pb and ^{210}Pb in peat. *Geochimica et Cosmochimica Acta* **54**, 3329-3346.

United States of America Corps of Army Engineers (USACAE) (1972) Charles river: main report and attachments. Waltham, MA. USA Cited In: Sather J.M. and Smith, R.D. (1984) *An Overview of Major Wetland Functions and Values.* Fish and Wildlife Service, FWS/OBS-84/18. 68 pp.

Usher, M.B. and Thompson, D.B.A. (1993) Variation in the upland heathlands of Great Britain: conservation importance. *Biological Conservation* **66**, 69-81.

Van de Noort, R. and Davies, P. (1993) *Wetland Heritage. An Archaeological Assessment of the Humber Wetlands.* Humber Wetlands Project, Hull.

Van den Bogaard, C., Dörfler W., Sandgren, P. and Schmicke, H-U. (1994) Correlating the Holocene records: Icelandic tephra found in Schleswig-Holstein (Northern Germany). *Naturwissenschaften* **81**, 554-556.

Van der Perk, N. (1989). Natuurbeeld-kritiek of maatschappijkritiek? Naar een 'ekologische bevrijdingstheologie'. In: Achterberg, W. (ed.) *Natuur:*

uitbuiting of respect? Natuurwaarden in discussiepp. Kok Agora, Kampen, pp. 160-176.

Van DeVeer, D. and Pierce, C. (eds) (1994). *The Environmental ethics and Policy Book: Philosophy, Ecology, Economics*. Wadsworth, Belmont.

Van Geel, B. and Middledorp, A.A. (1988) Vegetational history of Carberry Bog (County Kildare, Ireland) during the last 850 years and a test of the temperature indicator value of $^{2}H/^{1}H$ measurements of peat samples in relation to historical sources and meteorological data. *New Phytologist* **109**, 377-392.

Van Hoogstraten, H.D. (1993) *Geld en geest. Over milieu-ethiek*. Ten Have, Baarn.

Vijarnson, P. (1992) Utilisation in south-east Asia. *Workshop on Integrating, Planning and Management of Tropical Lowland Peatlands*, Cisarua, Indonesia.

Vogel, J.C., Casparie, W.A. and Munaut, A.V. (1969) Carbon-14 trends in subfossil pine stubs. *Science* **166**, 1143-1145.

Walker, D. and Walker, P.M. (1961) Stratigraphic evidence of regeneration in some Irish bogs. *Journal of Ecology* **49**, 169-85.

Waller, M. (1994) *The Fenland Project, Number 9: Flandrian Environmental Change in Fenland*. East Anglian Archaeology 70, Cambridgeshire County Council.

Warner, B.G. (1993) Palaeoecology of floating bogs and landscape change in the Great lakes drainage basin of North America. In: Chambers, F.M. (ed.) *Climate Change and human Impact on the Landscape*. Chapman and Hall, London, pp. 237-245.

Warner, B.G. and Charman, D.J. (1994) Holocene soil moisture changes on a peatland in north-western Ontario based on fossil testate amoebae (Protozoa) analysis. *Boreas* **23**, 270-279.

Warren, M.A. (1983) The rights of the nonhuman world. In: Elliot, R. and Gare, A. (eds) *Environmental Philosophy: A Collection of Readings* University of Queensland Press, St. Lucia, pp. 109-134.

Waton, P.V. and Barber, K.E. (1987) Rimsmoor, Dorset: biostratigraphy and chronology of an infilled doline. In: Barber, K.E. (ed.) *Wessex and the Isle of Wight Field Guide*. Quaternary Research Association, Cambridge, pp. 75-80.

Watson, A. and Hinge, M. (1989) *Natural Tree Regeneration on Open Moorland in Deeside and Donside*. Report to NCC (N.E. Scotland).

Watson, R.A. (1979) Self-consciousness and the rights of nonhuman animals and nature. *Environmental Ethics* **1**, 99 - 129.

Wayi, B.M. and Freyne, D.F. (1992) The distribution, characterisation, utilisation and management of peat soils in Papua New Guinea. *Proceedings of the International Symposium on Tropical Peatland*. Kuching, Sarawak, Malaysia, pp. 28-32.

Weber, C.A. (1902) *Über die Vegetation und Entstehung des Hochmoores von Augstumal im Memeldelta.* Paul Parey, Berlin, 252 pp.

Wein, R.W. (1973) Biological Flora of the British Isles: *Eriophorum vaginatum* L. *Journal of Ecology* **61**, 601-615.

Welch, D. (1975) Three Elizabethan documents concerning Milburn Fell. *Transactions of the Cumberland and Westmorland Antiquarian* and *Archaeological Society* **75**, 136-149.

Welch, D. (1984) Studies in the grazing of heather moorland in north-east Scotland. II. Response of heather. *Journal of Applied Ecology* **21**, 197-207.

Welch, D. and Rawes, M. (1966) The intensity of sheep grazing on high-level blanket bog in upper Teesdale. *Irish Journal of Agricultural Research* **5**, 185-196.

West, R. (1984) The future of Quaternary Research. *Quaternary Newsletter* **43**, 30-33.

Westgate, J.A. and Gorton, M.P. 1981. Correlation techniques in tephra studies. In: Self, S. and Sparks, R.S.J. (eds) *Tephra studies.* Reidel, Dordrecht, pp. 73-94.

Westhoff, V. (1989). Het natuurbeeld in schriftloze culturen en in hoger ontwikkelde oosterse wereldbeschouwingen. In: Achterberg, W. (ed.) *Natuur: Uitbuiting of Respect? Natuurwaarden in Discussie* Kok Agora, Kampen, pp. 177-198.

Wheeler, B.D. and Shaw, S.C. (1995a) *Monitoring Rehabilitation Work on Lowland Peatlands.* English Nature, Peterborough.

Wheeler, B.D. and Shaw, S.C. (1995b) *Restoration of Damaged Peatlands.* HMSO, London.

White, A. (1994a) *Border Mires Part One. The Spadeadam Mires: Redefinition of Hydrological Boundaries.* Northumberland Ecological Services for English Nature.

White, A. (1994b) *Border Mires Part Two. Assessment of the Full Extent Former Extent of Ombrotrophic Mire within the Border Mires Area.* Northumberland Ecological Services for English Nature.

White, A. (1995) *Ditch Survey and Mapping of the Northumbrian Border Mires and the Cumbrian Spadeadam Mires.* Report for Forest Enterprise.

White, B. and Wadsworth, R. (1994) A bioeconomic model of heather moorland management and conservation. *Ecological Economics* **9**, 167-177.

White, J. and Doyle, G.J. (1982) The vegetation of Ireland: a catalogue raisonné. *Journal of Life Sciences, Royal Dublin Society* **3**, 289-368.

White, L. (1967) The historical role of our ecological crisis. *Science* **155**, 1203-1207.

Whitehead, K.G. (1964) *The Deer of Great Britain and Ireland.* Routledge and Kegan Paul, London.

Whitehouse, N.J. (1996) Silent witnesses: an Urwald fossil insect assemblage from Thorne Moors. *Thorne and Hatfield Moors Papers* (in press).

Whitten, A.J., Damanik, S.J., Anwar, J. and Hisam, N. (1987) *The Ecology of Sumatra.* Gadjah Mada University Press, Yogyakarta, Indonesia.

Wildlife Trusts (1995) *Peat Campaign Review: After Five Years It's Not All Cut and Dried.* The Wildlife Trusts, Lincoln.

Wilford, G.E. (1960) Radiocarbon age determinations of quaternary sediments in Brunei and N.E. Sarawak. *British Borneo Geological Survey Annual Report,* 1959.

Wilkins, D.A. (1984) The Flandrian woods of Lewis (Scotland). *Journal of Ecology* **72**, 251-258.

Williams, M. (1990) Understanding Wetlands. In: Williams, M. (ed.) *Wetlands: A Threatened Landscape.* Blackwell, Oxford, pp. 1-41.

Williams, T.M. (1988) The geochemistry and magnetic mineralogy of Holocene strata in acid basins: case-study of granitic catchment in the Galloway and Rannoch moor regions of Scotland. PhD thesis, University of Edinburgh.

Wilson, E. (1995) *Integrated Ecological Monitoring on Lowland Raised Bogs: a Practical Guide.* Unpublished report. Scottish Wildlife Trust, Cramond House, Edinburgh.

Wind-Mulder, H.L., Rochefort, L. & Vitt, D.H. (1996) Water and peat chemistry comparisons of the natural and post-harvested peatlands across Canada and their relevance to peatland restoration. *Ecological Engineering* (in press).

Winkler, M.G. and de Witt, C.B. (1985) Environmental impacts of peat mining in the United States: documentation for wetland conservation. *Environmental Conservation*, **14** (4), 317-330.

Winpenney, J.T. (1991) *Values for the Environment. A Guide to Economic Appraisal.* HMSO, London.

Wood, T.F. and Foody, G.M. (1989) Analysis and representation of vegetation continua from Landsat Thematic Mapper data for lowland heaths. *International Journal of Remote Sensing* **10** (1), 181-191.

Woodin, S.J. and Lee, J.A. (1987) The effects of nitrate, ammonium and temperature on nitrate reductase activity in *Sphagnum* species. *New Phytologist* **105**, 103-115.

World Commission on Environment and Development (1987). *Our Common Future.* Oxford University Press, Oxford.

Zehetmayr, J.W.L. (1953) Experimental plantations in the far north of Scotland. *Scottish Forestry* 7, 71-78.

Zehetmayr, J.W.L. (1954) Experiments in tree planting on peat. *Forestry Commission Bulletin No. 24.* HMSO, London.

Zoltai, S.C. (1988) Wetland Environments and Classification. In: National Wetlands Working Group (eds) *Wetlands of Canada.* Ecological Land Classification Series, no. 24. Environment Canada, Ottawa, pp. 3-26.

Index

Page numbers in **bold** refer to key references, those in *italics* refer to tables/figures.